Private food law

Also available in the 'European Institute for Food Law series':

European Food Law Handbook
Bernd van der Meulen and Menno van der Velde
ISBN 978-90-8686-082-1
www.WageningenAcademic.com/foodlaw

Fed up with the right to food?
The Netherlands' policies and practices regarding the human right to adequate food
edited by: Otto Hospes and Bernd van der Meulen
ISBN 978-90-8686-107-1
www.WageningenAcademic.com/righttofood

Reconciling food law to competitiveness
Report on the regulatory environment of the European food and dairy sector
Bernd van der Meulen
ISBN 978-90-8686-098-2
www.WageningenAcademic.com/reconciling

Governing food security
Law, politics and the right to food
edited by: Otto Hospes, Irene Hadiprayitno
ISBN: 978-90-8686-157-6; e-book ISBN: 978-90-8686-713-4
www.WageningenAcademic.com/EIFL-05

efla

european food law association
association européenne de droit alimentaire

Private food law

Governing food chains through contract law, self-regulation, private standards, audits and certification schemes

edited by:

Bernd M.J. van der Meulen

Wageningen Academic Publishers

ISBN: 978-90-8686-176-7
e-ISBN: 978-90-8686-730-1
DOI: 10.3920/978-90-8686-730-1

ISSN 1871-3483

First published, 2011

Table of contents

About the authors

Ferdinando Albisinni

Ferdinando Albisinni is full professor of European Agricultural and Food Law, Tuscia University, Viterbo, Italy; Professor Jean Monnet of European Food and Agricultural Law; national secretary of the Italian Food Law Association (www.aida-ifla.it); director of the 'Rivista di diritto alimentare' on line (www.rivistadirittoalimentare.it); member of the Board of Directors of the European Food Law Association (www.efla-aeda.org); Member of the scientific committee of the Italian Ministry for Agriculture, Food and Forestry; Academician of the Academy of Georgofili, Florence; Member of the Bar of Rome, admitted before the Supreme Court; author of volumes and articles on agricultural law and food law in international and comparative perspective. E-mail: albisinni@unitus.it

Theo Appelhof

Theo Appelhof is a chemist (Utrecht University). In the period of 1975 to 2005 he was employed as a chemist and director at the Dutch Food Safety Authority (in Dutch: 'Keuringsdienst van Waren' or 'nVWA') in Nijmegen, Leeuwarden en Eindhoven, the Netherlands. Together with Precon Food Management BV he worked on various EU projects in Estonia and Lituania. In these projects, among others, local legislation was adjusted as preparation for EU accession. Since his retirement in 2005 he works as a senior consultant for Précon Food Managent BV in Bunnik, the Netherlands and is a member of the editorial board of the 'Praktijkgidsen Warenwet' (Dutch for 'practical guides on the Commodities Act'). In this position he writes publications on subjects such as claims and labelling related to foodstuffs. E-mail: tappelhof@precon-food.nl

Alessandro Artom

Alessandro Artom is attorney at law before the Court of Appeals Milan and the Superior Court Rome, Italy. He is founder and partner of '*Studio legale associato Artom – Papini*', a law firm dealing with the regulation of food and beverages under EU and national law and the EU and national law regulating production and sale of consumer goods. In this regard, the law firm assists the major national and international food/beverages companies operating in Italy. He is also a legal advisor of *Federalimentare*, the Italian federation of all the national associations of the producers of food and beverages, member of *CIAA*, as well as legal advisor of *Assobirra*, association of the Italian producers of beer and malt, member of Brewers of Europe. He is Vice President of EFLA, European Food Law Association, Brussels, Belgium; Member of the Council of *AIDA*, Italian Food Law Association, Rome, Italy; Member of the *CIBUS* Committee, International Food Fair taking place in Parma every two years; Member of the scientific Committee of the Italian Ministry of Industry and Commerce for the simplification and standardisation of national food and beverages legislation (*Codice di diritto alimentare*), in accordance to the Regulation (EC) No 178/2002. He has authored several voices of the legal Data Bank on Food Law '*Diritto alimentare. Mercato e sicurezza*' (www.leggiditaliaprofessionale.it) and is a Member of the Scientific Committee of the web legal journal '*Rivista di Diritto alimentare*' as well as author of several articles on themes of food and beverages law, especially in relation to commercial and fiscal problems (www.rivistadirittoalimentare.it). E-mail: artomlaw@iol.it

Harry Bremmers

Harry Bremmers graduated in Business Economics (Tilburg University, 1977) and Law (Erasmus University, 1993). He obtained a Phd in 1995 at Erasmus University with a study on the effects of environmental standards and legislation on financial accounting theory and practice. From 1993 on he joined Wageningen University, with subsequent appointments at the Business Administration and Law &Governance group. At present he is senior researcher and lecturer in the Law & Governance Group in the position of Associate Professor (Food Law and Economics). He focuses at green law (food law and environmental law) and its managerial and economic impact on the food industry. He published more than 100 papers, books and articles, mainly on the managerial and legal-economic impact of legal an sustainability requirements for the food industry. He participated in several research projects for the European Union, addressing the competitiveness and sustainability of the European food and drink industry, which at present are used to set out new policies. E-mail: harry.bremmers@wur.nl

Esther Brons-Stikkelbroeck

Lawyer Esther Brons-Stikkelbroeck heads the Retail and Franchise sector of the law firm Dijkstra Voermans. Praised as an outstanding franchise specialist in the 2010 Guide to the World's Leading Franchise Practitioners, she provides consultancy and litigation services to authoritative private and non-profit franchising organisations. Esther lectures on Franchise Law at the law training institute De Juridische Academie, is regularly published in the Dutch daily newspaper Het Financieele Dagblad, the business magazine Sprout and the journal Franchise Plus, and founded and writes the blog www.dekunstvanfranchise.nl. She is also an associate member of the Netherlands Franchise Association (Nederlandse Franchisevereniging) and a member of the Distribution, Franchise and Agency Law Association (Vereniging voor Distributie-, Franchise- en Agentuurrecht). E-mail: ester.brons-stikkelbroeck@dvan.nl

Lawrence Busch

Lawrence Busch is Professor of Standards and Society at the Centre for Economic and Social Aspects of Genomics (CESAGEN) at Lancaster University (UK) and University Distinguished Professor at the Center for the Study of Standards in Society at Michigan State University (USA). He is (co)author or (co)editor of twelve books including '*Standards: Recipes for Reality*' (MIT Press, forthcoming). He has also authored or co-authored more than 150 other publications. He is a fellow of the American Association for the Advancement of Science, a *Chevalier de l'Ordre du Mérite Agricole,* a member of the *Académie d'Agriculture de France,* and a *doctor honoris causa* of the Universidade Técnica of Lisboa, Portugal. E-mail: lbusch@msu.edu

Nicole Coutrelis

Nicole Coutrelis is a lawyer at the Paris Bar since 1985. Before starting her private practice, she was at the French 'Office National Interprofessionnel des Céréales' (Cereals Authority) where she was in charge of the European Affairs (1972-1978). She was then nominated at the French Permanent Representation to the EC in Brussels from 1978 to 1982 and joined afterwards the Legal Service of the European Commission where she was in charge of antitrust matters, particularly as applied in the Food and Agricultural sectors, from 1982 to1985. Nicole Coutrelis is a partner of Coutrelis & Associes, a French Law Firm with Offices in Paris and in Brussels, specializing in EU Law, mainly in Antitrust, Agriculture, Food and Drug regulations, Alcoholic Beverages, Free Movement of Goods, Customs and all matters related to the Single Market. She is particularly in charge of the Food Law and Customs Practice of the Firm, counseling, lobbying and litigating, in France, in other Member States of the EU and before the European Court of Justice and the Court of First Instance in Luxembourg. Currently, she is President of the European Food Law Association, as well as Vice President of the International Wine Law Association and a Member of the International Bar Association. Nicole Coutrelis is author of many articles and Conferences related to Food and Drug Law as well as, more generally, the European Law. E-mail: n.coutrelis@coutrelis.com

Tetty Havinga

Tetty Havinga is Associate Professor of Sociology of Law, Radboud University Nijmegen, the Netherlands. She published on the regulation of food safety, policy implementation and law enforcement, experiences of large companies with specialized courts, equal opportunities law, and migration and is particularly interested in relations between industry and law related to the public interest. Her current research deals with the development and effects of private regulation of food safety and experiences of European law in everyday life. She published in Agriculture and Human Values, 'Actors in Private Food Governance: The Legitimacy of Retail Standards and Multistakeholder Initiatives with Civil Society Participation.' (with Doris Fuchs and Agni Kalfagianni) and Law & Policy 'Private regulation of food safety by Supermarkets'. E-mail: t.havinga@jur.ru.nl

Spencer Henson

Spencer Henson is Professor, Department of Food, Agricultural and Resource Economics, University of Guelph, Canada and Professorial Fellow, Globalisation Team, Institute of Development Studies, UK. E-mail: s.henson@ids.ac.uk

Otto Hospes

Dr. Otto Hospes is an associate professor Public Administration and Policy at the Department of Social Sciences of Wageningen University and visiting professor at Nantes University. Using insights from legal anthropology, political sciences and economic sociology, he is doing research on plural legal orders for food security, multi-actor governance of biofuels, and private law making in the field of sustainable palm oil and soy production. He has visited, conducted evaluations and done research in many countries, including Indonesia, India, Sri Lanka, Zambia, Zimbabwe, Tanzania, Kenya and Brazil. E-mail: otto.hospes@wur.nl

Marinus P.C. Huige

Marinus P.C. Huige is working at the Ministry of Economic Affairs, Agriculture and Innovation in the Netherlands and was Chairman of the WTO SPS-Committee in the period 2007-2009. E-mail: m.p.c.huige@minlnv.nl

John Humphrey

John Humphrey is Professorial Fellow, Globalisation Team, Institute of Development Studies, UK. He has an interest in private governance of agri-food systems, and in particular the role of private standards, including impacts on developing countries. E-mail: j.humphrey@ids.ac.uk

Maria Litjens

Maria Litjens is PhD-researcher at Wageningen University, the Netherlands. Her PhD-research focuses on the tension between food law en competition law. Private regulation systems create on the one hand a high level of food safety and implement food law requirements, but on the other hand limit competition and finally may infringe on competition law. Previously she worked as an agronomist, including at Wageningen University as developer of research projects regarding agriculture and rural planning. E-mail: maria.litjens@wur.nl

Hanspeter Schmidt

Hanspeter Schmidt is the president of a law firm in Germany (Freiburg im Breisgau) specialising in the needs of the food industry. His practise covers unfair trade practises litigation, export import transactions and related fields, such as trademarks and food processing patents. He is a certified specialist of administrative law. He published of commentaries on EU food law, such as the Regulation (EC) No. 834/2007 on organic food. E-mail: hps@hpslex.de

Irene Scholten-Verheijen

Irene Scholten-Verheijen works as a lawyer in Amsterdam, the Netherlands. She specialises in regulated markets, with a focus on the pharma and food sectors. Irene is a member of the Dutch Food Law Association (NVLR) and the European Food Law Association (EFLA) and a board member of the European Institute for Food Law. She gives lectures on a regular basis and has several publications to her name (amongst other she is the co-author of the 'Landkaart levensmiddelenrecht', Praktijkgidsen Warenwet, Sdu 2010). E-mail: irene.scholten@dvan.nl

Fabian Stancke

Prof. Dr. Fabian Stancke is Professor of Law at the Brunswick European Law School of the Ostfalia University of Applied Sciences, Braunschweig/Wolfenbüttel, Germany and counsel at the Hamburg office of law firm Field Fisher Waterhouse LLP. He is specialised in EU and German competition and antitrust law and has extensive experience in the food and retail sector. Before his university career, he practiced competition law as competition counsel at the group legal department of a major blue chip company in Munich and as a lawyer at a leading international law firm. Fabian Stancke lectures and publishes regularly on topics concerning European and German competition law. He is member of the reputable Studienvereinigung Kartellrecht e.V. and listed in '*JUVE*' (German eq. to '*Chambers*') as a 'frequently recommended counsel'. E-mail: fstancke@ostfalia.de and fabian.stancke@lw.com

Ronald van den Heuvel

Ronald van den Heuvel graduated in food science and technology at Wageningen University, Wageningen, the Netherlands. Based from the United Kingdom he worked in many European countries as a business developper and market analist. In 2009 he started working as a consultant for Précon Food Management BV, Bunnik, the Netherlands providing legal and food safety management advice to food producing companies. He was a lecturer at the Food Law Academy on the topic of labelling. He is currently self employed and analyses food safety issues and international markets. He is also linked to the European Chair of Food Safety

Microbiology that is chaired at Wageningen University through the development of digital learning materials.

Bernd M.J. van der Meulen

B.M.J. van der Meulen is Professor of Law and Governance at Wageningen University, the Netherlands (www.law.wur.nl/UK/); chairman of the Dutch Food Law Association (www.NVLR.nl); member of the Board of Directors of the European Food Law Association (www.efla-aeda.org); member of the editorial board of the European Food and Feed Law Review (http://www.lexxion.de/en/effl-home) and director of the European Institute for Food Law (www.food-law.nl). He was responsible for the programme of the 18th international EFLA congress held in Amsterdam on 16 en 17 September 2010. E-mail: bernd.vandermeulen@wur.nl

Lomme C. van der Veer

Lomme C. van der Veer is attorney at law at Van der Veer Advocaten, Meppel, the Netherlands. He teaches law at Windesheim University of Applied Sciences in Zwolle. E-mail: lc@vanderveer.nl

Margret Will

Margret Will, is a freelance consultant specialised in agri-food value chain development, industry cluster promotion, food standards and development of agribusiness associations. She holds a MBA in international trade and a M.Sc. in horticultural economics and has got more than 20 years work experience in transition and developing countries in Eastern Europe, Africa and Latin America. She is (co)author or editor of several books and articles including 'Food Quality and Safety Standards as required by EU Law and the Private Industry – A Practitioners' Reference Book' (GTZ, 2007); 'Integrating Smallholders into Global Supply Chains: GlobalGAP Option 2 Smallholder Group Certification Generic Manual' (GTZ, 2010) and 'Public mandatory and private voluntary standards: challenge or opportunity for small scale farmers' participation in global supply chains?' (Edward Elgar, 2011). E-mail: margret.will@gmx.net

Foreword

Over the last decade, worldwide initiatives from the private sector have turned the legal and regulatory environment for food businesses upside down. Litigation is no longer solely framed by legislative requirements, but ever more by private standards such as GlobalGAP, BRC, IFS, SQF and ISO. Private standards incorporate public law requirements, thus embedding them in contractual relations and exporting them beyond the jurisdiction of public legislators. Private standards are used to remedy shortcomings in legislation, to reach higher levels of consumer protection than the ones chosen by the EU legislature, to impose new obligations on contracting parties, to manage risks and liability beyond the traditional limits of food businesses and finally to give substance to corporate social responsibility.

Private standards also play a role in defining specific markets of growing importance and in self-regulating the commercial communication/advertising for foods and beverages. Organic standards have found an interesting symbioses with public law. Halal standards express the demands of some two billion consumers worldwide. Food businesses are inspected more often by private auditors than by public inspectors. Effects in terms of receiving or being denied certification often far outweigh public law sanctions. In short, based on private law, an entire legal infrastructure for the food sector emerges, in parallel to, and sometimes complementing, the public law regulatory infrastructure.

The European Food Law Association (EFLA), in its 18th international congress held in Amsterdam in September 2010, explored this emerging private food law.

The congress looked into developments, backgrounds, structures, specific examples of varying nature, consequences for businesses and for the EU internal market; interplay with public law through accreditation, imposed self-regulation in the form of HACCP or hygiene codes, adoption of non-binding Codex Alimentarius standards in contractually binding private standards; possibilities for the food sector, limits due to competition law and all that needs to be known by lawyers, academics, quality managers, regulatory affairs officers, and civil servants active in the ever expanding world of European food law.

This book is not conference proceedings in the usual sense of the word. It does not simply bring the presentations made at the conference. It does not either reflect all the very rich debate that took place during the conference.

The conference has however inspired the speakers and some other experts to contribute to this first book in legal literature analysing private food law, a topic which clearly deserves great attention in our world. Many other topics could also have been discussed within that frame: Competition rules, Contract law,

International Commerce, Ethics, etcetera. Let us hope that the EFLA conference and this book will just be a start and inspire other works in the future.

I express my gratitude to Wageningen Academic Publishers and the European Institute for Food Law for the way they cooperated in the creation of this first 'EFLA-book' and for including it in the European Institute for Food Law series. This cooperation between EFLA and the European Institute for Food Law has proved to be extremely fruitful and promising. I thank all the authors for their beautiful contributions and, last but not least I thank professor Van der Meulen, who is also an active member of the Board of EFLA. He designed the conference topic, helped identify speakers and authors and edited the book. It is fair to say that he coined the concept 'private food law'.

Nicole Coutrelis
President of EFLA

Abbreviations

AB	Accreditation body
AFNOR	Association Française de Normalisation (French national organisation for standardisation)
AGCM	Autorità Garante della Concorrenza e del Mercato (Antitrust Authority)
AHC Europe	European Association of Halal Certifiers
AIJN	Association of the Industry of Juices and Nectars from Fruits and Vegetables of the EU
ASEAN	Association of Southeast Asian Nations
B2B	Business to business
B2C	Business to consumer
BER	Block exemption regulation
BERR	Department for Business, Enterprise and Regulatory Reform (United Kingdom)
BIPRO	Brand integrity programme
BMZ	Bundesministerium für wirtschaftliche Zusammenarbeit und Entwicklung (Federal Ministry for Economic Cooperation and Development)
BNN	Bundesverband Naturkost Naturwaren
BRC	British Retail Consortium
BSE	Bovine spongiform encephalopathy
CAC	Codex Alimentarius Commission
CAC/GL	CAC/Guideline
CAC/RCP	CAC/Recommended international code of practice
CB	Certification body
CBL	Centraal Bureau Levensmiddelenhandel (Dutch organisation for supermarkets)
CC	Civil code
CCP	Critical control point
CEC	Commission of the European Communities
CEN	Comité Européen de Normalisation (the European committee for standardisation
CIES	Comité International d'Entreprise à Succursales (The Food Business Forum)
CIPRO	Certification integrity programme
CISG	Convention for the International Sale of Goods
COLEACP	Comité de Liaison Europe-Afrique-Caraïbes-Pacifique
CPCC	Control points and compliance criteria
CQP	Critical quality points
CRC	Central rabbinical congress
CSR	Corporate social responsibility
CVUA	Chemischen und Veterinäruntersuchungsämter
CWG	Criteria working group of the RSPO

CZ	Coördinatiegroep Zelfreguleringsoverleg (Dutch Self-Regulation Coordination Group)
D. Leg.vo	Decreto legislativo (Legislative decree)
DDB	Dutch Dairymen Board
Defra	UK Department for Environment, Food and Rural Affairs
DfID	UK Department for International Development
DG	Directorate General of the European Commission's civil service
DNV	Det Norske Veritas
EC	European Commission (or European Community)
EDI	Electronic data interchange
EESC	European Economic and Social Committee
EFF	European Franchise Federation
EFIP	European Feed Ingredients Platform
EFLA	European Food Law Association
EFSA	European Food Safety Authority
EFTA	European Fair Trade Association
EJC	Court of Justice of the European Union
EMB	European Milk Board
EN	European norm
ETI	Ethical Trading Initiative
EU	European Union
EUREP	Euro-Retailer Produce Working Group
EurepGAP	European Retailers Protocol for Good Agricultural Practice
FAO	UN Food and Agriculture Organization
FCD	Fédération des entreprises du Commerce et de la Distribution
FCO	German Federal Competition Office
FDA	Food and Drug Administration of the United States of America
FDD	Franchise disclosure document
FEFAC	European Feed Manufacturers' Federation
FINE	Fairtrade Labelling Organizations International, International Fair Trade Association (now WFTO), Network of European Worldshops and the European Fair Trade Association
FLO	Fairtrade Labelling Organizations International
FMI	Food Marketing Institute
FOS	Friends of the Sea
FPEAK	Fresh Produce Exporters Association of Kenya
FSC	Forest Stewardship Council
FSEP	Food safety enhancement programme (Canada)
FSMS	Food safety management system
FSSC	Foundation for Food Safety Certification
FTF	Fair Trade Federation
FVO	Food and Veterinary Office
GAA	Global Aquaculture Alliace
GAP	Good agricultural practices

GAPKI	Gabungan pengusaha kelapa sawit indonesia (Association of Indonesian palm oil producers)
GATT	
GfRS	Gesellschaft für Ressourcenschutz
GFSI	Global Food Safety Initiative
GlobalGAP	Global Good Agricultural Practice
GMO	Genetically modified organism
GMP	Good manufacturing practice
GMP+	Good manufacturing practice + HACCP principles
GTZ	Gesellschaft für Technische Zusammenarbeit
HACCP	Hazard analysis and critical control points
HDE	Hauptverband des Deutschen Einzelhandels
HEII	Horticultural Export Investment Initiative
HIC	Halal International Control
HMSA	Humane Methods of Slaughter Act (USA)
IAF	International Accreditation Forum
IAP	Istituto dell'Autodisciplina Pubblicitaria (Institute of self regulation in Marketing Communication)
ICS	Internal control system
ICTSD	International Centre for Trade and Sustainable Development
IFA	Integrated Farm Assurance
IFANCA	Islamic Food and Nutrition Council of America
IFAT	International Fair Trade Association
IFIS	IFSA feed ingredient standard
IFOAM	International Federation of Organic Agriculture Movements
IFS	International Featured Standard
IFSA	International Feed Safety Alliance
IHIA	International Halal Integrity Alliance
IIED	International Institute for Environment and Development
IKB	Integrale ketenbeheersing (Integral chain control)
ILS	International logistic standard
IPOC	Indonesian Palm Oil Commission
IPPC	International Plant Protection Convention
IRMA	International Raw Material Assurance
ISC	Integrity Surveillance Committee
ISEAL	International Social and Environment Accreditations and Labelling
ISO	International Organization for Standardization
ISO/IEC	ISO/International Electrotechnical Commission
ISPO	Indonesian Sustainable Palm Oil
JAKIM	Jabatan Kemajuan Islam Malaysia (Department of Islamic Development Malaysia)
KKM	Keten Kwaliteit Melk (Chain Quality Milk)
KSA	Kosher supervision agency
L.	Legge (Law)

LCB	Licensed certification body
LOD	Limit of detection
LTO	Land- en Tuinbouw Organisatie (Dutch Organisation for Agriculture and Horticulture)
MEDA	Mesures d'accompagnement (Mediterranean Region Partnership Relations (Barcelona Agreement)
MEL-Japan	Marine Ecolabel Japan
MLA	Multilateral agreement or multilateral recognition arrangement
MP	Member of parliament
MPOA	Malaysian Palm Oil Association
MRL	Maximum residue level
MS	Member State(s)
MSC	Marine Stewardship Council
MUI	Majelis Ulama Indonesia (Indonesia Council of Ulama)
MVO	Dutch product board for margarines, fats and oils
NACMCF	National Advisory Committee on Microbiological Criteria for Foods
NEN	Netherlands normalisation institute or Nederlandse norm (Dutch standard)
NEWS!	Network of European Worldshops
NFV	Nederlandse Franchise Vereniging (Dutch franchise association)
NGO	Non governmental organisation
NMa	Nederlandse Mededingingsautoriteit (Dutch competition authority)
NMV	Nederlandse Melkveehouders Vakbond (Dutch dairy farmers union)
NNI	Stichting Nederlands Normalisatie Instituut (Dutch normalisation institute)
NRI	Natural Resources Institute
OECD	Organization for Economic Cooperation and Development
OIE	Office International des Epizooties (International Organisation for Animal Health)
OK	Organized Kashrus
OOS	Out of stock
OU	Kashruth Division of the Union of Orthodox Jewish Congregations of America
PDO	Products of Designated Origin
PDV	Productschap Diervoeder (Dutch product board animal feed)
PETA	People for the Ethical Treatment of Animals
PID	Pre-contractual information document
PMO	Produce Marketing Organisation or Primary Marketing Organisation
PPP	Public-private partnership
PVS	Private voluntary standards
QLIF	Quality low input food
QMS	Quality management system
QS	Qualität und Sicherheit für Lebensmittel vom Erzeuger bis zum Verbraucher

RA	Rainforest Alliance
RQCS	Regional quality control system
RRP	Recommended resale price
RSPO	Roundtable on Sustainable Palm Oil
RTRS	Round Table on Responsible Soy
SAI	Social Accountability International
SCM	Supply chain management
SGF	Schutzgemeinschaft der Fruchtsaftindustrie/Sure Global Fair
SGS	Société Générale de Surveillance
SPS	(Agreement on the Application of) Sanitary and Phytosanitary Measures (WTO)
SQF	Safe Quality Food
SQFI	SQF Institute
TBT	(Agreement on) Technical Barriers to Trade (WTO)
TFEU	Treaty on the Functioning of the European Union
TPC	Third party certification
TQS	Thai Quality Shrimp
TSR	Tripartite standards regime
UK	United Kingdom
UN	United Nations
UNCTAD	United Nations Conference on Trade and Development
US	United States (of America)
USA	United States of America
USAID	United States Agency for International Development
USDA	United States Department of Agriculture
VWA	Voedsel en Waren Autoriteit (Dutch food and consumer product safety authority)
WFTO	World Fair Trade Organization
WHO	World Health Organization
WIPO	World Intellectual Property Organization
WTO	World Trade Organisation
WWF	World Wide Fund for Nature (previously known as World Wildlife Fund)

1. Private food law

The emergence of a concept

Bernd van der Meulen

1.1 The first book on private food law

When I invited Lawrence Busch to contribute to the European Food Law Association (EFLA) conference on private food law, he remarked something to the effect of 'finally the lawyers are joining in the discussion'. For over a decade important developments take place in the field of food governance based on private law instruments, with hardly any serious participation of lawyers to the debate. Busch himself is a social scientist responsible for ground breaking empirical research on - in his vocabulary - grades and standards.[1]

The current book may well be the first book discussing the length and breadth of private food law for a law audience. It certainly is the first to present 'private food law' as an area of law in its own right. The expression 'law audience' here is to be taken in a broad sense. The world of food law, from the food business perspective also known as regulatory affairs, covers a wider audience than lawyers per se. Many come to the field from a food science background and pick up the legal part on the job.

While this book addresses the world of food law, its content is not strictly limited to legal scholarship but also includes empirical research and social science analyses of interest to the food law audience.[2]

This book is not a conference book in the customary meaning of the word. Its creation has been inspired by the presentations and discussions on 16 and 17 September 2010 at the 18th biannual scientific congress of the European Food Law Association (EFLA) held in Amsterdam.[3] The topic of this congress was 'Private food law; Non-regulatory dimensions of food law'. Some of the contributions to this

[1] See for example: Busch, L., 1997. Grades and standards in the social construction of safe food. Invited paper presented at a conference on The Social Construction of Safe Food and the Norwegian Technical University in Trondheim, Norway, April 1997; Reardon, T., Codron, J.-M., Busch, L., Bingen, J. and Harris, C., 1999. Global change in agrifood grades and standards: agribusiness strategic responses in developing countries. International Food and Agribusiness Management Review 2: 421-435; and most recently, Busch, L., 2010. Standards, law, and governance. Journal of Rural Social Sciences 25(3): 56-78. Generally see http://cs3.msu.edu/people/pubs/busch-lawrence/. See for a special issue of the Journal of Rural Social Sciences focused entirely on the work he and his team have done on standards: http://www.ag.auburn.edu/auxiliary/srsa/pages/TOCs/JRSS%20vol25-3.htm.
[2] See in particular Chapter 2 by Busch Lawrence, Chapter 7 by Otto Hospes, Chapter 8 by Margret Will, Chapter 12 by Tetty Havinga and Chapter 16 by Maria Litjens, Harry Bremmers and Bernd van der Meulen.
[3] On the congress programme and on EFLA in general see www.efla-aeda.org.

book are written by speakers at the congress elaborating on their contribution.[4] We have however completed the picture by inviting other scholars to update and rework previous publications[5] or write new ones[6] in order to cover as much as is possible at this moment in time of the emerging field of private food law. Finally we added the two most important policy documents from the European Commission as Appendices.

1.2 Private food law

1.2.1 Private

The label 'private food law' is meant to cover all applications to the food sector of rules and instruments generally labelled 'private' or 'civil'. These are rules found in civil codes and related legislation or case law. We have chosen the word 'private' to avoid possible confusion due to the double meaning of the word 'civil'. It can be used as synonymous to private to indicate the area of law that is not public law encompassing topics such as property, liability and contract. In comparative law, however, the word 'civil law' is used to indicate the continental European (Romano-Germanic) family of legal systems as opposed the Anglo-Saxon 'common law' family, encompassing both public and private law. Private food law, however, surpasses the borders between civil law and common law systems in the world.

Private food law may include topics such as (product) liability law where the relevant rules are in the civil code or related legislation. Indeed the contribution of Van der Veer goes into the specific legislative arrangements for distance sales as they apply to online purchases of food products. Also competition law has its place in this book.[7] It is public law 'par excellence' but an area mainly concerned with the regulation of private agreements and in this sense it may set a boundary to the expansion of private food law. Mainly, however, this book focuses on the elaborate structures of rules known as, self-regulation, private (voluntary) standards, codes of conduct or certification schemes. These structures have been created by private actors using private law instruments to regulate conduct of food businesses. The regulated businesses may be parties to contractual relations but also businesses further upstream the food chain and geographically remote. In this sense *private food law is food law privately made.*

[4] In particular Chapter 3 by myself, Chapter 2 by Lawrence Busch, Chapter 5 by Spencer Henson and John Humphrey, Chapter 6 by Marinus Huige, Chapter 8 by Margaret Will, Chapter 9 by Ferdinando Albisinni, Chapter 10 by Alessandro Arton, Chapter 13 Hanspeter Schmidt, Chapter 15 by Irene Scholten-Verheijenand Chapter 18 by Nicole Coutrelis.

[5] Chapter 4 by Theo Appelhof, Chapter 12 by Tetty Havinga, Chapter 16 by Maria Litjens, Harry Bremmers and myself and Chapter 17 by Fabian Stancke.

[6] Chapter 7 by Otto Hospes, Chapter 11 by Esther Brons-Stikkelbroeck and Chapter 14 by Lomme Van der Veer.

[7] See Chapter 16 by Maria Litjens, Harry Bremmers, Chapter 17 by Fabian Stancke and myself and Chapter 18 by Nicole Coutrelis as well as Appendix 1.

1.2.2 Law

At the EFLA congress' discussions the existence of such a thing as 'private food law' was contested. It was argued that businesses creating private standards never set out to create 'law' but merely to ensure compliance with regulatory food safety requirements. Partly this discussion relates to another language issue. In English language the word 'law' may refer to a specific piece of legislation. Obviously there does not exist any text deserving the label 'the private food law' in a similar way as in public law EU Regulation 178/2002 does deserve the name 'the general food law'. However, the word 'law' may also be used to refer to an entire rules based system or to an area of scholarly attention among lawyers.[8]

I believe that, indeed, in the food sector rules based systems emerge governing national and international food chains that are made by chain partners on the basis of private law instruments. This book may contribute to proving this point. And yes, I believe this to be an area worthy of scholarly attention. The research attention and the labels used by scholars to describe their findings not necessarily coincide with intentions and wordings chosen by the stakeholders subject to scrutiny.

Probably furthest in the direction of identifying private food law as a system of law, goes the chapter of Lawrence Busch, labelling private schemes in the form of tripartite standfards regimes 'quasi states'. After all is not a state (even quasi) very much more than 'just' a system of law? Not as explicit but also quite strong are Ferdinando Albisinni's labels: 'private regulatory law' and 'collective food law'.

The label 'private food law' is ambitious in that it designates private sector regulation of the food chain as an entire area in its own right, not just a detail within some other context.[9] It is modest in that it does not take a qualifying position. An expression such as 'self-regulation' creates an image of equality, mutuality and an inward direction in the sense that stakeholders organise their own behaviour. This label would be less appropriate for arguments that private food law in fact is used by dominant players in the chain to impose de facto duties upon dependent players. Such argument is indeed made by several authors in this book. For the same reason a label such as 'private *voluntary* standards' – while used by some of the authors in this book – may from the point of view of others be considered misleading in that the voluntariness may – as is argued be some authors in this book – be limited to legal theory but not apply to economic reality. The label 'private food law' is neutral in all these respects.

[8] In other languages the two meanings have distinct words, for example in Latin '*lex*' and '*ius*' respectively.
[9] In this sense the current book distinguishes itself from books discussing topics such as certification. While certification without a doubt is a core issue in private food, it is not a label claiming to cover a branch of law.

1.3 Cover

As law is often symbolised by Justitia[10] holding her paraphernalia: sword, blindfold and scales (or just the scales), private law is often represented by a hand shake (Figure 1.1).

The hand shake symbolises that private relations are self-made on the basis of equality and mutuality. For the cover of this book we have chosen a picture - gracefully provided by the European Union – that combines the two notions: law represented by Justitia, but self-made by private actors represented by the hand shake.

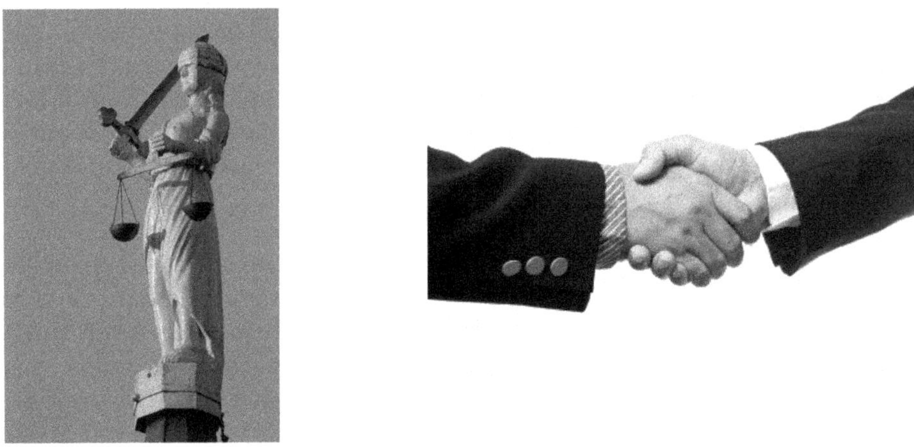

Figure 1.1. Symbols of Justitia (left) and Private law (right). Images found at: http://www.floridaadr.com/ and FreeDigitalPhotos.net (Nutdanai Apikhomboonwaroot). Used with the website's owners consent.

1.4 Food law

Private food law positions itself in different ways under the wider umbrella of 'food law'. In the following subsections I will discuss its place with regard to agricultural law and the different other parts of food law.

[10] Lady Justice; the Roman goddess of justice.

1.4.1 Agri-food law

The scope of the private schemes discussed in this book includes the food chain from farm to fork; from feed ingredients to retail. In particular schemes such as GlobalGAP and SQF 1,000 address primary production. Sometimes a distinction is made between 'food law' on the one hand and 'agricultural law' on the other. This distinction is not a watershed however. Food and agricultural law overlap in that the food chain (or agri-food chain) fully includes primary production of food. Thus this primary production of food is fully within the scope of food law. Agricultural law covers the legal aspects of everything related to the primary sector including the production of feed and food at the farm.

Many of the questions addressed in this book are just as relevant for the primary sector as they are for (industrial) production and retail. Our choice of the label 'private food law' by no means implies the claim that our topic is *exclusively* part of food law. It is just as well agricultural law and a case within regulatory law more in general.

1.4.2 Another level of food law?

The concept 'food law' is multi-level. It encompasses different topics and different levels. Among the topics corner stones are food safety, food security and food trade.[11] These topics are addressed by national law, regional law such as EU law and by international law. Arguably private food law is another level in addition to these three. Figure 1.2 attempts a graphic representation. It positions international

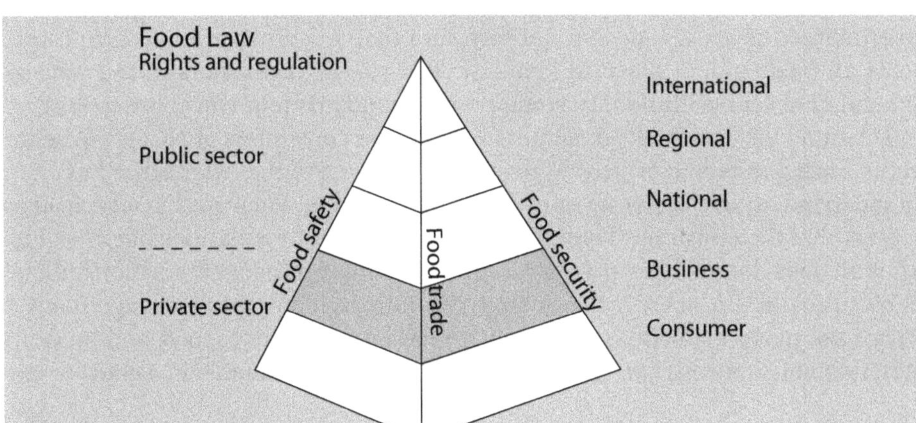

Figure 1.2. The pyramid of food law.

[11] I elaborated this opinion in Van der Meulen, B.M.J., 2004. The right to adequate food. Food law between the market and human rights. Elsevier, The Hague, the Netherlands.

food law at the top of a pyramid. International food law is the law on food as developed by the UN, the WTO, FAO, WHO, the Codex Alimentarius Commission and other international organisations. The WTO places emphasis on trade, the Codex Alimentarius on safety and FAO on food security. The three issues come together – therefore the pyramid shape – in the human right to adequate food, where the concept of adequacy combines availability with protection from harm, thus security via trade and otherwise with safety.[12]

International food law provides models and requirements for the other levels such as the regional level.[13] In the EU this regional level places requirements on the national level that are strict to the point of being mandatory for the national legislators. In this situation the distinction between regional and national is of limited significance. Should we for example consider food law of the United States – a federal state instead of a union of sovereign states – more akin to EU or to member state food law? Should we consider it regional or national? The national level in turn incorporates decentralised levels.

The figure goes on to place the private sector below these public law layers. This positioning only to a limited extent is hierarchical or geographically more focussed. For the public law layers one could argue in favour of the pyramid's tip to point downwards instead of upwards to indicate increased geographic scope. Private food law does not by its nature have a geographic orientation. In practice it can be very international and very independent but often it has to comply with mandatory requirements from several of the public law layers.

Figure 1.2 tentatively distinguishes the private sector in a business level and a consumer level. The topic of this book resides in the business level, the fourth layer marked in grey. I am not entirely sure that a consumer level (fifth layer) holds an independent place in food law. The role of consumers in the process of regulation seems limited. In some private standards retailers claim to express and channel the desires of consumers. Sometimes consumers or NGOs are heard in the setting of public or private regulations but not to the extent that I can see an independent consumer layer of food law emerging. Figure 1.2 can be read to suggest that the consumer layer, may not be a layer of regulations but of rights. After all does not food law revolve around consumer protection and consumers' rights? I doubt if it does. If I look from the European perspective, regrettably I do not see many countries committing to the human right to food to the extent that individual citizens can hold their governments accountable in a court of law

[12] On the human right to adequate food see: Hospes, O. and Hadiprayitno, I. (eds.), 2010. Governing food security. Law, politics and the right to food. Wageningen Academic Publishers, Wageningen, the Netherlands.

[13] On international food law see Van der Meulen, B.M.J., 2010. The global arena of food law: emerging contours of a meta-framework. Erasmus Law Review 3: 217-240. Available at: http://www.erasmuslawreview.nl/files/the_global_arena_of_food_law.

for living up to their obligations.[14] Also EU food law imposes many obligations upon businesses but hardly grants individual consumers any corresponding rights towards these businesses or participation rights towards the institutions.[15] Food law may *protect* consumers, it does not seem to *empower* them in any serious legal way.[16] At best it gives content to rights consumers can uphold in a court of law which rights they derive from other sources such as product liability law. A food unsafe by legal definition is very likely to qualify as 'defective' in the sense of product liability law.

So this figure of food law is tentative in several ways. I am convinced that food trade, food safety and food security are rightfully positioned as ribs marking the

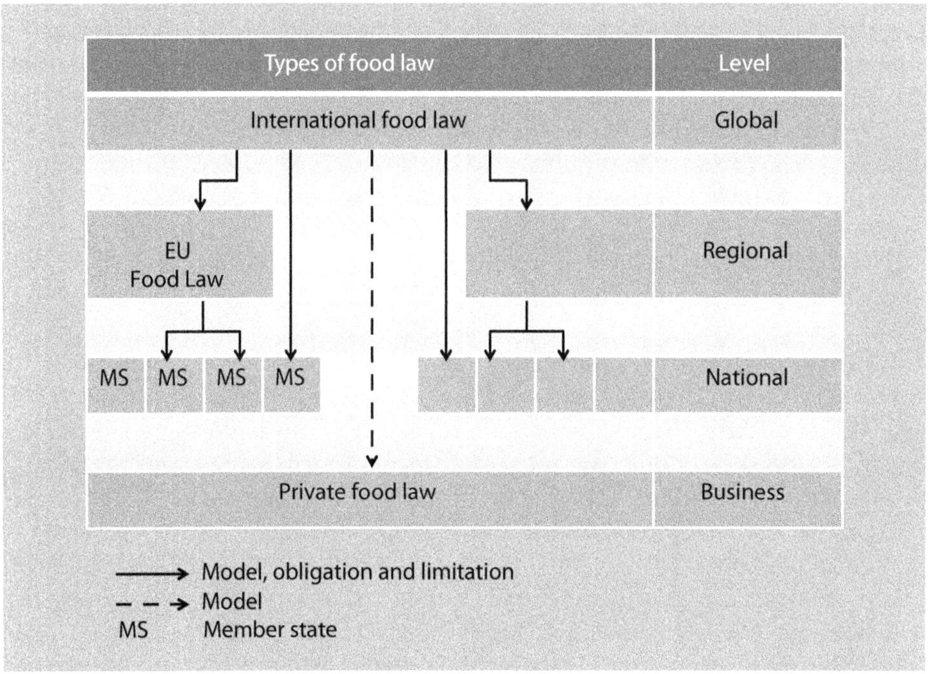

Figure 1.3. Multi-layered food law.

[14] On this issue see for example, Van der Meulen, B. and Hospes, O. (eds.), 2009. Fed up with the right to food. The Netherlands' policies and practices regarding the human right to adequate food. Wageningen Academic Publishers, Wageningen, the Netherlands.

[15] Article 9 of the General Food Law, Regulation (EC) 178/2002 only requires the public to be consulted. EU, 2002. Regulation (EC) No 178/2002 of the European Parliament and of the Council of 28 January 2002 laying down the general principles and requirements of food law, establishing the European Food Safety Authority and laying down procedures in matters of food safety. Official Journal of the European Union L 31, 1/2/2002: 1-24.

[16] It may empower in an economic way in that it endeavours to ensure availability of information that makes informed choice possible.

cornerstones. But, should it be a pyramid? If so should it taper above or below? Should there be four or five layers? Whatever the answers to these questions, I believe that this book contributes to proving that private food law provides a layer in its own right of a status comparable to the layers of public food law.

Maybe Figure 1.3 is less ambitious in depicting the relations and for this reason more acceptable. It shows international food law as inspiring both public and private food law and additionally making requirements on public food law (at regional and national level). In this book we see that private regulation adopts models from the Codex Alimentarius (mainly the HACCP concept) and thus is inspired by international food law. See Chapter 5 by Spencer Henson and John Humphrey. It seems to escape, however, the limits the WTO places on measures that may provide barriers to trade. See Chapter 6 by Marinus Huige. The influence private food law experiences from public (national and regional) law is much stronger. Just as the influence public food law experiences from international food law is stronger. The geographic scope of private food law may, however, be wider than the scope of national or regional food law. In this way it may connect different national systems of public food law and even export requirements from one system to businesses working within the jurisdiction of another.

So if we claim a certain level of autonomy for private food law, this in no way implies it to be disconnected from the other areas of food law.

1.4.3 International and national law

The effects of the global and regional level of food law may differ for countries depending on their membership to international organisations and ratification of international treaties. Most chapters in this book – but not all – take an EU perspective both on international and regional food law. For the global level this means membership of WTO and Codex Alimentarius, for the regional level this means applicability of EU food law. The national level differs among the contributions. We find connections to the Netherlands in Chapter 11 by Esther Brons-Stikkelbroeck, Chapter 12 by Tetty Havinga, Chapter 14 by Lomme van der Veer, Chapter 15 by Irene Scholten-Verheijen, Chapter 16 by Maria Litjens, Harry Bremmers and myself, and to a lesser extent Chapter 4 by Theo Appelhof and Chapter 7 by Otto Hospes. We find connections to Germany in the chapters by Hanspeter Schmidt (Chapter 13) and Fabian Stancke (Chapter 17), to Italy in the chapters by Ferdinando Albisinni (Chapter 9) and Alessandro Artom (Chapter 10), to the United States in Chapter 12 by Tetty Havinga, to Kenya, Ghana, Thailand and Macedonia in Chapter 8 by Margret Will and to Indonesia in Chapter 7 by Otto Hospes.

1.5 Classifications in private food law

Private food law revolves around private standards holding requirements with which businesses must comply to achieve directly or indirectly certain product characteristics as defined in the standard. These characteristics may relate to safety, sustainability, conformity to religious demands and many other aspects. These different aspects can be grouped under the heading 'quality' when quality is understood to mean any form of conformity to customers' desires. Criteria directly defining the desired product characteristics can be called 'outcome standards', criteria indicating the way to achieve them can be called 'process standards' (Henson and Humphrey, Chapter 5).

The standards are embedded in structures (called 'schemes' in some of the contributions) that ensure their development and fulfilment such as audits and third party certification. Maria Litjens in Chapter 16 proposes to label the entirety of standard plus scheme 'system'.

Different classifications of standards, schemes and systems have been proposed for different purposes. Some use as criterion 'who makes' the standard, other depart from the tool used to communicate compliance or the addressee of this communication, again others classify on the basis of the addressee of the requirements i.e. the regulated parties.

Spencer Henson and John Humphrey on the basis by whom standards are set distinguish individual firm standards and collective standards. Collective standards they further distinguish in national and international standards. This latter distinction is based on the location of the parties setting the standards. Within schemes they identify different functions: standard setting, adoption (which is requiring other businesses to implement the standard), implementation (fulfilling the requirements of the standard setter), conformity assessment (verify implementation) and enforcement in response to non-compliance.

Other authors place more emphasis on certification as basis of structures assessing and communicating compliance with standards. For example, the European Commission in its Best practice guidelines for voluntary certification schemes for agricultural products and foodstuffs (see Appendix 1 of this book) proposes a classification based on type of attestation, audience, object and content of requirements measured against the baseline of public law requirements (Table 1.1).

Maria Litjens, Harry Bremmers and myself (Chapter 16) on the basis of the requirements distinguishe vertical and horizontal standards. All these classifications have their limitations but may help to group and compare. It seems highly likely that with further research into private food law, nuance and precision will be added to the classifications used.

Table 1.1. Classification of voluntary certification systems for agricultural products and foodstuffs proposed by the European Commission

Type of attestation:	Self-declaration	Certification (third-party attestation)	
Audience:	B2C	B2C	B2B
Objects of specified requirements:	Products and processes	Mostly products (including services) and processes	Mostly management systems
Content of requirements:	Mostly above baseline	Mostly above baseline	Baseline and above baseline

1.6 Topics covered in this book

1.6.1 Introduction

This book provides a rich harvest of aspects of private food law. It enquires into the origin of private food law as it stands today, but also into the crafting of some individual schemes. Some are made by businesses exclusively. Others are the result of a meeting of stakeholders of varying background and interest. The book provides an overview of a wide variety of systems. Thus giving the flavour of the content of private food law. It attempts a legal theory as instrument for analysis. Then it continues to single out specific topics of significance. There is the relation to international food law. Several private standards draw inspiration and content from the Codex Alimentarius. On the other hand, private food law may enter into competition with the Codex as a means for global harmonisation of food law. Some fear that private food law may set trade barriers undoing some of the achievements of the WTO. This is partly a matter of WTO law and partly a matter regarding the makeup of private systems and measures taken for capacity building.

Many private systems place emphasis on product characteristics and on production processes. A classic is the organic standard. While organics are increasingly embedded in public law, new sustainability initiatives emerge. Another challenging area, is the area of religious standards. From a legal perspective they are public in some countries and private in others. From a religious perspective both qualifications may be inadequate as they make their legal significance depend on human agency. A majority of standards, however, addresses more mundane aspects of food quality in general and safety in particular. If this is the 'inside' of private food law, the 'outside' consists of the external relations of private systems to other private systems.

Apart from product and process, some systems address communication in labelling and advertisement. Public authorities seem reluctant to enforce the ban on misleading the consumers in situations where no more specific rules exist. Private initiatives can give meaning to what is good and proper in the presentation of products to consumers.

A topic of private regulation that does not have a parallel in public law is the format of businesses. Franchising provides requirements on how businesses are conducted and how they present themselves to the public.

The relation between private food law on the one hand and public (food) law and legislation on the other hand is complicated and divers. Private food law finds its legal bedrock in the national civil codes that provide contract law, sometimes adapted to new challenges such as the use of the Internet, (intellectual) property law, liability law and also in the national court systems to uphold private arrangements. Many private standards incorporate public food law requirements and some public rules require compliance with certain private standards. Public controls may take the performance of private systems into account. Can public authorities refer to private standards to express their desires in public procurement?

It is doubtful whether WTO law sets any limits to the expansion of private food law. The area of public law most likely to provide yardsticks for the legality and legitimacy of private systems seems to be competition law. Competition law makes strict requirements on agreements between businesses and unilateral conduct of dominant businesses that may restrict competition to the detriment of consumers. Private systems almost by definition come within the scope of agreements or unilateral conduct. Thus, from a competition law point of view the businesses concerned have to be aware of their systems' impact on competition and on consumers' interests.

In the 1980s, the EU chose a 'new approach'. European law would restrict itself to formulating the basic safety requirements, which would then be fleshed out by private standard setting organisations. Compliance with the private technical standard would then be considered to imply compliance with the European safety norm as well. In such a situation businesses complying with the technical standard are entitled to use CE-marking.[17] The food sector has explicitly been excluded from this new approach. In the final chapter of this book, Nicole Coutrelis argues against bringing the food sector within its ambit. It is my believe, however, that private food law *de facto* already has achieved what the new approach set out to do,

[17] See EU, 2002. Directive 2001/95/EC of the European Parliament and of the Council of 3 December 2001 on general product safety. Official Journal of the European Union L 11, 15/01/2002: 4-17.

namely to decide by private standard how to comply with the most fundamental norm in food law: the ban on unsafe food.[18]

The chapters in this book have been placed in a sequence that provides the reader with a storyline when reading them from the beginning to the end. However, they have been written in such a way that the reader can follow and understand the argument of each individual chapter without having to have read the previous chapters. In this sense the chapters each can stand alone like articles in a journal. The price we pay for this choice is that a little overlap and repetition between some of the chapters could not be avoided. The organisation of the chapters and appendices is represented in Table 1.2.

Table 1.2. The organisation of the chapters and appendices of this book.

Chapter / Topic	1	2	3	4	5	6	7	8	9	10	11	12	13	14	15	16	17	18	A1	A2
Background	x	x	x																x	
History	x	x	x																	x
Theory	x	x	x		x														x	
Concepts			x		x														x	
Overview			x	x																
International law					x	x														
Effect in 3rd world						x	x	x												
Individual schemes			x	x			x	x	x	x	x	x	x	x						
Standard setting							x													
Certification								x												
Advertisement									x	x										
Business format											x									
Religion												x								
Sustainability							x						x							x
Environment							x													
Organic													x							
Fair trade							x													x
Legislation														x			x	x		
Public sector													x		x		x	x		
Competition law																x	x	x	x	
EU policy																		x	x	x

[18] In the EU, Article 14(1) of EU, 2002. Regulation (EC) No 178/2002 of the European Parliament and of the Council of 28 January 2002 laying down the general principles and requirements of food law, establishing the European Food Safety Authority and laying down procedures in matters of food safety. Official Journal of the European Union L 31, 1/2/2002: 1-24.

1.6.2 The origin of private food law

After this introductory chapter, the book continues with Chapter 2 'Quasi-states? The unexpected rise of private food law' by Lawrence Busch. In his opinion much of the global economy has been transformed as trade in food and agricultural products has burgeoned, many supermarket chains have begun to operate across national boundaries, and Supply Chain Management has eclipsed business to business exchanges on spot markets. One important aspect of this transformation has been the rise of what he calls the 'Tripartite Standards Regime', a largely private regime of standards, certifications, and accreditations that parallels formal legal regimes and is dependent on it (e.g. with respect to contract, intellectual property, and criminal law). The neoliberal project of limiting the role of the state has led to the unexpected rise a wide range of 'quasi-states' consisting of individual firms, industry groups, and private voluntary organisations, each pursuing their own aims and interests through the production of private codes, laws, rules, and regulations. Whether some or all of these quasi-states are able to achieve legitimacy and develop democratic modes of governance remains to be seen.

1.6.3 The anatomy of private food law

In Chapter 3 'The anatomy of private food law' [19] I attempt a legal theory of private food law by identifying the legal instruments used and by unravelling the structure of the creation, the binding character, control and enforcement of private food law. In this way this chapter maps the legal structure of private food law. Using the basic instruments from civil law such as property, intellectual property, corporate law, labour law but mainly general contract law food businesses have set up systems of private regulation of the food chain. These systems include standard setting, auditing, accreditation, enforcement and sometimes also conflict resolution.

The chapter goes on to present summary descriptions of the currently most important systems to provide an impression of the topics most prominently regulated today in private food safety law.

1.6.4 Inventory of private food law

In Chapter 4 Theo Appelhof and Ronald van den Heuvel present a more encompassing overview of standards and schemes in the food and feed sectors. Where several of the other authors consider private standards to be at least de facto binding, this chapter sees them as voluntary in more than name only.

---

[19] This chapter elaborates on the chapter 'Private food law' in Van der Meulen, B. and Van der Velde, M., 2009. European Food Law Handbook, 2^e edition. Wageningen Academic Publishers, Wageningen, the Netherlands.

The chapter starts from the EU imposition of self-regulation in the HACCP system and the possibility to make hygiene codes as a collective alternative. These codes can be recognised by national or EU authorities as proper implementation of the hygiene obligation of the businesses. The authors continue to set out the Dutch policy of supervision of controls. This is public authorities taking account of trustworthy private systems in prioritising official controls.

They then visit one by one the relevant sub-sectors of the food sector such as animal feed, primary production, manufacturing, packaging and transport to identify the most important standards used and to provide a summary of their content. It turns out that a forest of schemes has grown almost overnight.

1.6.5 The Codex Alimentarius in private food law

Chapter 5 'Codex Alimentarius and private standards' by Spencer Henson and John Humphrey further elaborates on a legal theory by providing categorisations of private and mixed standards and by distinguishing roles of stakeholders setting and using such standards. It then goes on to discuss how the Codex Alimentarius Commission addresses the issue of private standards and how private standard setting bodies are clients of the Codex Alimentarius in that they turn legally non-binding Codex standards and codes of practice into contractually binding requirements by including them in their standards.

In Henson's and Humphrey's view private standards have become a much more prevalent part of the governance of global agri-food value chains in the last 10 to 15 years. Private firms and standards-setting coalitions, including companies and NGOs, have created and adopted standards for food safety, as well as for food quality and environmental and social aspects of agri-food production. This has raised profound questions about the role of public and private institutions in establishing and enforcing food safety norms. Such discussions have, however, been hampered by a failure to recognise the diversity of agri-food private standards with respect to their institutional and administrative characteristics, scope or functions, and also the often tight inter-relationships between private standards and public regulation. Private food safety standards are predominantly directed at the management of risk of food safety failures, many of which are defined by regulatory requirements. While there are instances where private standards lay down requirements that are beyond regulations, in many cases their function is to establish systems for more reliable and cost-effective regulatory compliance. Although there are concerns that private food safety standards are undermining the standards, guidelines and recommendations promulgated by the Codex Alimentarius Commission, private standards can be seen as substantively packaging multiple Codex norms and national legislation. At the same time, private standards fill 'voids' where international standards are missing. Certainly the rise of private standards presents challenges for Codex, including the need to reflect on its client base, review its procedures and to examine where international norms are needed in a world where private governance is taking on an increasing role.

1.6.6 International versus private food law

In Chapter 6 'Private retail standards and the law of the World Trade Organisation' Marinus Huige sets out the discussion within the WTO SPS Committee on the question whether private standards constitute barriers to international trade and whether WTO members have a responsibility in this regard.

The argument against private standards in general and EurepGAP/GlobalGAP in particular has been based on Article 13 of the WTO Agreement on the Application of Sanitary and Phytosanitary Measures (SPS). Huige does not believe that the argument holds but calls for improved communication between the players in international food law and the players in global private food law.

1.6.7 The making of private food law

Public law is made on the basis of well-defined procedures ensuring democratic input and legitimacy. Private food law can be made whichever way the standard setting organisations have agreed upon. In Chapter 7 'Private law making at the round table on sustainable palm oil' Otto Hospes presents a case study on the setting of private standards for the sustainable production of palm oil for food and other uses.

The 1990s and even more so the 2000s marked the rise of different, world-wide initiatives of non-governmental organisations and multinational companies to develop private law for the sustainable production of global commodities. One of these global initiatives has been the Round Table for Sustainable Palm Oil (RSPO). The chapter by Hospes analyses private law making at the RSPO: it describes how the normative contents, actors and instruments of the RSPO have evolved in relation to market power, public law and state authorities. A key question is whether the law making process at the RSPO has contributed to the development of new public standards on sustainable palm oil. The focus is on the different ways in which the governments of Indonesia and the Netherlands have each responded to the establishment of the RSPO principles and criteria.

1.6.8 Capacity building in private food law

Margret Will, in Chapter 8 'GlobalGAP smallholder certification. Challenge and opportunity for smallholder inclusion into global value chains', addresses the core topic in private food law: certification. In particular a form of certification developed within GlobalGAP to overcome financial and technical barriers to certification for small producers in third world countries. This chapter particularly shows private food law as a form of global governance, the fairness of which can – probably – only be ensured from the inside as a global government that could counterbalance the global business world does not exist. The chapter is based on empirical research in African, Asian and Eastern-European countries. It is optimistic about the potential of private food law to provide producers in developing countries access to western markets.

1.6.9 Private food advertisement law

In Chapter 9, 'Towards the self-regulation code on beer advertising in Italy: Steps on the long lasting path of competition/co-operation of public and private food law' Ferdinando Albisinni describes the history of the Italian Self-Regulation Code on Beer Advertising. Both the legislator and the private sector realised the importance of consumer protection in the context of sales of and publicity for alcoholic beverages. The Self-Regulation Code, discussed in more detail in Chapter 10, has found an interesting balance between the different interests involved.

Allesandro Artom, in Chapter 10 'Self-Regulation Code on Beer Advertising' elaborates on the previous chapter, presenting a case study on the 'Self-Regulation Code on Beer Advertising' in Italy. It shows communication from businesses to consumers as topic of a private standard and it shows a structure within the private scheme providing third parties – consumers – with a procedure to address cases of (perceived) non-compliance. The case thus provides a model that can prove valuable for private food law in general. One of the weak spots in private food law seems to be the legal protection of interested parties. It suffers, as Busch puts it in Chapter 2, from inadequate appeals mechanisms. Artom shows that this can be remedied within the private schemes.

1.6.10 Business format regulation

In Chapter 11 'Franchising strengthens the use of private food standards' Esther Brons-Stikkelbroeck adds the topic 'franchising' to the discussion of private food law. This topic is of twofold relevance. Franchising in itself is a form of private food law. A form that on private law basis – franchising contracts – imposes strict requirements upon participating businesses on how to present themselves to the public and how to conduct business. A large proportion of the world's population will recognise a McDonald's restaurant from the outside and know what to expect on the inside. The margin for the owner of an individual outlet to do things differently is very small.

The chapter goes on to discuss how within the franchise framework private standards as understood in the other chapters play a role. It appears that in franchising we encounter two forms of private food law mutually reinforcing each other.

1.6.11 Religious standards

Tetty Havinga in Chapter 12 'On the borderline between state law and religious law: regulatory arrangements connected to kosher and halal foods in the Netherlands and the United States' addresses religious standards. The Netherlands, like other Western countries, is a growing market for halal food products, that is, food products that comply with Islamic food laws. Halal food is becoming more visible as Dutch supermarkets, hospitals and schools decide to include halal food in their supply. Havinga compares the regulation of halal food in the Netherlands to the regulation of kosher food in the Netherlands and the United States. She analyses the division of roles between state actors, the food industry, certification agencies and religious authorities in these regulatory arrangements. Contrary to expectation, the regulatory arrangements are rather state-centred in several US states (liberal market economy), whereas the Dutch corporatist welfare state plays a limited role by allowing religious slaughter and leaving the issue of halal and kosher certification entirely to commercial and religious organisations.

1.6.12 Organics

Organic agriculture is based on the philosophy that farms should function like organisms do with as little external inputs as possible and no use of chemical fertilisers and pesticides.[20] In Chapter 13 'Organic food: a private concept's take-over by government and the continued leading role of the private sector', Hanspeter Schmidt discusses two developments. One is how the initially private initiative of organic agriculture in many countries was included in public legislation. This is what he calls 'a friendly takeover'. This takeover provides status and protection to organic agriculture. Infringement on organic standards has become a matter of public law enforcement. Audits and certification in many countries remain with private organisations. The downside of the friendly takeover is that the legislator now holds the position to change with one stroke of a pen the meaning of the concept of organic.

The second development discussed by Schmidt is that the takeover is not complete. Whenever gaps emerge, private initiative fills them in. He argues this point at the example of chemical contamination. Standards laid down in legislation ban certain chemical fertilisers and pesticides from *use* in organic agriculture. They do not deal with *presence* of their residues in organic food due to external contamination. In Germany private norms have been developed to distinguish residues that do and from those that do not raise a suspicion of unauthorised use. In this way private norms continue to complement public organic food law.

[20] It bears repeating that the label 'organic' in no way implies any suggestion that other foods would be 'inorganic'.

1.6.13 Digital food law

In Chapter 14 'Food online: reconnaissance into a consumer protection no-man's land between food law and the Civil Code' Lomme van der Veer takes us into cyberspace. In the Netherlands sale of food via the internet takes place in a contested area between Civil Code requirements on distance selling and public law requirements on food labelling. Due to the characteristics of food, the consumer usually cannot enjoy much of the additional protection the legislator provides her/him regarding distance contracts. Due to the characteristics of internet sales consumers cannot enjoy many of the rights conferred upon them in food labelling law to make informed choices. Private food law so far does not seem to be closing the gap.

The situation is not as bleak as it would seem at first sight. The disadvantages facing the consumer of food online, awaken the potentials of the general contract law requirements of conformity. If anywhere they apply to their fullest, it is in the specific context of food stores along the digital highway.

1.6.14 The public side of private food law

Chapter 15, 'National public sector and private standards: cases in the Netherlands' by Irene Scholten-Verheijen presents cases on current discussions in the Netherlands on the interplay between private standards and public law. What happens to the private law character of private standards if reference is made to them in legislation? What happens if public authorities refer to private standards in calls for tender under EU procurement law? And, to what extent can enforcement bodies in setting their priorities rely on the performance of private quality management systems? The answer to these questions is contested in Dutch case law and practices. It goes without saying that the relevance of these question reaches far beyond the jurisdiction of the Netherlands. Some courts took the position that reference in legislation to private standards endows them with public law status which cannot be done unless they are publicly available the same way as legislation. This position, however, has not been upheld in appeal. In public procurement of coffee provision to civil servants the Dutch version of Fair Trade certification (Max Havelaar) was required. This was contested by coffee suppliers certified against a different sustainability standard. Dutch courts so far accepted the use of private standards in public procurement, but the European Commission started an infringement procedure against the Netherlands. On this topic see also Appendix 2.

1.6.15 The outside of private food law

Chapter 16 'The outside of private food law: the case of braided private regulation in Dutch dairy viewed in the light of competition law' by Maria Litjens, Harry Bremmers and myself presents a case study of the dairy sector in the Netherlands

– stretching upstream to include the chains supplying feed to dairy farms. The study shows that private regulatory systems externally (on their 'outside') relate to each other with such intensity that private food law can be seen as an inter-regulated system of entangled systems. Individual systems consist of private standards fleshed out with private schemes for their operation and enforcement. They usually regulate practices of businesses upstream from the businesses setting the standards and operating the schemes.

The interconnectedness of systems strongly increases their coercive effect on regulated businesses. Initially the Netherlands' competition authority set a limit to this development with a view to its exclusion of non-participating businesses from the market. While, if anything, the web of private systems has only become more tightly knitted, the competition authority has not continued its involvement. The analysis speculates on a likely explanation. Is the ultimate cause a change in competition law, a change in the formal policy or just loss of interest from the side the competition authority?

1.6.16 The limit of private food law

As indicated above, competition law likely presents the most important limit to private food law. In Chapter 17 'The limit of private food law: competition law in the food sector' Fabian Stancke brings this limit to the lime light.

Despite intense competition, the food sector in Germany is currently faced with challenges from both competition authorities and courts. Dawn-raids conducted by the German Federal Competition Office (FCO) took place only recently and three-digit million Euro fines were imposed. Among the companies concerned were sugar producers, fruit importers, sweets producers, coffee roasters, traders in spices, dairy producers and retail chains. The European Commission and the FCO have announced that they intend to further intensify competition law enforcement in the food sector. Both agencies are led by a concern for free competition for the benefit of consumers.

1.6.17 New approach

In the last chapter of the book, Chapter 18 'EU 'new approach' also for food law?', Nicole Coutrelis takes stock of the findings in this book. 'New approach' is a label applied to a policy dating back to the 1980s. According to Coutrelis this new approach can be summarised as follows:
- EU legislation should be limited to the adoption of essential requirements, regarding specifically safety or other requirements of general interest;
- The task of drawing up the technical specifications of products, in conformity with the essential requirements laid down in the legislation, is entrusted to organisations which are competent in the standardisation area;

- These technical specifications are not mandatory;
- However, products conforming to the standards are presumed to conform to the essential requirements.

The food sector has been excluded from this new approach. Coutrelis now askes the question in the light of the findings in this book – that private standards in minute detail elaborate safety requirements – if it has become possible or desirable to now fully apply this approach to the food sector. She answers this question to the negative. Not because private standards in the food sector would not be suitable for the task, but because public food law has meanwhile developed too far from being limited to the adoption of essential requirements, regarding specifically safety or other requirements of general interest leaving the details to the private sector.

Like rails on a railway track public food law has changed its course as much as private food law has and never the twain shall meet.

1.6.18 EU involvement

The book does not end there. Two communications from the European Commission addressing private food law have been added as Appendices. Appendix 1, the 'EU best practice guidelines for voluntary certification schemes for agricultural products and foodstuffs' from 2010 provides definitions and best practices as seen by the Commission.[21] It elaborates on the limits from competition law. Appendix 2, is a communication from 2009 'Contributing to Sustainable Development: The role of Fair Trade and nongovernmental trade-related sustainability assurance schemes'. It is rich in information on private systems of a predominantly ethical nature.

1.7 Law and governance

In social sciences, a concept of governance has been introduced to account for the fact that the management of society no longer is a matter of government alone. The notion of 'governance' shifts attention from 'who' is managing society to 'what' is being done in social management. Legal theory often applies a distinction in public law and private law based on involvement of some form of government. Topics such as self-regulation are approached from the perspective of government. Do they provide alternatives? Should the government leave it to the market, etcetera. This approach may increasingly lose adequacy when – as Busch puts it in Chapter 2 – the distinction between public state agencies and non-governmental organisations becomes blurred.

[21] EU, 2010. Commission communication – EU best practice guidelines for voluntary certification schemes for agricultural products and foodstuffs. Official Journal of the European Union C 341, 16/12/2010: 5-11.

This book may contribute to the realisation that private forms of governance may emerge outside the scope of government initiative or control. Even in areas where there is no or no effective government to manage stakeholders' actions such as in weak states with weak regulatory compliance and enforcement structures and in international trade. International trading businesses are confronted by national governments covering only a part of the operations and therefore sometimes unable to effectively influence what goes on. There is no such as an international government. The international organisations such as the WTO set requirements on the role the national governments play in this context, not on the trading businesses. The authority of national governments largely ends at their borders. The only set of rules that trading partners at opposite sides of the world have in common, are the rules they created for themselves by contract including the private systems they include in their relation. Private law may bring forth more truly global law than public international law. It may introduce some form of rule of law where no government exists.

Legal scholarship may be well advised to consider including a governance concept in its analytical toolbox. For the authors from Wageningen University participating in this book it is a contribution to the development of legal theory in the Netherlands Institute for Law and Governance.[22]

1.8 Last but not least

A book is nothing if it is not read. We, the authors, believe that we have created a product worthy of your – the reader's – attention. We hope you will enjoy it and benefit from it. We invite you to feel part of a global community developing and studying private food law. Your comments are welcome: bernd.vandermeulen@wur.nl.

Acknowledgements

This book would not have come into being without the contributions of the authors and the publisher. At the side of the publisher Marijn van der Gaag and Mike Jacobs in particular put a lot effort into the creation of this book. Some others have also made important but less visible contributions. Jessica van Wijngaarden designed the cover of the book. Geronda Klop, one of the best students I ever had in Comparative Food Law, applied her skills in power point to ensure that several diagrams in this book express what the authors wanted them to express. Robin Kaplan, also a student at Wageningen University, gave me a helping hand with the index. Finally we are grateful to all who contributed by reviewing and commenting on draft chapters.

[22] See http://nilg.nl/en/.

References

Busch, L., 1997. Grades and standards in the social construction of safe food. Invited paper presented at a conference on The Social Construction of Safe Food and the Norwegian Technical University in Trondheim, Norway, April 1997.

Busch, L., 2010. Standards, law, and governance. Journal of Rural Social Sciences 25(3): 56-78.

EU, 2002. Directive 2001/95/EC of the European Parliament and of the Council of 3 December 2001 on general product safety. Official Journal of the European Union L 11, 15/01/2002: 4-17.

EU, 2002. Regulation (EC) No 178/2002 of the European Parliament and of the Council of 28 January 2002 laying down the general principles and requirements of food law, establishing the European Food Safety Authority and laying down procedures in matters of food safety. Official Journal of the European Union L 31, 1/2/2002: 1-24.

EU, 2010. Commission communication – EU best practice guidelines for voluntary certification schemes for agricultural products and foodstuffs. Official Journal of the European Union C 341, 16/12/2010: 5-11.

Hospes, O. and Hadiprayitno, I. (eds.), 2010. Governing food security. Law, politics and the right to food. Wageningen Academic Publishers, Wageningen, the Netherlands.

Reardon, T., Codron, J.-M., Busch, L., Bingen, J. and Harris, C., 1999. Global change in agrifood grades and standards: agribusiness strategic responses in developing countries. International Food and Agribusiness Management Review 2: 421-435.

Van der Meulen, B.M.J., 2004. The right to adequate food. Food law between the market and human rights. Elsevier, The Hague, the Netherlands.

Van der Meulen, B.M.J., 2010. The global arena of food law: emerging contours of a meta-framework. Erasmus Law Review 3: 217-240. Available at: http://www.erasmuslawreview.nl/files/the_global_arena_of_food_law.

Van der Meulen, B. and Hospes, O. (eds.), 2009. Fed up with the right to food. The Netherlands' policies and practices regarding the human right to adequate food. Wageningen Academic Publishers, Wageningen, the Netherlands.

Van der Meulen, B. and Van der Velde, M., 2009. European Food Law Handbook, 2e edition. Wageningen Academic Publishers, Wageningen, the Netherlands.

2. Quasi-states? The unexpected rise of private food law

Lawrence Busch

2.1 Introduction

> Ther ys the veray and true commyn wele; ther ys the most prosperouse and perfayt state, that in any cuntrey, cyte, or towne, by pollycy and wysdom, may be stablyschyd and set.[23]

When Thomas Starkey wrote the above description of the state in 1538, the earliest usage of the state in this context known in English, he clearly saw it as a singular phenomenon. Each country would have a singular, stabilised state that would serve the common weal, i.e. that would provide for common prosperity. The study and implementation of food law has been premised on that assumption. Hence, either one studied/implemented food law in a given nation, or one studied/implemented the transnational arrangements and agreements facilitating trade in food. However, in recent decades, just as numerous pundits have discussed the 'hollowing out of the state',[24] the state has seemingly proliferated.

In this chapter I shall argue that what is variously called the 'hollowing out of the state', the shift from government to governance, or the devolution of the state, has actually resulted in the creation of a range of quasi- or even pseudo-states. This, in turn, has resulted in the proliferation of standards, rules, regulations, good practices, and other means of ordering and organising the production, processing, transport, retailing, and testing of food.

First, I examine several critical events associated with the rise of neoliberalism. Then, I relate the transformation of the global food economy. In the next section I examine the rise of what has been called the 'Tripartite Standards Regime'. Finally, I ask whether the governance of food can, in fact, be plural.

2.2 Building neoliberalism

Under the theory and practice of classical liberalism a space was to be carved out and reserved for the market. Practitioners might debate the size of that space, what should be in it, and how it was to be organised, but its central feature was

[23] Starkey, T., 1878. England in the reign of Henry the Eighth, a dialogue between Cardinal Pole and Thomas Lupset. N, Trubner & Co, London, UK,. for the Early English Text Society, 1878 [1538] as cited in the Oxford English Dictionary.
[24] See e.g. Peters, G. and Pierre, J., 1998. Governance without government? Rethinking public administration. Journal of Public Administration Research and Theory 8(2): 223-243.

that it was to be a *laissez-faire* space, a space to be left alone, where markets could develop without the interference of the state. Thus, for the classical liberals markets were spaces of liberty, places where tyrannical governments could be excluded and business could be conducted in an untrammelled manner. This view was arguably best articulated in the seminal works of Adam Smith.[25]

By the 1930s this vision was clearly under attack. On the one hand, the forces of Soviet communism, Nazism, and fascism controlled a considerable part of Europe. On the other hand, the welfare state had grown considerably in size in the United States as well as in the remaining democratically governed parts of Europe. Concerned about these developments, a French philosopher of mathematics, Louis Rougier, invited a selected group of intellectuals from around the world to the *Colloque Walter Lippmann* in Paris.[26] Over the course of a week in 1938 the participants were to analyse the ills of contemporary societies and to propose the renovation of liberalism, the creation of what Rougier called 'neoliberalism'.

What came to be known as neoliberalism may be said to have had several founding documents. Walter Lippmann's, *The Good Society*, the centrepiece of the *Colloque*, made a case for a more active form of liberalism.[27] In that best-selling work, Lippmann voiced his concerns about totalitarian regimes as well as the New Deal in the USA. In particular, Lippmann was sceptical about the uncritical acceptance of central planning. He argued that, given the complexities of the modern state, such planning would always fall short of its promise. In contrast, the uncountable transactions that took place in markets, as defined in classical liberal philosophy, appeared to be a means by which the complexities of modern life might be better addressed. However, Lippmann was also concerned that the classical liberal principle of *laissez-faire* was simply too negative. It needed to be replaced by a more positive, active promotion of markets to be accomplished by a program of legal reform.

A much shorter work by University of Chicago economist Henry C. Simons, that curiously proposed a seemingly oxymoronic 'positive program for *laissez-faire*' was also important.[28] Simons proposed nothing less than a complete overhaul of the economic system so as to reduce the ever-expanding role of the state. However,

---

[25] Smith, A., 1994 [1776]. An inquiry into the nature and causes of the wealth of nations. Modern Library, New York, NY, USA.
Smith, A., 1982 [1759]. The theory of moral sentiments. Liberty Fund, Indianapolis, IN, USA.

[26] Travaux du Centre International d'Etudes Pour la Rénovation du Libéralisme (ed.), 1939. Le Colloque Walter Lippmann. Cahier no. 1, Librairie de Médicis, Paris, France.

[27] Lippmann, W., 1938. The good society. George Allen and Unwin Ltd, London, UK.

[28] Simons, H.C., 1948 [1934]. A positive program for Laissez Faire: some proposals for a liberal economic policy, In economic policy for a free society. University of Chicago Press, Chicago, IL, USA, pp. 40-77. Simons's perspective deviated from that of many other neoliberals in that he favoured a sharp reduction of corporate power, a progressive tax system, and limits on advertising. For a more detailed discussion, see Foucault, M., 2008. The birth of biopolitics: lectures at the Collège de France, 1978-79. Palgrave Macmillan, New York, NY, USA.

the program of work outlined at the *Colloque,* was not to be carried out as war soon intervened.

It was left to F.A. Hayek and Milton Friedman to spell out the details of the neoliberal program. Hayek's best selling, *The road to serfdom*, became a kind of manifesto for neoliberal policies.[29] In particular, in it Hayek argued that there was a 'slippery slope' from the welfare state to totalitarianism. Planning was to be avoided unless it was aimed at enhancing competition and markets; in those instances it was to be encouraged. Moreover, other forms of state planning were to be restricted by creating a set of international organisations that would specifically limit state power.

After the war ended, Rougier's project was resurrected by F.A. Hayek and others in the Mont Pelerin Society.[30] The Society has since served to promote neoliberal ideals. But it was Hayek's later trilogy and Friedman's influential *Capitalism and freedom* that spelled out more clearly a neoliberal vision of the future.[31] Space here does not permit a detailed review of even these works, let alone all the many other works generated in the last 70 years. But one can distil (admittedly with some distortions) a set of positions widely held among neoliberals and now largely enshrined in national and international law:

1. Markets should be actively promoted, since they allow goods and services to be distributed without recourse to a central authority. Indeed, governments must work to make markets fully competitive, i.e. to make them conform as much as possible to the mathematical model of the market.[32]
2. Governmental regulation of social, political, and economic affairs should itself be limited to support of the market mechanism whenever possible. Hence, public services and enterprises should be privatised as much as possible.
3. Individuals should be encouraged to be entrepreneurs of themselves.
4. Free trade should be promoted along with competitive exchange rates and market determination of prices. Moreover, global institutions that ensure free trade should also serve to limit state sovereignty.

All of this remained rather theoretical and far from practice (with the partial exception of Chile) until the elections of Margaret Thatcher and Ronald Reagan in Great Britain and the United states, respectively. Both leaders were familiar

[29] Hayek, F.A., 2007 [1944]. The road to serfdom. University of Chicago Press, Chicago, IL, USA.
[30] Mont Pelerin Society, http://www.montpelerin.org/. For a history, see: Mirowski, P. and Plehwe, D. (eds.), 2009. The road from Mont Pèlerin: the making of the neoliberal thought collective. Harvard University Press, Cambridge, MA, USA.
[31] Friedman, M., 1962. Capitalism and freedom. University of Chicago Press, Chicago, IL, USA; Hayek, F.A., 1973-1979.. Law, legislation and liberty, 3 vols. University of Chicago Press, Chicago, IL, USA.
[32] For Hayek, markets are forms of self-generating or 'spontaneous order'. Hayek's mentor, Ludwig von Mises, went further arguing that economic science is not empirical, but a priori. Put differently, the functioning of markets is prior to and independent of experience. See: von Mises, L., 1978 [1933]. Epistemological problems of economics. Ludwig von Mises Institute, Auburn, AL, USA.

with and committed to (something akin to) neoliberal values and proceeded to put them into practice. In particular, they railed against what they saw as unnecessary regulation and the excesses of the welfare state.

Furthermore, commensurate with this new approach to governance, over the next several decades the 'big four' international governance institutions came into being or were greatly strengthened. The World Bank, initially created to help Europe rebuild after World War II, was reinvented after the debt crisis as a means of remaking governments in poor countries along neoliberal lines.[33] The International Monetary Fund,[34] working together with the bank, ensured that nations not following the recently invented rules of the 'Washington Consensus' would be forced into default.[35] The World International Property Organization (WIPO) created a global forum where intellectual property rules could be extended and unified. But the major new global actor was the World Trade Organization (WTO). By limiting its mandate to trade, and avoiding thorny questions of environment, labour, and the like, the signatories to the WTO very nearly instantiated Hayek's imaginary of 1944. Put differently, the WTO limited the role of member nations, put trade issues squarely at the centre of international negotiations, and promoted a market-led form of economic development. Indeed, when WIPO's one-nation, one-vote approach to strengthening intellectual property rights proved too slow for some, shifting the venue to the Trade-Related Intellectual Property Rights (TRIPs) agreement at the WTO proved more effective. Similarly, technoscientific issues were subordinated to trade through the Technical Barriers to Trade (TBT) agreement, while sanitary and phytosanitary issues were subordinated through the eponymous SPS agreement.

In sum, the last 30-40 years may be seen as a marked change in the form(s) of global governance. It involved a shift from (welfare) state-centred governance to market-led governance. Often food safety agencies and related government oversight of food and agriculture has been weakened or eliminated, to be replaced – according to its advocates – by private sector incentives of various sorts.[36] Although performed

[33] For a critical review of bank policies, see: Goldman, M., 2005. Imperial nature: The World Bank and struggles for justice in the age of globalization. Yale University Press, New Haven, CT, USA.

[34] Insight into IMF decision making can be found in: Harper, R., 2000. The social organization of the IMF's mission work. In: Strathern, M. (ed.). Audit cultures: anthropological studies in accountability, ethics, and the academy. Routledge, New York, NY, USA, pp. 21-53.

[35] Williamson, J., 2005. The strange history of the Washington consensus. Journal of Post Keynesian Economics 27(2): 195-206.

[36] For a brief review of the US scene, see: Fortin, N.D., 2009. United States Food Law: consumers, controversies, current issues. European Journal of Consumer Law 2009(1): 15-42. A comprehensive review can be found in: Fortin, N.D., 2009. Food regulation: law, science, policy, and practice. John Wiley, Hoboken, NJ, USA.

For the EU, see: Allemanno, A., 2009. Solving the problem of scale: The European approach to import safety and security concerns. In: Coglianese, C., Finkel, A.M. and Zaring, D. (eds.). Import safety: regulatory governance in the global economy. University of Pennsylvania Press, Philadelphia, PA, USA, pp. 171-189.

differently in different nations, the neoliberal approach to governance advocated by Hayek, Friedman, and others was slowly but surely put into practice, eclipsing other forms of governance to varying degrees.

2.3 Transformation of the global economy

In large part as a result of the neoliberal reforms, the last one-third of the twentieth century marked a sea change in the global economy. Trade in food and agricultural products expanded at a rapid pace. Supermarkets began to operate across national boundaries. And, supply chain management (SCM) eclipsed business to business exchanges in spot markets. Let us examine each of these phenomena in turn.

2.3.1 Trade in food and agricultural products

The last 50 years has seen a rapid, indeed astonishing, growth in global trade in food and agricultural products. From 1961 to 2007 the value of global food imports rose from US$ 34,696 million to US$ 903,431 million. Similarly, the value of global food exports rose from US$ 32,118 million to US$ 876,410 million (Figure 2.1).[37]

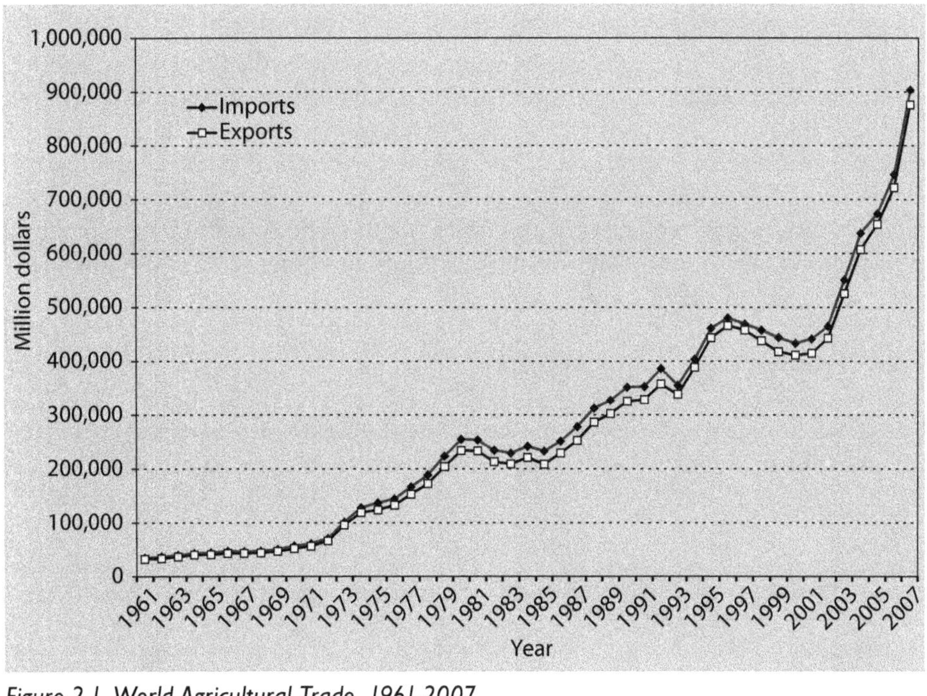

Figure 2.1. World Agricultural Trade, 1961-2007.

[37] Food and Agriculture Organization, 2009. FAO Statistical Yearbook 2009, FAO, Statistics Division, Rome, Italy.

This extraordinary growth can be directly linked to the opening of world markets under the GATT and later the WTO.

2.3.2 The rise of transnational supermarket chains

The legal restructuring of global food and agricultural markets through the GATT and the World Trade Organization, and especially the SPS and TBT agreements, had another important consequence: They opened the door for the larger supermarket chains to penetrate previously closed markets.

Prior to that time, few supermarket chains operated across national boundaries. If they did, their operations in one nation were largely divorced from that in other nations, since tariffs and quotas, as well as different local laws and regulations on food products, made global sourcing and distribution expensive, difficult, or impossible.

However, this all changed with the new trade regime. Global sourcing of most fresh and processed products became possible. Moreover, different chains employed different strategies with respect to store format (e.g. big box, convenience store, small supermarket, etc.). More importantly, the various chains took different strategies with respect to regional penetration. On the one hand, some chains invaded the national territory of other chains, either by direct competition or by purchasing smaller national chains. For example, recently the British supermarket chain, Tesco, entered the US market by opening a group of upscale small format stores on the west coast under the name *Fresh and easy*. On the other hand, some chains ventured into the nations of the former Soviet bloc and/or the middle and even lower income nations of Africa, Asia, and Latin America where supermarket chains were either weak or non-existent.[38] And, at the same time, major firms consolidated their market shares in their home markets, such that a few firms dominate in most industrialised nations.[39]

2.3.3 Supply chain management

The growth in scale and geographical reach of supermarket chains was paralleled by a growth in their market power. While much of the twentieth century was marked by the dominance of large food manufacturers (e.g. Coca-Cola, Kraft,

[38] Dries, L., Reardon, T. and Swinnen, J.F.M., 2004. The rapid rise of supermarkets in central and eastern Europe: implications for the agrifood sector and rural development. Development Policy Review 22(5): 525-556.
Reardon, T., Timmer, C.P., Barrett, C.B. and Berdegue, J., 2003. The rise of supermarkets in Africa, Asia, and Latin America. American Journal of Agricultural Economics 85(5): 1140-1146.
Weatherspoon, D.D. and Reardon, T., 2003. The rise of supermarkets in Africa: implications for agrifood systems and the rural poor. Development Policy Review 21(3): 1-17.

[39] For trends in concentration see: Busch, L. and Bain, C., 2004. New! Improved? The transformation of the global agrifood system. Rural Sociology 69(3): 321-346.

Nestlé), the current century has so far been marked by the growing dominance of food retailers. This is a consequence of several parallel developments. First, the sheer size of some of the larger retailers dwarfs anything found in the food processing sector. Wal-Mart, for example, now has annual sales of over $375 billion, making it the largest single company in the world.[40] This can be compared to the annual sales of Nestlé, currently the world's largest food processor. Its annual sales for 2008 were ca. $100 billion.[41]

Second, supermarkets have been at the forefront of organisational innovation. In particular, they have shifted their buying practices dramatically. As late as the 1960s, supermarkets tended to buy whatever products were available on the wholesale market, bringing them in through the back door, stacking them on shelves, and sending them out the front. Even the invention ofbar codes had little effect on these practices. In addition to seasonal variations for fresh products, there were regular shortages of packaged products, the result of disruptions in shipping or availability of raw materials. This meant that supermarkets had to have large on-site storerooms where additional items for sale were kept 'just in case', or they had to have rather large numbers of 'out of stock' (OOS) items. Both were costly. Warehousing products for sale later required extra space (that might have otherwise been used for sales floors), while OOS items meant empty shelves and lost sales.

The favoured solution to this problem proved to be Supply Chain Management (SCM). Two somewhat different versions of the origins of SCM can be identified.[42] One places its origin in the systems literature of the 1950s; the other places it in the transformation of the Japanese automobile industry, especially at Toyota, in the 1970s.[43] Regardless of its origins, SCM transformed the food industry. SCM was simultaneously a *conceptual* and a *practical* innovation. It involved the organisation of an entire supply chain such that goods would be delivered 'just in time', and that shelves would always be full. This involved paying much greater attention to all the details of logistics (i.e. the flow of goods, resources, and information within and across firms), as well as to the development of commensurate handling practices all the way through the chain. Moreover, it involved vertical coordination of most or all the actors in a supply chain by a 'supply chain captain', a company with sufficient marketing clout to discipline the other companies in the chain and to ensure that goods flowed smoothly through the entire chain.

[40] Wal-Mart, 2008. Annual Report. Available at: http://walmartstores.com/sites/annualreport/2008/message_from_lee_scott.html.

[41] Nestlé, 2008. Management Report 2008. Nestlé, Geneva, Switzerland.

[42] Croom, S., Romano, P. and Giannakis, M., 2000. Supply chain management: an analytical framework for critical literature review. European Journal of Purchasing & Supply Management 6: 67-83.

[43] Importantly, Toyota was greatly influenced by the post-war work on quality control of J. Edwards Deming. See: Edwards Deming, W., 1982. Out of the crisis. Massachusetts Institute of Technology, Center for Advanced Engineering Study, Cambridge, MA, USA.

What this has meant in practice is that large retailers have been able to specify standards for a wide range of product qualities and supplier practices. However, relations between suppliers and buyers in supply chains vary considerably. Kaplinsky suggests that there are:

> '... two contrasting paths of value chain standards in corporate-driven value chains. One involves close and high-trust relations along the chain, with cost-reduction an outcome of largely cooperative efforts between lead-buyers and their tiers of suppliers. The second involves the use of standards to promote much more conflictive, arms-length relations along the chain.'[44]

At least with respect to the food industry, this distinction may be overdrawn. Instead, at one end of a rough continuum we find companies that employ the 'Toyota model', directly working with suppliers to ensure that the qualities of what suppliers produce meet their quality standards in the most efficient and profitable manner. This may involve provision of various forms of advice (which is nearly impossible to ignore), as well as supplier investments (i.e. lock-in costs) in specialised capital equipment (e.g. packaging, varieties, delivery arrangements) in return for increased profits and long-term relationships. At the other end of the continuum are buyers who employ standards, but who remain at arms-length from their suppliers. As one auditor put it, 'The objective is to push the responsibility for food safety and quality back down to the suppliers. Unfortunately, this is the reason why most companies are audited.'[45] This allows those demanding audits to profit from competition among suppliers, while avoiding the costs incurred by working closely in long-term relationships. In point of fact, however, supermarket chains and fast food restaurants use a wide variety of approaches to link suppliers to buyers.

Thus, on the one hand, retailers have been able to specify the sizes and shapes of food packages, and, especially in the case of private label (also known as own brand, or store brand) products, the specific ingredients and processing techniques for the product. On the other hand, retailers have been able to intervene in supplier practices, including setting of standards for on-farm growing and handling, and even reorganising suppliers' businesses so as to make them more efficient (thereby lowering costs for retailers). Some firms, such as Walmart, also link to their suppliers through Electronic Data Interchange (EDI); put differently, as products are sold at retail, orders are automatically sent electronically to suppliers, specifying timing and volume of replenishments.[46]

[44] Kaplinsky, R., 2010. The role of standards in global value chains. The World Bank, International Trade Department, Washington, DC, USA, p. 5.
[45] Stier, R.F., 2009. Third party audits: what the food industry really needs. Food Safety Magazine 15(5): 44.
[46] See Walmart Corporate, Requirements: http://walmartstores.com/Suppliers/248.aspx.

SCM has been spurred on by the neoliberal transformations of the market noted above. In particular, retailers have enjoyed and defended the 'freedom to operate' provided by relatively weakly regulated markets. But they have also been frightened by the *'horror vacui'* of weakly regulated markets. SCM, combined with private standards, provided a solution to the problem. Put differently, SCM provided a means for developing and imposing private standards on suppliers. Yet, paradoxically, even as it avoided additional regulation, it depended heavily on nation-states' legal systems for the enforcement of contract and criminal law.[47]

2.4 Rise of the Tripartite Standards Regime (TSR)

Another aspect of the growth in global trade was a major shift in its governance. While neoliberal and other proponents of 'free trade' were of the opinion that merely reducing and eventually eliminating tariffs and quotas would be sufficient to create a trade explosion, this left questions about specifications, standards, quality and safety unanswered. In short, even the most scrupulous firms engaged in trade needed to know, among other things, that (a) the terms they used in contracts (even those in a single language) had 'the same' meanings for both parties, (b) that a set of standards was shared by both seller and buyer, (c) that standards were adhered to by the seller, (d) that various legally-mandated quality and safety measures were taken, and (e) some recourse for breach of contract was available. Given that all contracts are incomplete,[48] these were hardly matters to be taken lightly. The result was the creation of a new governance structure, what we call the Tripartite Standards Regime (TSR)[49] consisting of standards, certifications, and accreditations. Let us examine each of these in turn.

2.4.1 Standards development organisations

Standards development organisations (SDOs) began to form in the late nineteenth century. Many were initially highly specialised, focused on specific problems found in a given industry. In addition, by the 1930s nearly every industrial nation had a general national SDO, partially the result of mishaps during World War I that were attributable to non-standard parts. However, in the last third of the 20th century, the number and scope of SDOs appears to have expanded, such that nearly every nation on the planet now has a national general SDO and many specialised ones. In the agro-food sector specifications and standards [(a) and (b) above] were delegated to this ever-expanding list of SDOs. These included both specialised and general international organisations established in mid-century [e.g. Codex

[47] Contract law is also essential to the 'success' of large-scale food animal processing companies in shifting risks and dictating management styles to their suppliers. It has been arguably most successfully used in the broiler industry.

[48] Williamson, O.E., 1975. Markets and hierarchies. Free Press, New York, NY, USA.

[49] Loconto, A. and Busch, L., 2010. Standards, techno-economic networks, and playing fields: performing the global market economy. Review of International Political Economy 17(3): 507-536.

Alimentarius, International Organization for Standardization (ISO), Organization for Economic Cooperation and Development (OECD)], a few individual buyers large enough to set their own standards,[50] as well as a wide range of newly established SDOs including GlobalGAP, COLEACP, the International Federation of Organic Agriculture Movements (IFOAM), the British Retail Consortium (BRC), and Safe Quality Food (SQF). Importantly, these SDOs include industry consortia, private voluntary associations, and quasi-public organisations.[51] In short, they blur the distinction between public, state agencies and non-governmental organisations.

2.4.2 Third party certifiers

In the past it was commonplace for relations between buyer and seller to be based largely on the buyer's trust in the seller and the seller's expectations for future sales. Moreover, in those instances where the trust was found to be misplaced, it was possible for the buyer to take the seller to court. However, the massive increase in global food trade combined with the withdrawal of the state from direct action posed a new set of problems. A buyer might be in the position of dealing with hundreds of sellers, who could offer goods at lower prices than domestic producers, but all of whom were located in other nations, had considerably different expectations as to quality than the buyer, and were – for reasons of cost, time, and/or inadequate means for legal redress in that jurisdiction – impossible to bring to court in cases of non-compliance with the terms of a contract. The most common solution to this problem was the introduction of various forms of certification. This appeared to address points (c) and (d), and to remove the need for (e) above.

There are essentially three forms of certification.[52] First party certification occurs when the producer certifies that products meet the standards in question. This works well in the case of products bearing the brand of the first party, since there are reputational issues at play. Second party certification involves the continuous inspection of the delivered product by the buyer. In short, it requires that the buyer police the seller continuously. Finally, third party certification involves the use of an allegedly neutral third party, i.e. not a party to the exchange in question,

[50] For example, Wal-Mart is sufficiently large that it can and does specify package sizes for many of the products it stocks.

[51] Fulponi, L., 2006. Private voluntary standards in the food system: the perspective of major food retailers. Food Policy 31(1): 1-13.

[52] See, for example: Busch, L., Thiagarajan, D., Hatanaka, M., Bain, C., Flores, L.G. and Frahm, M., 2005. The relationship of third-party certification (TPC) to sanitary/phytosanitary (SPS) measures and the international agri-food trade: Final report. Development Alternatives, Inc., Washington, DC, USA.
Deaton, B., 2004. A theoretical framework for examining the role of third-party certifiers. Food Control 15: 615-619.
Gibbon, P. and Ponte, S., 2008. Global value chains: from governance to governmentality? Economy and Society 37(3): 365-392.
Hatanaka, M., Bain, C. and Busch, L., 2005. Third-party certification in the global agrifood system. Food Policy 30(3): 354-369.

whose job it is to engage in 'conformity assessment,' i.e. to determine to what extent the seller's products conform to the standards upon which an agreement had been reached. Given that the costs of this last form of certification can be imposed on the seller, and that it separated the policing role from that of the seller, it has become the most favoured form of certification worldwide.

It has also given rise to a vast, largely new, industry – that of certification. While certifiers were initially focused on particular industries and many still are, today many certifiers are more than happy to certify across industries. Hence, the largest certifying firms, e.g. Société Générale de Surveillance (SGS), and Det Norske Veritas (DNV), certify everything from food products to steelworks. In contrast, medium and small certifying firms tend to stick with one domain such as food products (e.g. Quality Assurance International, NSF International).

However, the certification industry can hardly be said to be disinterested. Although there is no evidence of widespread fraud or deceit, certifiers find themselves between the proverbial rock and a hard place. Demand for certification services almost always originates with buyers of food products, but it is the firm that is to be certified that pays the certifier. Thus, failing to give a firm a passing score on a certification is tantamount to losing a customer. I return to this dilemma in section 2.5.2 below.

2.4.3 Accreditors

The increased interest by retailers in purchasing only goods certified to a given standard opened a global market for certification services. Initially, virtually anyone could hang out a sign that said that they were a certifier. Indeed, even today in most nations nothing prohibits one from so doing. However, it soon became apparent that some certifiers were more trustworthy than others. To resolve this problem, numerous national accreditation bodies were created; these national bodies range from public agencies to public-private partnerships, to private entities.[53] Furthermore, two non-governmental organisations were developed that now accredit the national accreditors. Put differently, they certify (through peer evaluation) that accreditors can be trusted to certify that certifiers do their job correctly.

The International Laboratory Accreditation Cooperation (ILAC) is, as the name suggests, a cooperation among national laboratory accreditors.[54] It attempts to ensure that national accreditors employ the same standards of testing, specification,

[53] For a recent list, see Donaldson, J.L., 2005. Directory of National Accreditation Bodies. National Institute of Standards and Technology, Gaithersburg, MD, USA.
[54] International Laboratory Accreditation Cooperation, Welcome to ILAC, http://www.ilac.org/home.html.

accuracy, and precision in certifying the competence of national testing laboratories, ensuring that they meet minimum standards. The International Accreditation Forum (IAF), as it notes on its website, 'is the world association of Conformity Assessment Accreditation Bodies in the fields of management systems, products, services, personnel and other similar programmes of conformity assessment'.[55] Importantly, neither of these organisations is a governmental body, although there is little doubt that they engage in governance activities.

2.4.4 The new private governance: TSRs

Together, SDOs, certifiers, and accreditors have formed a TSR, or perhaps more accurately, an assortment of TSRs, which – through markets – collectively govern global production and trade in agrifood and other goods and services. TSRs are perhaps best seen as a plethora of new law-making and enforcing entities. Although participation in a given TSR is nominally voluntary, each TSR produces/enforces its own private (i.e. non-state) codes, laws, rules, standards, specifications, and regulations that are binding on adherents to the TSR. Failure to comply is generally dealt with by market sanctions, by barring access to a particular (and usually lucrative) market. Moreover, TSRs extend beyond the borders of nation-states; indeed, they are not, at least in principle, geographical in scope. Hence, a given TSR may involve growers in Guatemala, Mexico, and Ghana, supermarket chains operating in Thailand, France, and Canada, and transporters shipping goods from growers' farms to those supermarkets.

At the same time, it is important to note that all TSRs are ultimately dependent on the state. States provide vital services to TSRs including, most obviously, (1) corporate, contract, intellectual property, and anti-fraud laws, (2) appropriate enforcement mechanisms (beyond market sanctions) such as civil and criminal courts, police, and prosecutors, and (3) international agreements and organisations (e.g. the WTO). Without these state-sponsored and supported institutions, TSRs would be powerless to discipline errant participants.

2.5 Can governance be plural? Legitimacy and markets revisited

Nation-states took centuries to gain legitimacy. Even now, that legitimacy is still contested by various sub-national groups (e.g. Basque separatists, Scottish nationalists, the Michigan Militia). Yet, with the exception of a few rare jurisdictions with dual legal regimes, and a few places where states have lost control over both territory *and* population (e.g. Somalia) nation-states now cover the land area of the entire planet. To my knowledge, there are no places where citizens may choose among many legal regimes until they find one to their liking. Yet, this is precisely

[55] International Accreditation Forum, International Accreditation Forum, Inc., http://www.iaf.nu/.

what one finds with respect to the private systems of governance described above. Consider the advantages and disadvantages of TSRs.

2.5.1 Advantages of TSRs

TSRs have several advantages for supply chain captains. First, and arguably most importantly, they are quite flexible, far more flexible than state-sponsored law. They are easily changed if (some) parties to the TSR decide that their current form is unacceptable for whatever reason. Furthermore, they are, or can be, quite proactive. Hence, in a time of short harvests, quality standards can be relaxed to ensure that processors and retailers are supplied; conversely, in a time of abundant harvests, standards can be more rigidly enforced, so as to exclude non-conforming goods. Third, TSRs can be tailored to specific situations to a much greater degree than legal regimes. Fourth, TSRs generally have an opt-out provision that is lacking in formal law. Indeed, it is these added flexibilities that attracted private actors - both corporate and private voluntary organisations - to TSRs in the first place.

At the same time, TSRs have the potential to provide advantages to other groups. Perhaps most importantly, TSRs may allow participation in (some) markets that would otherwise be closed to certain participants, especially those in developing nations.[56] Moreover, TSRs may help such producers to improve the effectiveness and efficiency of their production practices as well. Finally, TSRs may be effective means of grappling with complex transnational environmental, labour, and human rights issues on which to date international agreement has been either lacking or unenforced.[57]

2.5.2 Problems associated with TSRs

At the same time, there are several major problems associated with TSRs. Perhaps the area of greatest concern with respect to TSRs is that they tend to behave like states, or perhaps better, like quasi-states. They are like states in that each TSR issues its own set of rules and regulations, and each has its own enforcement mechanism, generally relying on denial of market access to violators of the rules.[58] While technically TSRs are voluntary, in many instances they are *de facto* mandatory, as no market actors outside the TSR exist in a given locale. But TSRs are unlike states

[56] In some instances this is a zero-sum game. Some get to participate as a result of the TSR, even as others are excluded.
[57] For a positive view of the role of private voluntary organisations in promoting TSRs, see: Conroy, M.E., 2007. Branded! How the 'certification revolution' is transforming global corporations. New Society Publishers, Gabriola Island, Canada; for a more critical view, see Gereffi, G., Garcia-Johnson, R. and Sasser, E., 2001. The NGO-industrial complex. Foreign Relations 125(July-August): 56-65.
[58] Denial of market access is arguably less consequential in industrialized nations, where other less discriminating buyers may be available. In contrast, in many poor nations, there are often local monopolies, making denial of market access tantamount to business failure.

in that they generally have few or inadequate appeals mechanisms. They are also unlike states in that they are not territorial in their authority but extraterritorial. And, and as noted above, they depend on states for enforcement of certain legal obligations, but at the same time, they are often able to enforce rules far beyond the boundaries of any given state.

Moreover, they are certainly not like *democratic* states. They generally have little separation of powers, rarely represent more than a small fraction of those in a given supply chain (generally those who are most powerful), and they may apply their rules in a highly capricious manner. Some are reminiscent of George Orwell's *Animal Farm*,[59] where all animals are equal, but some animals are more equal than others.

The state-like character of TSRs, as well as their democratic deficit, poses a number of other problems as well. These include questions about accountability, effectiveness, transparency, innovation, fairness, and legitimacy.[60] Let us consider each of these briefly in turn.

Accountability. It is not clear to whom the participants in TSRs are accountable. TSRs are designed such that they promote (a form of) trust among the private actors within a given supply chain.[61] However, at the same time, TSRs perform functions traditionally reserved to states (e.g. food safety). As such, they are at least partially responsible for certain public goods. Moreover, as one observer has noted, '... most private safety regulation currently faces northward. It protects developed country interests, and has only haltingly and partially incorporated the voices and interests of developing country producers and publics'.[62] In addition, the very plurality of standards tends to dilute accountability.

Effectiveness. As one observer notes, '... whether private standards benefit consumers and society ultimately depends on the actual improvement that they generate with respect to the previous situation'.[63] The effectiveness of the now widespread

[59] Orwell, G., 1984 [1945]. Animal farm. Folio Society, London, UK.

[60] In addition, there is some concern about the compatibility of private standards with the WTO and especially the SPS and TBT agreements. I leave it to others to address this issue. See, e.g. Schroder, H.Z., 2009. Definition of the concept 'international standard' in the TBT Agreement. Journal of World Trade 43(6): 1223-1254; see also Chapter 6 by Marinus Huige.

[61] The IAF claims that promoting trust is central to its aims: International Accreditation Forum, Inc., http://www.iaf.nu/. Cf. O'Neill, O., 2002. A question of trust. Cambridge University Press, Cambridge, UK.

[62] Meidinger, E., 2009. Private import safety regulation and transnational new governance. In: Coglianese, C., Finkel, A.M. and Zaring, D. (eds.). Import safety: regulatory governance in the global economy, University of Pennsylvania Press, Philadelphia, PA, USA, p. 233.

[63] Liu, P., 2009. Private standards in international trade: issues, opportunities and long-term prospects. In: Sarris, A. and Morrison, J. (eds.). The evolving structure of world agricultural trade: implications for trade policy and trade agreements. Food and Agriculture Organization of The United Nations, Trade and Markets Division, Rome, Italy, p. 218.

use of HACCP, ISO 9000, and other management-oriented schemes assumes that the participants are competent and want to improve the quality and safety of their businesses. Clearly, if the control points chosen do not include all those of relevance, then HACCP is of little value. Indeed, in some quarters the initials HACCP are spelled out as 'Have another cup of coffee and pray.'

One can make similar observations with respect to certification. The recent problems with salmonella in peanut butter in the United States provide a rare glimpse of the nature of the problem. The now-defunct Peanut Corporation of America was an ingredient supplier to a number of food manufacturers. It was inspected by Eugene A. Hatfield, who worked for the well-known and long-established food certifier, AIB International. According to its website, AIB International is accredited by ANSI and UKAS, the US and UK national accreditors.[64] The company knew when the audit would take place and had plenty of time to prepare. Moreover, Mr. Hatfield had less than one day to review the large plant which processed several million pounds of peanuts each month.

Mr. Hatfield was hardly a novice, but at age 66 a seasoned plant inspector. Nevertheless, his expertise was in the fresh produce sector, not in groundnuts. He gave the plant a rating of 'Superior' in the certification process. Yet,

'Federal investigators later discovered that the dilapidated plant was ravaged by *Salmonella* and had been shipping tainted peanuts and paste for at least nine months. But they were too late to prevent what has become one of the nation's worst known outbreaks of food-borne disease in recent years, in which nine are believed to have died and an estimated 22,500 were sickened.'[65]

Firms buying the ingredients were also forced to recall potentially dangerous products. Although there was no evidence of any form of collusion or fraud, clearly the audit was woefully inadequate. But the problem goes further than this would suggest: 'If the certifier is a for-profit company, it may have an interest in not interpreting the standard in too strict a manner, lest some clients switch to competitors who have a more flexible interpretation. Also, withdrawing certification in case of non-compliance means losing a customer.'[66]

Indeed, all of this took place in an industrial nation with a strong state and generally high relations of trust among both buyers and the general public. In

[64] AIB International, https://www.aibonline.org/.

[65] The New York Times, 6 March 2009, A1. See also: Stier, R.F., 2009. Third party audits: what the food industry really needs. Food Safety Magazine 15(5): 43-44, 46, 48-49, 60.

[66] Liu, P., 2009. Private standards in international trade: issues, opportunities and long-term prospects. In: Sarris, A. and Morrison, J. (eds.). The evolving structure of world agricultural trade: implications for trade policy and trade agreements. Food and Agriculture Organization of The United Nations, Trade and Markets Division, Rome, Italy, p. 224.

nations with weak states and widespread distrust, the problem may be far greater. For example, one of the few studies of the effectiveness of a TSR in a weak state, the 'safe vegetable production system' in Vietnam, concludes that it is largely ineffective.[67]

It should also be noted that the very obverse situation may prevail, where – usually for reasons of non-price competition and/or due diligence – buyers will not only demand that high quality standards be met, but that safety standards exceed those deemed necessary by health authorities. This is especially problematic for producers of fresh fruits and vegetables, since they have little or no control over the (apparently harmless) small quantities of microorganisms that are commonly found on such items. Moreover, these are perishable products that must be sold before they lose value. For example, one study found that blueberry buyers were demanding seemingly arbitrary bacterial counts for organisms not implicated in disease outbreaks; moreover, in one instance one buyer's threshold was 20 times higher than another.[68] This and similar demands certainly do not make the delivery of public or private goods more effective, but they do add to costs.

Transparency. While Kafkaesque legal regimes certainly exist, in most states, most commercial law is publicly available and known (or knowable) to all. But many standards regimes are deliberately made quite opaque.[69] That opacity has several aspects. First, most SDOs sell their standards; indeed, that is their major source of income. Hence, they have a vested interest in limiting circulation to paying customers. Second, in some instances standards may actually be a trade secret, such that using them requires signing a non-disclosure agreement. Finally, the standards may regularly change based (usually) on the changing demands of the buyer. This lack of transparency is particularly troublesome.

Innovation. TSRs can and often do block innovation. This is particularly true when they are based on process standards. For example, many current TSRs in the agrifood sector demand that producers follow rigidly defined calendars for planting, spraying, fertilising, and harvesting crops, or similarly rigid rules for livestock management. While such rules certainly help to produce uniform products according to a fixed schedule, they are also quite inflexible with respect to innovation. They presume, often wrongly, that all innovation must come from those in charge of the supply chain; conversely, they limit the managerial capacities of suppliers. Similar problems may exist in the processing sector.

---

[67] Pham Van Hoi, Mol, A.P.J. and Oosterveer, P.J.M., 2009. Market governance for safe food in developing countries: the case of low-pesticide vegetables in Vietnam. Journal of Environmental Management 91(2): 380-388.
[68] Bain, C. and Busch, L., 2004. Standards and strategies in the Michigan blueberry industry. Michigan Agricultural Experiment Station, East Lansing, MI, USA.
[69] Kaplinsky, R., 2010. The role of standards in global value chains. The World Bank, International Trade Department, Washington, DC, USA.

Fairness. One problem frequently noted in the literature is the negative impacts on small producers. Indeed, it is worth noting that both ends of the continuum are biased against small producers. On the one hand, small producers may lack the funds necessary to invest in specialised equipment; on the other hand, without technical help, they may find themselves squeezed out of the market. Many aid agencies, both public and private, have attempted to grapple with these issues, with a mixed record of success. In some instances producer cooperatives and similar organisations have been able to successfully meet buyer requirements,[70] while in other situations this goal has been unachievable. Furthermore, when the standards change rapidly or are rather opaque, they would seem to require the permanent presence of some advisory group to maintain fairness. This seems a highly unlikely prospect.

Legitimacy. TSRs pose a number of legitimacy problems. First, legitimacy is undermined when suppliers are required to meet ever higher specifications.[71] This is often the case as downstream actors usually have the ability to squeeze upstream actors.[72] It is possible that in the future climate change and high oil prices will change this situation, but that remains to be seen. Moreover, it might have the unintended consequence of provoking food riots in poor nations.

Second, even so-called ethical standards (e.g. fair trade, sustainability) pose legitimacy problems. While the private voluntary organisations that promulgate these standards doubtless have noble goals in mind, they are often imposed on producers who have little say in their use and enforcement.[73]

Third, each TSR or each supply chain develops its own rules, and creates its own market with its own outcomes. These outcomes not only include those persons within a given TSR, but may also include those connected to, but not quite part of the TSR. Put differently, a TSR may (deliberately or inadvertently) create two classes of 'citizens' – those who are within the domain of the TSR and those who are excluded and hence ignored. For example, Bain has shown how contract labour is systematically excluded from labour protections in a Chilean table grape TSR.[74] Moreover, on a national scale, as standards pluralise law, they also make it more and more difficult to calculate, channel, or forecast the behaviour of markets. As such, they undermine national and even international policies.

---

[70] In this book, see for example Chapter 8 by Margret Will.

[71] This problem is exacerbated when there is little or no price premium for higher standards.

[72] This point was officially recognised in Great Britain by the Competition Commission, 2000. Supermarkets: a report on the supply of groceries from multiple stores in the United Kingdom. Available at: http://www.competition-commission.org.uk/rep_pub/reports/2000/446super.htm. It noted that even as consumers benefitted from low prices, suppliers were often squeezed by supermarkets.

[73] Gereffi, G., Garcia-Johnson, R. and Sasser, E., The NGO-industrial complex. Foreign Relations 125(July-August): 56-65.

[74] Bain, C., 2010. Structuring the flexible and feminized labor market: Globalgap standards for agricultural labor in Chile. Signs: Journal of Women in Culture and Society 35(2): 343-370.

Finally, the proliferation of TSRs seems to have given state regulators the mistaken notion that they can now sit back and relax. The lack of adequate regulation of new food technologies, e.g. nanotechnologies, suggests that we now have (at least in some areas) a *laissez-faire state*, in which the normally reactive character of law is taken to the point of leaving things alone – or not having the wherewithal to act – even in the face of genuine concerns.[75] Likely it will take a serious nano disaster to prod state agencies into serious action. In short, there are now legitimacy problems for both states and TSRs.

2.6 Conclusions

The ostensibly voluntary character of standards suggests that the resolution of the problems associated with standards, certifications, and accreditations lie neither in voice or loyalty, but in exit.[76] Put differently, one might argue that those unable or unwilling to participate in marketing to the industrial world might still be able to sell their products to the middle and low income nations whose standards are lower. Doubtless, there are such opportunities at the present. And, given the rising purchasing power of nations such as India, China, and Brazil, one might see this as a reasonable alternative. Yet, both buyers and governments in these nations also appear to be rapidly adopting higher quality and safety standards because (1) these nations are also suppliers to the world market and are aware of the economic and safety advantages of higher quality products, and (2) articulate and well-educated middle class consumers in these nations are demanding protection from unscrupulous local producers in the light of various food scares.[77]

Of course, it would be naïve to think of state-led governance as always perfect. It is far from that. But plural governance by various quasi-states raises fundamental issues for democracy. It appears to me that there are, *grosso modo*, four options:

---

[75] Busch, L., 2008. Nanotechnologies, food, and agriculture: next big thing or flash in the pan? Agriculture and Human Values 25(2): 215-218; House of Lords, 2010. Nanotechnologies and food: evidence. House of Lords, Committee on Science and Technology, London, UK; Institute for Food and Agricultural Standards, 2007. An issues landscape for nanotechnology standards: report of a workshop. Institute for Food and Agricultural Standards, Michigan State University, East Lansing, MI, USA; Taylor, M., 2006. Regulating the products of nanotechnology: Does FDA have the tools it needs? Woodrow Wilson Center for International Scholars, Project on Emerging Nanotechnologies, Washington, DC, USA.

[76] Hirschman, A.O.,1970. Exit, voice, and loyalty: responses to decline in firms, organizations, and states. Harvard University Press, Cambridge, MA, USA.

[77] In China quasi-private certification schemes are already being explored. See: Yamei, Q., Zhihua, Y., Weijun, Z., Heshan, T., Hongping, F. and Busch, L., 2008. Third-party certification of agro-products in China: a study of agro-product producers in Guangzhou, Shenzhen, Hangzhou and Qingdao. Food Protection Trends 28(11): 765-770. Moreover, the melamine scandal has caused the state to enhance its food safety laws and to encourage certification schemes.

2.6.1 Return to unitary governance

Many would doubtless like a return to unitary governance within the confines of the system of nation-states. This would mean putting severe restrictions on the ability of non-state organisations to impose quality standards and the dismantling of the TSRs. Alternatively, it might mean a strengthening of international governance, e.g. by strengthening the WTO and the extension of the SPS and TBT agreements to many if not most otherwise private transactions. Indeed, some observers have suggested that these agreements should extend to the private sector.[78] And, the expanding use of private standards to further state policies would seem to shift the policy scene in that direction. However, it appears unlikely that WTO members would agree on a reigning in of private standards.

2.6.2 Continued proliferation

Far more likely is a continuing proliferation of standards and of TSRs, especially as encouraged by the advantages for larger firms; these include non-price competition for final consumers, supplier lock-in, shifting of risks to suppliers, and just-in-time deliveries. However, this proliferation may be counteracted by both the growing concentration in the agrifood sector and decisions not to compete on certain types of standards. For example, the Global Food Safety Initiative is premised on the notion that food safety is an area to be excluded from competition.[79]

2.6.3 A new feudalism

A third possibility is a new form of feudalism in which the large food retailers act more and more like latter-day feudal lords, using their market power both to enact favourable state policies and to bind suppliers to their particular supply chains. Indeed, Thurman Arnold noted many years ago that corporations are the last vestiges of feudal institutions.[80] More recently, Vladimir Shlapentokh has noted the feudal tendencies in contemporary society.[81]

---

[78] See, e.g. Dankers, C., 2003. Environmental and social standards, certification and labelling for cash crops. Food and Agriculture Organization of the United Nations, Rome, Italy.

[79] Fulponi, L., 2006. Private voluntary standards in the food system: the perspective of major food retailers. Food Policy 31(1): 1-13; Global Food Safety Initiative, http://www.mygfsi.com/.

[80] Arnold, T., 1937. The folklore of capitalism. Yale University Press, New Haven, CT, USA.

[81] Shlapentokh, V. and Woods, J., 2007. Contemporary Russia as a feudal society: a new perspective on the post-Soviet era. Palgrave Macmillan, New York, NY, USA; Shlapentokh, V. and Woods, J., 2011. The Middle Ages in America: feudal elements in contemporary society. The Pennsylvania State University Press, University Park, PA, USA.

2.6.4 A hybrid system

Finally, it is possible that some sort of hybrid arrangement will emerge. In such a system there would be state-run, fully private, and state-private standards and TSRs of various sorts. However, likely some would dominate in some substantive domains and other forms in other domains. For example, it might be decided that most aspects of food safety should be left to the state, while cosmetic qualities should be entirely left to the buyers' discretion. Process standards for sellers might be demarcated by states, but administered by non-state actors. There are a nearly infinite number of such arrangements possible; which ones are instituted will be a matter of prolonged political, legal, and economic debate and conflict. The outcome of those debates and conflicts will affect nearly everyone on the planet.

Acknowledgements

The support of the Economic and Social Research Council (ESRC) is gratefully acknowledged. This work is part of the Research Programme of the ESRC Genomics Network at Cesagen (ESRC Centre for Economic and Social Aspects of Genomics). Allison Loconto and Carolina Maciel made helpful comments on an earlier draft of this chapter.

References

Allemanno, A., 2009. Solving the problem of scale: The European approach to import safety and security concerns. In: Coglianese, C., Finkel, A.M. and Zaring, D. (eds.). Import safety: regulatory governance in the global economy. University of Pennsylvania Press, Philadelphia, PA, USA, pp. 171-189.

Arnold, T., 1937. The folklore of capitalism. Yale University Press, New Haven, CT, USA.

Bain, C., 2010. Structuring the flexible and feminized labor market: Globalgap standards for agricultural labor in Chile. Signs: Journal of Women in Culture and Society 35(2): 343-370.

Bain, C. and Busch, L., 2004. Standards and strategies in the Michigan blueberry industry. Michigan Agricultural Experiment Station, East Lansing, MI, USA.

Busch, L., 2008. Nanotechnologies, food, and agriculture: next big thing or flash in the pan? Agriculture and Human Values 25(2): 215-218.

Busch, L. and Bain, C., 2004. New! Improved? The transformation of the global agrifood system. Rural Sociology 69(3): 321-346.

Busch, L., Thiagarajan, D., Hatanaka, M., Bain, C., Flores, L.G. and Frahm, M., 2005. The relationship of third-party certification (TPC) to sanitary/phytosanitary (SPS) measures and the international agri-food trade: Final report. Development Alternatives, Inc., Washington, DC, USA.

Competition Commission, 2000. Supermarkets: a report on the supply of groceries from multiple stores in the United Kingdom. Available at: http://www.competition-commission.org.uk/rep_pub/reports/2000/446super.htm.

Conroy, M.E., 2007. Branded! How the 'certification revolution' is transforming global corporations. New Society Publishers, Gabriola Island, Canada.

Croom, S., Romano, P. and Giannakis, M., 2000. Supply chain management: an analytical framework for critical literature review. European Journal of Purchasing & Supply Management 6: 67-83.

Dankers, C., 2003. Environmental and social standards, certification and labelling for cash crops. Food and Agriculture Organization of the United Nations, Rome, Italy.

Deaton, B., 2004. A theoretical framework for examining the role of third-party certifiers. Food Control 15: 615-619.

Deming, W.E., 1982. Out of the crisis. Massachusetts Institute of Technology, Center for Advanced Engineering Study, Cambridge, MA, USA.

Donaldson, J.L., 2005. Directory of National Accreditation Bodies. National Institute of Standards and Technology, Gaithersburg, MD, USA.

Dries, L., Reardon, T. and Swinnen, J.F.M., 2004. The rapid rise of supermarkets in central and eastern Europe: implications for the agrifood sector and rural development. Development Policy Review 22(5): 525-556.

Food and Agriculture Organization, 2009. FAO Statistical Yearbook 2009, FAO, Statistics Division, Rome, Italy.

Fortin, N.D., 2009a. Food regulation: law, science, policy, and practice. John Wiley, Hoboken, NJ, USA.

Fortin, N.D., 2009b. United States Food Law: consumers, controversies, current issues. European Journal of Consumer Law 2009(1): 15-42.

Foucault, M., 2008. The birth of biopolitics: lectures at the Collège de France, 1978-79. Palgrave Macmillan, New York, NY, USA.

Friedman, M., 1962. Capitalism and freedom. University of Chicago Press, Chicago, IL, USA.

Fulponi, L., 2006. Private voluntary standards in the food system: the perspective of major food retailers. Food Policy 31(1): 1-13.

Gereffi, G., Garcia-Johnson, R. and Sasser, E., 2001. The NGO-industrial complex. Foreign Relations 125(July-August): 56-65.

Gibbon, P. and Ponte, S., 2008. Global value chains: from governance to governmentality? Economy and Society 37(3): 365-392.

Goldman, M., 2005. Imperial nature: The World Bank and struggles for justice in the age of globalization. Yale University Press, New Haven, CT, USA.

Harper, R., 2000. The social organization of the IMF's mission work. In: Strathern, M. (ed.). Audit cultures: anthropological studies in accountability, ethics, and the academy. Routledge, New York, NY, USA, pp. 21-53.

Hatanaka, M., Bain, C. and Busch, L., 2005. Third-party certification in the global agrifood system. Food Policy 30(3): 354-369.

Hayek, F.A., 1973-1979. Law, legislation and liberty, 3 vols. University of Chicago Press, Chicago, IL, USA.

Hayek, F.A., 2007 [1944]. The road to serfdom. University of Chicago Press, Chicago, IL, USA.

Hirschman, A.O.,1970. Exit, voice, and loyalty: responses to decline in firms, organizations, and states. Harvard University Press, Cambridge, MA, USA.

House of Lords, 2010. Nanotechnologies and food: evidence. House of Lords, Committee on Science and Technology, London, UK.

Institute for Food and Agricultural Standards, 2007. An issues landscape for nanotechnology standards: report of a workshop. Institute for Food and Agricultural Standards, Michigan State University, East Lansing, MI, USA.

Kaplinsky, R., 2010. The role of standards in global value chains. The World Bank, International Trade Department, Washington, DC, USA.

Lippmann, W., 1938. The good society. George Allen and Unwin Ltd, London, UK.

Liu, P., 2009. Private standards in international trade: issues, opportunities and long-term prospects. In: Sarris, A. and Morrison, J. (eds.). The evolving structure of world agricultural trade: implications for trade policy and trade agreements. Food and Agriculture Organization of The United Nations, Trade and Markets Division, Rome, Italy, pp. 205-235.

Loconto, A. and Busch, L., 2010. Standards, techno-economic networks, and playing fields: performing the global market economy. Review of International Political Economy 17(3): 507-536.

Meidinger, E., 2009. Private import safety regulation and transnational new governance. In: Coglianese, C., Finkel, A.M. and Zaring, D. (eds.). Import safety: regulatory governance in the global economy, University of Pennsylvania Press, Philadelphia, PA, USA, p. 233.

Mirowski, P. and Plehwe, D. (eds.), 2009. The road from Mont Pèlerin: the making of the neoliberal thought collective. Harvard University Press, Cambridge, MA, USA.

Nestlé, 2008. Management Report 2008. Nestlé, Geneva, Switzerland.

O'Neill, O., 2002. A question of trust. Cambridge University Press, Cambridge, UK.

Orwell, G., 1984 [1945]. Animal farm. Folio Society, London, UK.

Peters, G. and Pierre, J., 1998. Governance without government? Rethinking public administration. Journal of Public Administration Research and Theory 8(2): 223-243.

Pham Van Hoi, Mol, A.P.J. and Oosterveer, P.J.M., 2009. Market governance for safe food in developing countries: the case of low-pesticide vegetables in Vietnam. Journal of Environmental Management 91(2): 380-388.

Reardon, T., Timmer, C.P., Barrett, C.B. and Berdegue, J., 2003. The rise of supermarkets in Africa, Asia, and Latin America. American Journal of Agricultural Economics 85(5): 1140-1146.

Schroder, H.Z., 2009. Definition of the concept 'international standard' in the TBT Agreement. Journal of World Trade 43(6): 1223-1254.

Shlapentokh, V. and Woods, J., 2007. Contemporary Russia as a feudal society: a new perspective on the post-Soviet era. Palgrave Macmillan, New York, NY, USA.

Shlapentokh, V. and Woods, J., 2011. The Middle Ages in America: feudal elements in contemporary society. The Pennsylvania State University Press, University Park, PA, USA.

Simons, H.C., 1948 [1934]. A positive program for Laissez Faire: some proposals for a liberal economic policy, In economic policy for a free society. University of Chicago Press, Chicago, IL, USA, pp. 40-77.

Smith, A., 1982 [1759]. The theory of moral sentiments. Liberty Fund, Indianapolis, IN, USA.

Smith, A., 1994 [1776]. An inquiry into the nature and causes of the wealth of nations. Modern Library, New York, NY, USA.

Starkey, T., 1878. England in the reign of Henry the Eighth, a dialogue between Cardinal Pole and Thomas Lupset. N, Trubner & Co, London, UK,. for the Early English Text Society, 1878 [1538] as cited in the Oxford English Dictionary.

Stier, R.F., 2009. Third party audits: what the food industry really needs. Food Safety Magazine 15(5): 43-44, 46, 48-49, 60.

Taylor, M., 2006. Regulating the products of nanotechnology: Does FDA have the tools it needs? Woodrow Wilson Center for International Scholars, Project on Emerging Nanotechnologies, Washington, DC, USA.

Travaux du Centre International d'Etudes Pour la Rénovation du Libéralisme (ed.), 1939. Le Colloque Walter Lippmann. Cahier no. 1, Librairie de Médicis, Paris, France.

Von Mises, L., 1978 [1933]. Epistemological problems of economics. Ludwig von Mises Institute, Auburn, AL, USA.

Wal-Mart, 2008. Annual Report. Available at: http://walmartstores.com/sites/annualreport/2008/message_from_lee_scott.html.

Weatherspoon, D.D. and Reardon, T., 2003. The rise of supermarkets in Africa: implications for agrifood systems and the rural poor. Development Policy Review 21(3): 1-17.

Williamson, J., 2005. The strange history of the Washington consensus. Journal of Post Keynesian Economics 27(2): 195-206.

Williamson, O.E., 1975. Markets and hierarchies. Free Press, New York, NY, USA.

Yamei, Q., Zhihua, Y., Weijun, Z., Heshan, T., Hongping, F. and Busch, L., 2008. Third-party certification of agro-products in China: a study of agro-product producers in Guangzhou, Shenzhen, Hangzhou and Qingdao. Food Protection Trends 28(11): 765-770.

3. The anatomy of private food law

Bernd van der Meulen[82]

3.1 Introduction

3.1.1 Plan

This chapter elaborates the legal structure of private food law, its anatomy so to speak. To this end the next section (section 3.2) paints a picture of the way private food law has or might have developed. Sections 3.3 till 3.11 discuss the elements that together make up the structure of private food law: forms of chain orchestration (3.3), ownership of private schemes (3.4), enforcement of private food law (3.5), adjudication and conflict resolution (3.6), the role of audits (3.7), certification (3.8), accreditation (3.9), the emergence of private alternatives to accreditation (3.10) and standard setting (3.11). Section 3.12 summarises the findings in graphic form, which may well be seen as the skeleton of private food law. Sections 3.13 and 3.14 discuss connections among private schemes (3.13) and between private schemes and public law (3.14). Section 3.15 goes into motives underlying private food law. Sections 3.16 till 3.22 describe the content of the currently most important examples of private regulation: the underlying concepts of good agricultural practices, good manufacturing practices and HACCP (3.17); GlobalGAP (3.18), BRC (3.19), IFS (3.20), SQF (3.21) and ISO 22.000 (3.22). Section 3.23 addresses the attempts at harmonisation of private food law through the Global Food Safety Initiative. Sections 3.24 and 3.25 analyse the relevance of EU and WTO law respectively for private food law. This chapter ends with some concluding remarks in section 3.26.

3.1.2 Voluntary rules

In food law, one encounters several types of rules to which no legal obligation applies to comply with them; they are in other words not binding or at least not made binding by legislation. This is true for example for hygiene codes, for the codes of conduct elaborated by the Codex Alimentarius Commission and also for quite a lot of other texts on food.

A food business may choose to apply such rules thinking this is a good way to comply with the mandatory requirements from the law (e.g. apply hygiene codes to comply with HACCP), that this may improve the safety or other quality aspects of the product or for many other reasons.

---

[82] This contribution is based on Chapter 17 'Private food law' of Van der Meulen, B. and Van der Velde, M., 2008. European Food Law Handbook. Wageningen Academic Publishers, Wageningen, the Netherlands. Many thanks to Maria Litjens for ideas and inspiring discussion and to Geronda Klop for designing the figures. Comments are welcome at: bernd.vandermeulen@wur.nl.

If the owner of a business decides to apply non-binding rules, this may establish an internal obligation: the premises owned by the concern and the people working there may come under the obligation to apply these rules. The legal mechanism behind such obligations are not in the rules themselves, but may be found in property law (the owner has a say over his business), corporate law (the articles of association bind the partners in the company) or labour law (the labour contract gives the employer a certain power to impose duties upon the employee).

If a business communicates to its trading partners – its customers in particular – that it applies certain rules in the production of its products,[83] this may become part of the offer it makes in the market: a guarantee that the product meets a certain level of quality as expressed in the rules applied to it(s production). If this feature is important to the customer, it may become part of the contract between the producer and his customer. A contract is binding upon the parties to it. Therefore, if a set of rules not binding by themselves is included in a contract, it thus becomes binding for the parties to the contract.

Customers may require their suppliers to apply certain rules/meet certain standards.[84] If the suppliers agree, again this becomes part of the contract and thus a binding obligation. A contract requires mutual agreement ('a meeting of minds'). The specific content may be suggested by either party.

As long as there is no contract, there is no obligation to apply the non-binding rules. If most of the purchasers on the market (or a very powerful purchaser) make the application of certain rules a strict condition to enter into contractual relations, there still is no legal obligation for producers to apply these rules. They will know, however, that they can only acquire contracts if they are willing to accept the obligation. In a way the obligation is in the air. It is hovering over the contracts yet to be concluded. Legally there is no obligation to apply the rules, but if you do not apply them, you are out of business. In such a situation we may call the rules concerned 'de facto' binding. No obligation from the law (*de jure*) to apply, but such a necessity to apply from the facts in case (*de facto*) that in practice it feels like an obligation.

The combination of these elements – the power to create obligations by means of contracts and the power of certain players to dictate the terms of contract to such an extent that they have to be fulfilled even before the contract is concluded – forms the legal basis of a development where the private sector creates norms that apply to the food sector in addition to and even in competition with food law

[83] For example the rules on organic production. See Chapter 13 by Hanspeter Schmidt.
[84] Such as Fair Trade.

found in legislation e.g. in public law. Contract is a part of private (civil) law.[85] For this reason we have labelled the entirety of rules that in this way confront the food sector: 'private food law'.

3.2 The *history* of private standards

Most accounts on public European food law start with an account of the BSE crisis and the measures that have since then been taken. Many aspects of food law can best be understood if we take into account how it has developed historically. Also for private food law it helps to take its development into consideration. This development, however, is scattered and as yet little recorded. For these reasons and for simplicity's sake, in this section I will make a composition of developments that have taken place at divers times and places or that could have taken place and weld this into a 'story' of private food law that is only partly history, but that may help to understand in a similar way as history does.

One can imagine a development of contracts in the food chain from simple to complex. A simple contract states the identity, the quantity and price of the product to be sold by the vendor to the purchaser. The contract becomes more complex if it defines safety or other quality requirements like levels of contamination. Experience has shown that the safety of the product is influenced by the way it is handled (hygiene). The contract may therefore go beyond the product as such and include aspects of product handling, etcetera, etcetera.

Now one can imagine a second development. A company that has invested time and resources (maybe to pay for legal and technical advice) in elaborating a complex contract and is happy with the results, will try to use the same provisions again. If this company has the bargaining power to impose the contractual provisions on its partners, it may start to use this contract as a model or even as general provisions to all the contracts it concludes.

It takes a lot of expertise to develop a good model. Consultants may come to the market or businesses may cooperate to develop model contracts of ever increasing quality and complexity. And isn't it much easier instead of exchanging huge stacks

[85] In this connection often the expression 'self-regulation is applied'. EU, 2010. European Parliament, Council and Commission, The interinstitutional agreement on better law-making, Official Journal of the European Union C 321, 31/12/2003: 1-5, for example defines self-regulation as: 'the possibility for economic operators, the social partners, non-governmental organisations or associations to adopt amongst themselves and for themselves common guidelines at European level (particularly codes of practice or sectoral agreements).' While this definitions applies to most of the private schemes discussed in this chapter, I avoid the expression 'self-regulation'. In my opinion the word as well as the definition convey too much of an impression of harmony and mutual understanding while – as we will see – in practice the creation of private schemes may well be based on market power of some businesses and dependence of others. In such situations legal theory still recognises equality, but in common speech this word is likely to create an incorrect image.

of paper every time a contract is concluded to simply indicate the model that applies? Today it is sufficient to indicate in the contract the identity, quantity and price of the product and to stipulate that it conforms to ISO 22000, BRC or SQF.[86]

3.3 Chain orchestration

3.3.1 Contracts

The impact of the private scheme can go beyond the immediate contractual relation. A contract is a relation between two parties. The safety and quality of a food product supplied by one business to the next, may largely depend, however, on the way it has been dealt with earlier up in the chain. So the customer depends on the agreements reached between her or his supplier and the previous supplier. S/he is, however, an outsider in this agreement. By demanding that in this relation a certain standard applies, a purchaser can exercise considerable influence on contractual relations upstream. A core issue, for example, in the Fair Trade scheme[87] is that the workers at the very beginning of the chain (employees and smallholders in third world countries) receive reasonable remuneration for their efforts. The businesses expressing such wish often are not the ones who have financial relations with these people. For this reason they demand their suppliers to provide proof (through certification) of applying fair conditions.

In this way private standards can be used as an instrument for what is called 'chain orchestration'.

3.3.2 Vertical integration

Another approach to ensuring performance upstream in the chain – to which chain orchestration on the basis of contracts is the modern alternative – is through vertical integration. Businesses are vertically integrated if the different links of the chain are part of one concern. This can be achieved by setting up or acquiring businesses taking care of supply or to enter into a joint venture with a supplier. In such situations, the legal instrument of governance is found in property and/ or associations law.

3.4 Owning a standard

The models described above (and below) represent a certain value. Sometimes the texts are available for free.[88] It can be in the interest of the businesses that

[86] The meaning of these abbreviations is explained hereafter.
[87] See: http://fairtrade.nl/EN/MainContent/Home.aspx.
[88] Like GlobalGAP. This conforms to Article 5.1(4) of EU, 2010. Commission communication – EU best practice guidelines for voluntary certification schemes for agricultural products and foodstuffs, Official Journal of the European Union C 341, 16/12/2010: 5-11.

apply it if they are easily accessible. It is also possible that access to the standard is given only for payment.[89] The legal instrument that makes it possible to claim ownership of the text of the standard is copyright. The copyright holder is entitled to decide on the conditions and price for circulation.

3.5 Enforcement

Generally speaking, legislation creates obligations towards the public authorities representing society as a whole. In case of non-compliance, sanctions may come as a consequence.[90]

Contracts create obligations between parties and provide these parties with instruments to deal with non-compliance by the other parties. In case of non-compliance liability for damages arises, contractual relations may be ended and all kinds of consequences may arise that have been agreed upon in the contract (like contractual fines). The difference in consequence of non-conformity with public law and non-conformity with private law can make it sensible to create obligations by contract that are similar or even identical to obligations that are already present in public law. We see for instance that all the major private food law standards include the obligation to apply the Hazard Analysis and Critical Control Point system (HACCP). This does not create an obligation for the supplier that s/he does not already have, but it turns it from an obligation towards society into an obligation towards the purchaser and provides the purchaser with civil law instruments to enforce this obligation.

3.6 Adjudication

Contractual rights and duties can be invoked in civil courts of law. However, the scheme at issue may also provide for other dispute settlement structures such as arbitration. In case of requirements in the interest of third parties such as consumers, complaint mechanisms for these third parties may be included.[91]

3.7 Audits

In the contractual relation it can be agreed that the customer has the right to visit the suppliers' premises to check if the agreed practices are being applied with the agreed results. Such inspection visits are generally known as 'audits'. It has been recorded that big production companies had special divisions where several people

[89] Like ISO.
[90] Indeed according to Article 6.2 of EU, 2010. Commission communication – EU best practice guidelines for voluntary certification schemes for agricultural products and foodstuffs, Official Journal of the European Union C 341, 16/12/2010: 5-11, unsatisfactory inspection results *should* lead to appropriate action.
[91] For an example, see Chapter 10 by Alessandro Artom on the Italian Self-Regulation Code on Beer Advertising.

had full time jobs receiving auditors. In due course it was noted that all these auditors where checking for similar requirements. Along with the development described above where standard contracts were welded into harmonised and commonly used models, another development took place.

Instead of every customer of a certain producer auditing the producer, it became accepted that one player would audit on behalf of all.[92] If the auditor considered the requirements of the applicable standard to be met, s/he would provide a certificate to the audited business to give it an instrument to communicate its compliance to its customers.

In case the audit shows non-compliance, no certification is provided and/or the right to use the mark representing the certification is withdrawn. By consequence the company can no longer do business with customers that demand certification.

In this way the formulation of norms (standard), the inspection of their fulfilment and the proof of their fulfilment in the form of a certificate and the sanction on non-compliance in the form of withdrawal of certification, developed into structures known as certification schemes (Figure 3.1). If the customers whom the certification communicates are other businesses, the certification scheme is called B2B (business to business). If the customers are the final consumers, the certification scheme is called B2C (business to consumer).

3.8 Certification mark

Connected to a certification scheme is often a symbol that is owned by the owner(s) of the scheme. This ownership usually takes the form of a trademark. The owner of a trademark has the right to allow or forbid its use by others. On this basis the owner (both of the scheme and the related trademark) has the legal power to impose compliance with the standard, to reward compliance or to sanction for non-compliance (Textbox 3.1).

3.9 Accreditation

It is considered of vital importance that certifying bodies can be trusted. To ensure this, schemes have been set up to certify the certifiers. Such schemes can be strictly private, but in many countries public authorities take an interest as well.

[92] This player would be an independent third party trusted by all parties concerned. Three types of audits can be distinguished: one (or first) party audits (self-controls or internal audits), second party audits (audits by the costumer) and third party audits (audits by auditors independent from both other parties). Generally in legalese, the expression 'third' is used to indicate parties outside the relation at issue. This applies regardless if the 'we' relation consists of two parties or more. Talking about a contractual relation, players outside this relation are 'third'. Talking about the EU, states who are not members are 'third'. Talking about the Cold War, the world outside the opposing political blocs is 'third'.

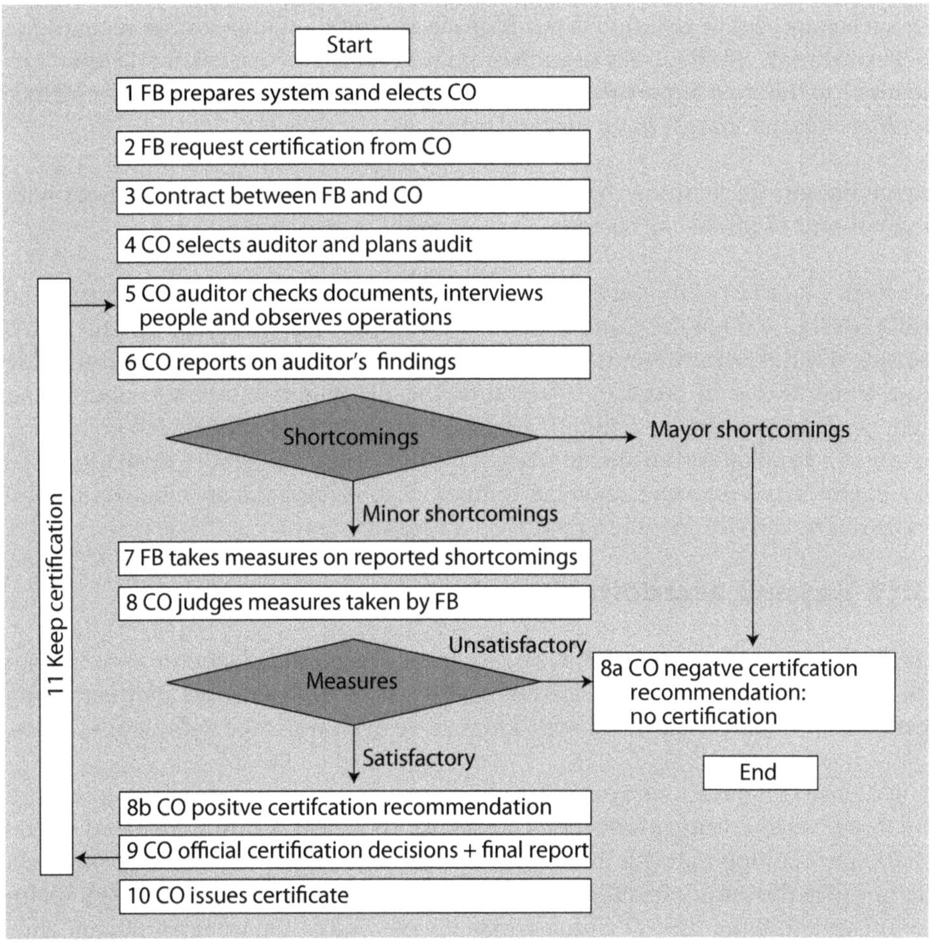

Figure 3.1. Steps in the certification process. FB = food business. CO = certifying organisation. (This figure has been adapted from Bergsma, N., 2010. Voedselveiligheid: certificatie en overheidstoezicht. Praktijkgids Warenwet, Sdu, The Hague, the Netherlands, p. 40.)

Textbox 3.1. SQF on use of certification trade mark.

3 Conditions for using the SQF 2000 Certification Trade Mark

3.1 A producer shall, for the duration of its certification, prove to the satisfaction of SQFI[1] or the LCB[2] that its quality system satisfies the requirements set forth in the current edition of the SQF 2000 Code; and

3.2 A producer must only use the SQF 2000 Certification Trade Mark in accordance with its Certificate of Registration and these rules.

[1] SQF Institute (footnote added).

[2] Licensed certification Body (footnote added).

In such countries, legislation is put in place setting standards for the recognition of certification. On the basis of such standards, certifiers can ask for *accreditation* to prove to their customers that the product they deliver (e.g. proof of compliance with private standards) is up to standard.

Accreditation therefore is the official certification of certifiers often by or with the consent of public authorities.

As from 1 January 2010, an EU framework for accreditation is place. Regulation (EC) No 765/2008 of the European Parliament and of the Council of 9 July 2008 setting out the requirements for accreditation and market surveillance relating to the marketing of products,[93] requires the EU member states to appoint an independent national accreditation body. The national accreditation body shall issue an accreditation certificate when it evaluates that - what the regulation calls - a conformity assessment body is competent to carry out a specific conformity assessment activity (Article 5).

3.10 Beyond accreditation

Accreditation takes place independently from schemes and scheme owners. Not all scheme owners are fully satisfied that accreditation ensures that certifying bodies only certify businesses with a high level of compliance with their scheme.

For GlobalGAP questions arose regarding the efficacy of accreditation to ensure up to standards certification when GlobalGAP certified products appeared on the market not complying with the applicable maximum residue levels for pesticides. To improve the output of audits and the quality of certification, GlobalGAP set up an Integrity programme. The programme is governed by the Integrity Surveillance Committee (ISC) which was made operational in 2009.[94] The Integrity programme consists of a Brand Integrity Programme (BIPRO) and a Certification Integrity Programme (CIPRO). BIPRO provides an online database of certified businesses with a publicly available search site. This enables businesses to ensure themselves if a claim to GlobalGAP certification is valid.

The aim of CIPRO is to ensure that each certified producer meets the same level and to make sure that the control of these producers has been done consistently and each certification body applied the GlobalGAP rules the same way. In addition to accreditation, certification bodies need approval. Results of the CIPRO assessment

[93] EU, 2008. Regulation (EC) No 765/2008 of the European Parliament and of the Council of 9 July 2008 setting out the requirements for accreditation and market surveillance relating to the marketing of products and repealing Regulation (EEC) No 339/93. Official Journal of the European Union L 218, 13/8/2008: 30-47.

[94] See: http://www.globalgap.org/cms/front_content.php?idcat=129.

of certification bodies are available to accreditation bodies. Ultimately, the license of certification bodies can be cancelled if they do not perform up to CIPRO standard.

3.11 Standard setting

In the story described above the initiative to set up private schemes is placed with businesses holding a dominant position in the market. Initially these businesses were mainly the famous brand holders requiring B2B certification to be able to ensure a constant product quality in order to uphold the reputation of the brand. Such brand related quality schemes still play an important role. Of late, however, retailers brands (the so-called private labels) have acquired a position on the marked that has made them leading in the creation of private food law.[95] Most of the examples discussed below originate in retailer initiative.

Private standards, however, are not always based on market power. They may also result from agreements reached with various stakeholders, for example in round tables.[96] In particular schemes dealing with general interests like protection of the environment and corporate social responsibility take account of opinions of third parties (third from the point of view of the parties to the contract to which the scheme applies) including NGOs. Some schemes – such as GlobalGAP originated from a power base and are moving to more representative forms including other stakeholders in decision making.

3.12 Structure of private food law

The structure of private food law as set out in the above, is summarised in Figure 3.2. Rules have been agreed upon in the context of the standard setting organisation. This organisation is the owner (copy right holder) of these rules. The standard setting organisation can consist of businesses using (or imposing) the standard, but it can also be independent. Connected to the rules can be a certification symbol protected as a trade mark. The standard setting organisation agrees (maybe through a license contract or – as far as the rules are concerned – by putting them in the public domain) to the use of these rules and the trademark by auditors and food businesses. One of the actors in the chain decides on applicability of the scheme. It includes the scheme in its contractual requirements and requires it to be included in contracts earlier in the chain as well. The auditor checks for

[95] On the power of private labels, see Bunte, F., Van Galen, M., De Winter, M., Dobson, P., Bergès-Sennou, F., Monier-Dilhan, S., Juhász, A., Moro, D., Sckokai, P., Soregaroli, C., Van der Meulen, B. and Szajkowska, A., 2011. The impact of private labels on the competitiveness of the European food supply chain. Publications Office of the European Union, Luxembourg, Luxembourg.

[96] See Chapter 7 by Otto Hospes. See also Hospes, O., 2009. Regulating biofuels in the name of sustainability or the right to food? In: Hospes, O. and Van der Meulen, B. (eds.) Fed up with the right to food? The Netherlands' policies and practices regarding the human right to adequate food, Wageningen Academic Publishers, Wageningen, the Netherlands, pp. 121-135.

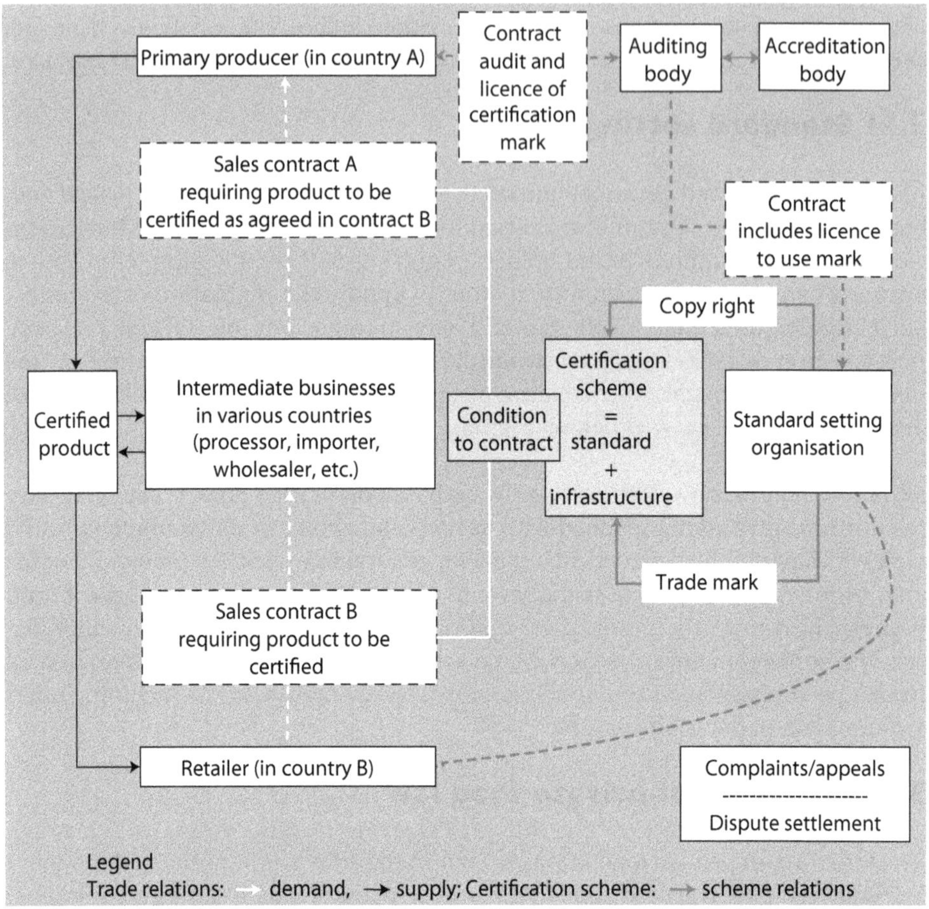

Figure 3.2. The structure of private food law.

compliance with the scheme, certifies (or refuses to certify) the audited business and decides to licence (or withhold) the right to use the certification mark. The quality of the auditor in turn is ensured through accreditation.[97]

3.13 Interconnected private schemes

Figure 3.2 shows two contractual relations (representing a longer chain) where in each link the same private scheme applies. In practice also examples are found where the scheme applied in one link, requires another link to apply another scheme. In the Dutch dairy sector for example, dairy producing businesses require dairy farmers to be certified (Textbox 3.2). The applicable certification scheme requires them to use as feed for milk producing animals only products obtained

[97] Most private schemes require accreditation as laid down in the International Standard ISO/IEC Guide 65.

Textbox 3.2. Dutch HACCP on application of other schemes.

HACCP Certification Regulations 2006
Article 4
Where these regulations do not stipulate any other requirements with respect to the HACCP certification process, the certification bodies have to apply the procedures set in force for the certification of quality systems, that are accredited on the basis of EN 45012 or ISO/IEC Guide 62, excepting article 3.3. of ISO/IEC Guide 62. This article is replaced by article 5.3 of ISO/IEC Guide 66:1998.
With the information/application phase of the certification process, these regulations have to be submitted by the certification body to anyone requesting HACCP certification in accordance with clause 3.1.1.1 of EN 45012:1998 and ISO/IEC Guide 62:1996.

from businesses certified against a certain feed quality scheme.[98] In this way private food law obtains a web-like structure of interrelated schemes.

3.14 Public – private interconnections

Private schemes are not only interconnected among themselves, but also with public law. The vast majority of private certification schemes refers to public law requirements that have to be complied with. Less common but also existing is the inverse where public law provisions require compliance with private schemes (Textbox 3.3 and 3.4).[99] Legislation on community reference laboratories or on methods of sampling refer to private law technical standards.

Textbox 3.3. The official controls regulation referring to private CEN standards.

Regulation 882/2004 *Article 11*
Methods of sampling and analysis
1. Sampling and analysis methods used in the context of official controls shall comply with relevant Community rules or,
(a) if no such rules exist, with internationally recognised rules or protocols, for example those that the European Committee for Standardisation (CEN) has accepted or those agreed in national legislation;

[98] For a detailed analysis of private regulation of the Dutch dairy sector see Chapter 16 by Maria Litjens, Harry Bremmers and myself.
[99] On this issue see Chapter 15 by Irene Scholten-Verheijen, National public sector and private standards.

Textbox 3.4. Regulation 2073/2005 on microbiological criteria, in Annex I referring to private EN/ISO standards[1].

Chapter 1. Food safety criteria

Food category	Micro-organisms/ their toxins, metabolites	Sampling-plan		Limits		Analytical reference method	Stage where the criterion applies
		n	c	m	M		
1.1. Ready-to-eat foods intended for infants and ready-to-eat foods for special medical purposes	*Listeria monocytogenes*	10	0	Absence in 25 g		EN/ISO 11290-1	Products placed on the market during their shelf-life
1.2. Ready-to-eat foods able to support the growth of *L. monocytogenes*, other than those intended for infants and for special medical purposes	*Listeria monocytogenes*	5	0	100 cfu/g		EN/ISO 11290-2	Products placed on the market during their shelf-life
		5	0	Absence in 25 g		EN/ISO 11290-1	Before the food has left the immediate control of the food business operator, who has produced it
1.3. Ready-to-eat foods unable to support the growth of *L. monocytogenes*, other than those intended for infants and for special medical purposes	*Listeria monocytogenes*	5	0	100 cfu/g		EN/ISO 11290-2	Products placed on the market during their shelf-life

[1] EC, 2005. Commission Regulation (EC) No 2073/2005 of 15 November 2005 on microbiological criteria for foodstuffs. Official Journal of the European Union L 338, 22/12/2005: 1-26.

Yet another public law approach to private regulation is where legislation imposes upon stakeholders the duty to regulate for themselves. The most notorious example is the HACCP requirement in Regulation 852/2004. The very essence of HACCP is that a business sets up rules for its own processes. In the case of HACCP it is not a voluntary choice but a public law obligation that will be enforced by public authorities. Such situations are known as imposed self-regulation or enforced self-regulation. An alternative for the application of HACCP, is the application of a hygiene code. The expression hygiene code is used to refer to the national or community guides of good practice.[100] Member states approve the national guides. In this way this form of private regulation acquires status under public law. Compliance with the private standard is deemed to imply compliance with the legal HACCP requirement.

Some food safety inspection agencies acting on the basis of risk based policies reduce the intensity of inspections for businesses operating under private schemes that are trusted to provide good results in ensuring food safety.[101] They mainly limit themselves to an assessment of the quality of the private scheme. Such controls of the quality of private control schemes are known as meta-controls.[102]

Finally we find examples where public authorities partake in private standard setting to achieve objectives in foreign countries that could not be achieved by means of public law instruments. Hospes for example analyses principles formulated by Dutch authorities for the sustainable production of biomass (in countries such

[100] Articles 8 and 9 of EU, 2004. Regulation (EC) No 852/2004 of the European Parliament and of the Council of 29 April 2004 on the hygiene of foodstuffs, Official Journal of the European Union L 139, 30/4/2004: 1-54.

[101] The official controls regulation (882/2004) takes such practices into account in Article 27(6): When, in view of own-check and tracing systems implemented by the feed or food business as well as of the level of compliance found during official controls, for a certain type of feed or food or activities, official controls are carried out with a reduced frequency or to take account of the criteria referred to in paragraph 5(b) to (d), Member States may set the official control fee below the minimum rates referred to in paragraph 4(b) (...).
EU, 2004. Regulation (EC) No 882/2004 of the European Parliament and of the Council of 29 April 2004 on official controls performed to ensure the verification of compliance with feed and food law, animal health and animal welfare rules. Official Journal of the European Union L 65, 30/4/2004: 1-141.

[102] The Dutch ministry of agriculture is strongly in favour of such policies. Publications on this topic with a summary in English language are for example: Bondt, N., Deneux, S.D.C., Van Dijke, I., De Jong, O., Smelt, A., Splinter, G., Tromp, S.O. and De Vlieger, J.J., 2006. Voedselveiligheid, ketens en toezicht op controle. Rapport 5.06.01, LEI, The Hague, the Netherlands; Beekman, V., Kornelis, M., Pronk, B., Smelt, A. en Teeuw, J., 2006. Stimulering eigen verantwoordelijkheid. Zorgen dat producenten en consumenten zorgen voor voedselveiligheid, Rapport 5.06.05. LEI, The Hague, the Netherlands; De Bakker, E., Backus, G., Selnes, T., Meeusen, M., Ingenbleek, P. and Van Wagenberg C., 2010. Nieuwe rollen, nieuwe kansen? Een programmeringsstudie voor toezicht op controle in het agro-foodcomplex. Rapport 6.07.08. LEI, The Hague, the Netherlands.

as Brazil).[103] These principles are operationalised through private certification schemes. The Netherlands consider requiring certification as condition for the import of biofuels.

3.15 Motives

What are the driving forces behind private food law? Probably they are too numerous to provide an exhaustive overview and motives may differ from stakeholder to stakeholder, but at least some points can be identified.

The motive mentioned in most private food schemes is food safety. Food safety is important for the protection of consumers, to comply with consumers wishes and also to comply with public law requirements.

Compliance with public law requirements can be a motive in itself.[104] To comply with their own legal obligations, businesses depend on how the product has been dealt with upstream. Therefore they may want to ensure themselves with private law instruments that legal obligations are being complied with, or to impose these obligations on producers working in countries where different legal requirements apply, thus using private law to bridge the gap between different legal systems. Connected to compliance is liability. On the one hand businesses may try to pass on liability to other links in chain (some require insurance and guarantees from producers), on the other hand explicit agreements are a way to show that everything possible has been done to avoid non-compliance. In civil and criminal cases this may be used in what – in the UK – is called a due diligence defence.

Further private law is used to discourage the legislator from taking charge. If businesses solve problems there will be less urgency for the legislator to intervene. Businesses prefer private law to public law as it reflects their own wishes better and it is easier to change if the need arises. Also private law can be used to supplement or repair public law. For example public law on traceability (Article 18 of Regulation 178/2002) is lacking in certainty whether internal traceability

[103] Hospes, O., 2009. Regulating biofuels in the name of sustainability or the right to food? In: Hospes, O. and Van der Meulen, B. (eds.) Fed up with the right to food? The Netherlands' policies and practices regarding the human right to adequate food. Wageningen Academic Publishers, Wageningen, the Netherlands, pp. 121-135. See for another example of the use of private schemes by public authorities to influence behaviour abroad, Chapter 7 in this book.

[104] Empirical research shows that private standards are helpful in complying with public law requirements. The explaining factor is probably the embeddedness of private standards in audit schemes that provide feedback on performance. See Van der Meulen, B., 2009. Reconciling food law to competitiveness. Wageningen Academic Publishers, Wageningen, the Netherlands.

(within a business) is required.[105] ISO 22.000 explicitly requires internal traceability. While EU legislation exempts the primary sector from HACCP, private schemes such as GlobalGAP impose it on this sector as well.

On the basis of Regulation 882/204, official controls should be risk based and control intensity can be related to compliance history. As discussed above, certification may be an instrument to convince inspection agencies that the level of compliance is high and therefore the urgency for official controls low.[106]

Private standards that go beyond compliance, that is to say apply higher safety and/or quality standards than public law,[107] may help a business to distinguish itself on the market and acquire a share of the (top end) market. Or, to phrase a similar thought differently, raising standards may be used to protect markets from competitors.

Finally moral considerations such as religion and corporate social responsibility is a driving factor of private regulation. With a view to showing their commitment to contributing their part to sustainable development, businesses bind themselves to private schemes that elaborate on these interests.

3.16 Examples

In the next sections some examples are presented of the private schemes that currently seem to be leading on the market.[108] The discussion of the examples focuses mainly on the content of the standards and less on the governance and certification structure of the schemes. The objective of these examples is to make, in addition to the structure set out above, the content of private food law more concrete. The examples have all been taken from the area of private food *safety*

[105] The Standing Committee on the Food Chain and Animal Health has published a more or less official interpretation of the General Food Law, where they argue that 'the Regulation does not expressly compel operators to establish a link (so called internal traceability) between incoming and outgoing products. Nor is there any requirement for records to be kept identifying how batches are split and combined within a business to create particular products or new batches'. See EU, 2010. Guidance on the Implementation of Articles 11, 12, 14, 17, 18, 19 and 20 of Regulation (EC) N° 178/2002 on General Food Law. Conclusions of the standing committee on the food chain and animal health. Available at: http://ec.europa.eu/food/food/foodlaw/guidance/guidance_rev_8_en.pdf.

[106] On the stacking of private and official controls see: Van der Meulen, B.M.J. and Freriks, A.A., 2006. Millefeuille. The emergence of a multi-layered controls system in the European food sector. Utrecht Law Review 2(1): 156-176.

[107] According to Article 5.2(3) of the EU best practice guidelines for voluntary certification schemes for agricultural products and foodstuffs, it should be clearly indicated where this is the case. EU, 2010. Commission Communication – EU best practice guidelines for voluntary certification schemes for agricultural products and foodstuffs. Official Journal of the European Union C 341, 16/12/2010: 5-11.

[108] The overview is based on the information provided by the owners of these schemes. No critical comparison is intended at this stage of the research. A more elaborate overview can be found in Chapter 4 by Theo Appelhof and Ronald van den Heuvel.

law. As we have seen above, many other areas of concern are covered by private food law as well.

Public law requirements on food businesses can be distinguished[109] in rules regarding the *product* (vertical standards, market approval requirements for certain ingredients and safety objectives in the form of maximum levels of contaminants and residues), rules regarding the *process* (hygiene, traceability and incident management), rules regarding the *presentation* (labelling and advertisement) and public powers (enforcement and incident management by authorities). In private food law we find similar ingredients, but in a different mix (see Figures 3.3 and 3.4). *Product* related rules mainly concern safety and quality objectives. Rules on the *process* (hygiene, traceability and risk management) are the core. *Labelling* provisions will usually be limited to the use of the certification mark. The place public powers of inspection and enforcement hold in public food law, in private food law is taken by the powers granted by the businesses themselves to the auditors and certifiers. The most important additions to the types of food rules we have encountered in public law, are provisions governing business organisation and management systems including management commitment and provisions on information sharing within the food chain (so-called chain transparency).[110]

My first impression in comparing private standards to legislation is that the drafting is done sloppily. It seems that lawyers experienced in drafting legislation are not involved. Nevertheless, stakeholders seem to understand the meaning of private standards better than they do legislation.[111] Maybe this can be explained by a different attitude towards private standards than towards legislation. Private standards have their place within a business relation that stakeholders intend to continue. This is a strong motivator to understand private standards the way they are meant. In case of legislation by contrast, misunderstanding may justify non-compliance to what the legislator envisaged. Lawyers are trained to search

[109] For this approach see: in general Van der Meulen, B. and Van der Velde, M., 2008. European Food Law Handbook. Wageningen Academic Publishers, Wageningen, the Netherlands; and in particular: Van der, Meulen, B.M.J., 2009. The system of food law in the European Union. Deakin Law Review 14(2): 305-339. See also: Van der Meulen, B. and Van der Velde, M., 2010. The general food law and EU food legislation. In: Oskam, A., Meester, G. and Silvis, H. (eds.) EU Policy for agriculture, food and rural areas. Wageningen Academic Publishers, Wageningen, the Netherlands, pp. 211-224.

[110] On private schemes, see: Van Plaggenhoef, W., Batterink, M. and Trienekens, J.H., 2003. International trade and food safety. Overview of legislation and standards. Wageningen University, Wageningen, the Netherlands; Chia-Hui Lee, G., 2006. Private food standards and their impacts on developing countries. European Commission, DG Trade Unit G2; Will, M. and Guenther, D., 2007. Food quality and safety standards, as required by EU law and the private industry with special reference to the MEDA countries' exports of fresh and processed fruits & vegetables, herbs & spices. A practitioners' reference book. 2nd edition GTZ 2007; OECD, 2006. Working party on agricultural policies and markets, final report on private standards and the shaping of the agro-food system.

[111] On this topic see: Van der Meulen, B., 2009. Reconciling food law to competitiveness. Report on the regulatory environment of the European food and dairy sector. Wageningen Academic Publishers, Wageningen, the Netherlands.

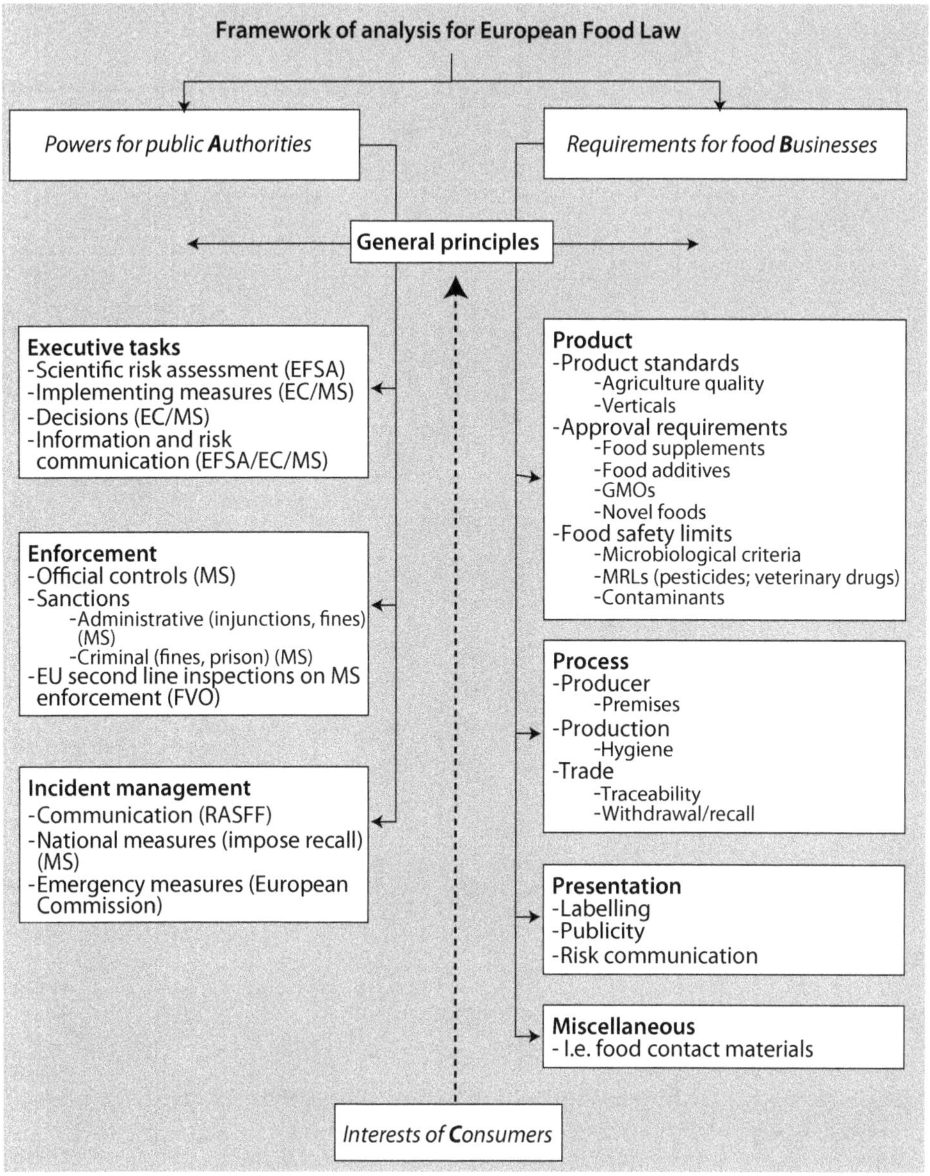

Figure 3.3. Possible framework for the analysis of EU (public) food law.

for alternative meanings within the wording of the law, an attitude that would mean the end of the business relation if it were applied to contractual provisions prior to a conflict.

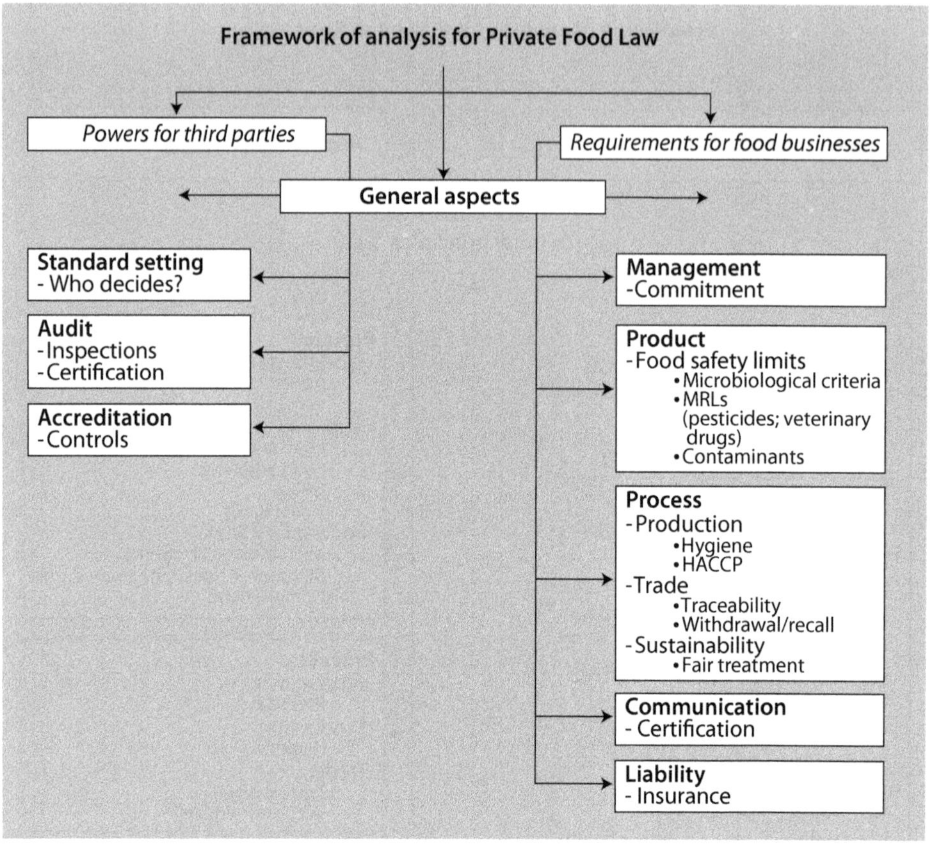

Figure 3.4. Possible framework for the analysis private food law.

3.17 Underlying concepts

3.17.1 GAP/GMP

Many private food safety systems are based on HACCP and some form of good practices, like good agricultural practices (GAP) or good manufacturing practices (GMP). In the words of the UN Food and Agriculture Organization FAO,[112] the concept of Good Agricultural Practices has evolved in recent years in the context of a rapidly changing and globalising food economy and as a result of the concerns and commitments of a wide range of stakeholders about food production and security, food safety and quality, and the environmental sustainability of agriculture. These stakeholders include governments, food processing and retailing industries, farmers, and consumers, who seek to meet specific objectives of food security, food quality,

[112] FAO Committee on Agriculture, Development of a Framework for Good Agricultural Practices, Seventeenth Session, Rome, 31 March-4 April 2003.

production efficiency, livelihoods and environmental benefits in both the medium and long term. GAP offers a means to help reach those objectives. Broadly defined, GAP applies available knowledge to addressing environmental, economic and social sustainability for on-farm production and post-production processes resulting in safe and healthy food (and non-food agricultural products). Many farmers in developed and developing countries already apply GAP through sustainable agricultural methods such as integrated pest management, integrated nutrient management and conservation agriculture. These methods are applied in a range of farming systems and scales of production units, including as a contribution to food security, facilitated by supportive government policies and programmes.

GAP represents the state of the art in sustainable agriculture. Something similar is true for GMP. For this reason, its content in terms of do's and don'ts is continuously developing and cannot be caught in one lasting description. It is not law in itself,[113] but through private schemes it can acquire legal relevance.

3.17.2 HACCP

The version of HACCP mandatory for food business operators, is the one codified in Regulation 852/2004 of the European Union. The version referred to in private schemes is often the one laid down in the Codex Alimentarius.[114] The former is based on the latter but it is not as elaborate. Through their inclusion in private schemes, the non-binding Codex acquires a measure of legal effect. Inclusion of HACCP in private schemes adds for European businesses applicability in sectors exempted from Regulation 852/2004 (the primary sector in particular), enforcement through private law instruments and visibility through certification.[115] For non-European businesses it brings an obligation that may not or not in the same way follow from their own public law system.

3.18 EurepGAP/GlobalGAP[116]

EurepGAP started in 1997 as an initiative by retailers belonging to the Euro-Retailer Produce Working Group (EUREP). British retailers in conjunction with supermarkets in continental Europe were the driving forces. They reacted to

[113] Luning, P.A., Marcelis, W.J. and Jongen, W.M.F., 2009, Food Quality Management. A techno-managerial approach, Wageningen Academic Publishers, Wageningen, the Netherlands, p. 225 refer to Good Practices as 'self-discipline'.

[114] The Codex Basic Text on hygiene, 2009, is a standard work comprising of: Recommended International Code of Practice General Principles of Food Hygiene CAC/RCP 1-1969, Rev. 4 (2003); Hazard Analysis and Critical Control Point (HACCP) System and Guidelines for its Application; Principles for the Establishment and Application of Microbiological Criteria for Foods CAC/GL 21 - 1997, and Principles and Guidelines for the Conduct of Microbiological Risk Assessment CAC/GL-30 (1999).

[115] See for example the Dutch HACCP certification scheme (based in Apeldoorn, the Netherlands). See: http://www.foodsafetymanagement.info.

[116] Information from the GlobalGAP website http://www.globalgap.org.

growing concerns of consumers regarding product safety, environmental and labour standards and decided to harmonise their own often very different standards for agricultural products.

The development of common certification schemes was considered to also be in the interest of producers. Those with contractual relations to several retailers explained that they had to undergo multiple audits against different criteria every year. With this is in mind, EUREP started working on harmonised standards and procedures for the development of Good Agricultural Practices[117] in conventional agriculture including highlighting the importance of Integrated Crop Management and a responsible approach to worker welfare. This resulted in what was initially called the European Retailers Protocol for Good Agricultural Practice (EurepGAP).

Over the next ten years a growing number of producers and retailers around the globe joined in with the idea as this matched the emerging pattern of globalised trading: EurepGAP began to gain in global significance. To align EurepGAP's name with the proposition as the pre-eminent international GAP-standard and to prevent confusion with its growing range of public sector and civil society stakeholders, the Eurep Board decided to undertake the step to re-brand. It was considered a natural path and evolution that led EurepGAP to become GlobalGAP. The decision was announced in September 2007 at the 8th global conference in Bangkok.

GlobalGAP has established itself as a key reference for Good Agricultural Practices in the global market-place, by translating consumer requirements into agricultural production in a rapidly growing list of countries – currently more than 80 on every continent (Figure 3.5).

GlobalGAP is a private sector body that sets standards for the certification of agricultural products around the globe. The aim is to establish one single standard (the Integrated Farm Assurance (IFA) Standard) for Good Agricultural Practice with different product applications capable of fitting to the whole of global agriculture. Governance is by a Board whose decisions are based on a structured consultation process. In the board retailers and suppliers are represented. Sector specific interests and multi-stakeholder input are consolidated to ensure global acceptance. Sector Committees discuss and decide upon product and sector specific issues. All committees have 50% retailer and 50% producer/supplier representation.

GlobalGAP is a pre-farm-gate standard, which means that the certificate covers the process of the certified product from farm inputs like feed or seedlings and all the farming activities until the product leaves the farm. GlobalGAP is a business-to-business (B2B) label and is therefore not directly visible to consumers. Its certification is carried out by more than 100 independent and accredited certification bodies in

[117] Hence the GAP part in the name.

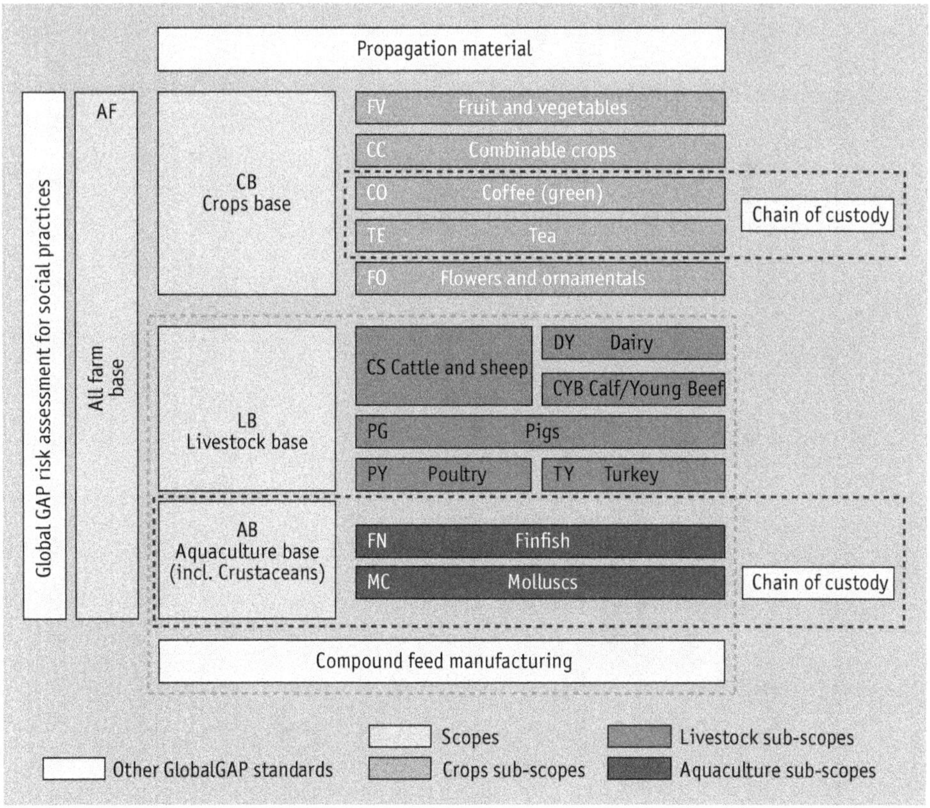

Figure 3.5. GlobalGAP structure of the Integrated Farm Assurance Standard.(Source: http://www.globalgap.org/cms/front_content.php?idcat=176.)

more than 80 countries. It is open to all producers worldwide. GlobalGAP includes annual inspections of the producers and additional unannounced inspections.

GlobalGAP consists of a set of normative documents. These documents cover the GlobalGAP General Regulations, the GlobalGAP Control Points (Textbox 3.5) and Compliance Criteria and the GlobalGAP Checklist.

As many other on-farm assurance systems have been in place for some time prior to the existence of GlobalGAP, a way had to be found to encourage the development of regionally adjusted management systems and so to prevent farmers from having to undergo multiple audits. Existing national or regional farm assurance schemes can seek recognition as equivalent to GlobalGAP through independent benchmarking.

The GlobalGAP standard is subject to a three year revision cycle of continuous improvement to take into account technological and market developments.

TextBox 3.5. GlobalGAP on traceability and record keeping (Source: GlobalGAP Control Points and Compliance Criteria Plant Propagation Material March 2008).

N° b	Control point	Compliance criteria	Level
PM 1	Traceability		
PM 1.1	Is GLOBALGAP (EUREPGAP) registered product traceable back to and trackable from the registered nursery (and other relevant registered areas) where it has been grown?	There is a documented identification and traceability system that allows GLOBALGAP (EUREPGAP) registered plants to be traced back by individual batch numbers which relate to customer orders, on a per batch basis to inputs (such as Seed Lot / Growing Media Batch / Growing or Germination Temperature Regimes / Crop protection materials applied / plant movements within the nursery) to the registered nursery, and tracked forward to the immediate customer. No N/A.	Major must
PM 1.2	Do all propagators have a documented procedure to manage the withdrawal of registered products from the market?	All propagators must have access to documented procedures which identify the type of event that may result in a withdrawal, persons responsible for taking decisions on the possible withdrawal of product, the mechanism for notifying customers and the GLOBALGAP (EUREPGAP) CB (if a sanction was not issued by the CB and the propagator or group recalled the products out of free will) and methods of reconciling stock. The procedures must be tested annually to ensure that it is sufficient. Procedure must be demonstrated.	Major must
PM 2	Record keeping and internal self-assessment		
PM 2.1	Are all records requested during the external inspection accessible and kept for a minimum period of time of two years, unless a longer requirement is stated in specific control points?	Propagators keep up to date records for a minimum of two years from the date of first inspection, unless legally required to do so for a longer period. No N/A.	Minor must
PM 2.2	Does the propagator take responsibility to undertake a minimum of one internal self-assessment per year against the GLOBALGAP (EUREPGAP) Standard?	There is documentary evidence that the GLOBALGAP (EUREPGAP) or benchmarked standard internal self-assessment under responsibility of the propagator has been carried out and are recorded annually. No N/A.	Major must
PM 2.3	Are effective corrective actions taken as a result of non-conformances detected during the internal self-assessment?	Effective corrective actions are documented and have been implemented. No N/A	Major must

3.19 BRC[118]

In 1998 the British Retail Consortium (BRC based in London), responding to industry needs, developed and introduced the BRC Food Technical Standard to be used to evaluate manufacturers of retailers own brand food products. In BRC the British supermarkets Tesco, Sainsbury, Safeway and Summerfield participate. In early days each retailer inspected his own suppliers. These common efforts to inspect suppliers have huge cost advantages for retailers, because a supplier fulfils the requirements of all British retailers once.

BRC is designed to be used as a pillar to help retailers and brand owners with their 'due diligence' defence, should they be subject to a prosecution by the enforcement authorities. Under EU food law, retailers and brand owners have a legal responsibility for their brands.[119]

In a short space of time, the BRC Standard became invaluable to other organisations across the sector. It is regarded as a benchmark for best practice in the food industry. This and its use outside the UK has seen it evolve into a global standard used not just to assess retailer suppliers, but as a framework upon which many companies have based their supplier assessment programmes and manufacture of some branded products.

The majority of UK, and many continental European and global retailers, and brand owners will only consider business with suppliers who have gained certification to the appropriate BRC Global Standard (Textbox 3.6).

Following the success and widespread acceptance of the Global Standard - Food, the BRC published the first issue of the Packaging Standard in 2002, followed by Consumer Products Standard in August 2003, and by the BRC Global Standard - Storage and Distribution in August 2006. In 2009, the BRC partnered with the Retail Industry Leaders Association (RILA) to develop the Global Standard for Consumer Products North America edition. Each of these Standards is regularly reviewed and each standard is fully revised and updated at least every 3 years after extensive consultation with a wide range of stakeholders.

[118] Information from BRC website http://www.brc.org.uk.

[119] See for example Article 17(1) of Regulation 178/2002 (the General Food Law) and Directive 85/374 on product liability. EU, 2002. Regulation (EC) No 178/2002 of the European Parliament and of the Council of 28 January 2002 laying down the general principles and requirements of food law, establishing the European Food Safety Authority and laying down procedures in matters of food safety. Official Journal of the European Union L 31, 1/2/2002: 1-24; EU, 1985. Council Directive 85/374/EEC of 25 July 1985 on the approximation of the laws, regulations and administrative provisions of the Member States concerning liability for defective products. Official Journal of the European Union L 210, 7/8/1985: 29-33.

Textbox 3.6. BRC principle on management commitment (From BRC Global Standard for Food Safety, issue 5 2008, section 2.1 Principles of the Global Standard for Food Safety).

2.1.1 Senior Management Commitment
Within a food business, food safety must be seen as a cross-functional responsibility, including activities that draw on many departments using different skills and levels of management expertise in the organisation. Effective food safety management extends beyond technical departments and must involve commitment form production operations, engineering, distribution management, procurement of raw materials, customer feed back and human resource activity such as training. The starting point for an effective food safety plan is the commitment of senior management to the development of an all-encompassing policy as a means to guide the activities that collectively assure food safety. The Global Standard for Food Safety places a high priority on clear evidence of senior management commitment.

3.20 IFS[120]

In 2002, in order to create a common food safety standard, German food retailers from the HDE (Hauptverband des Deutschen Einzelhandels) have developed a common audit standard called International Food Standard or IFS. In 2003, French food retailers (and wholesalers) from the FCD (Fédération des entreprises du Commerce et de la Distribution) have joined the IFS Working Group.

The aim of the IFS (now based in Paris) is to create a consistent evaluation system for all companies supplying retailer branded food products with uniform formulations, uniform audit procedures and mutual acceptance of audits, which will create a high level of transparency throughout the supply chain. Its scope is now beyond the food sector alone: 'IFS' has become to mean: International Featured Standard. Among its standards IFS food still holds a prominent position.

The IFS food defines requirements in content, procedure and evaluation of audits and a requirement profile for the certification bodies and auditors. The IFS food standard (the so-called catalogue of requirements) consists of five parts called chapters:

- senior management responsibility;
- quality management system;
- resource management;
- product process;
- measurements, analyses and improvements.

[120] The information in this section has been taken from the IFS website http://www.ifs-certification.com.

The auditor will audit against the IFS food standard which is divided into two levels plus recommendation on higher level. The chapter 'Senior Management Responsibility' deals with the responsibility of the management, the management commitment, the management review and the customer focus. In the chapter 'Quality Management System' requirements concerning the HACCP system, the HACCP team and the HACCP analysis are defined. It also contains rules for a quality manual to be applied and the obligation to keep reports and documents. The chapter 'Resource Management' addresses personnel issues (hygiene, medical screening) and staff facilities. The chapter 'Product Process' is the most extensive one. It considers topics about e.g. specifications for products, factory environment, pest control, maintenance, traceability; GMOs and allergens. The last chapter, 'Measurements, Analyses and Improvements', deals with e.g. internal audit, all kind of controls during production steps, product analysis and corrective actions.

The requirements for auditors and the certification bodies are strictly regulated. All certification bodies shall have an accreditation against EN 45011 on IFS food.[121] Only authorised auditors who have passed a written and oral examination can audit against the standard. The auditor shall have professional knowledge of the IFS food. The auditors can only audit against their competence in a certain sector (at least 2 years professional experience in the specific sector or at least 10 audits in this sector). Finally, auditors who comply with these requirements shall only work for one IFS certification body accredited for auditing against the IFS food.[122]

3.21 SQF[123]

Safe Quality Food (SQF; now based in Arlington, USA) is an Australian initiative. Taking over this system seems to have been the American answer to the mainly European initiatives described above.

Besides food safety, SQF focuses on product quality and stimulation of improvement strategies. The main goal of SQF is to control the whole chain. However, SQF believes that one standard does not work for all companies in the chain and that most other standards only work for big companies. Most procedures associated with the standards are considered too elaborate and laborious for small companies. So SQF developed two different norms, the SQF 1000 and the SQF 2000. The SQF 2000 Code was developed in consultation with food industry and quality professionals. HACCP guidelines, as developed by the Codex Alimentarius Commission, form the basis of the Code. Unlike other well-recognised quality systems like BRC, HACCP

121 Again an example of interconnected private schemes as discussed above.

122 The question if such requirement is compatible with competition law (Article 81 EC Treaty) is outside the scope of this chapter.

123 The information for this section has been taken from the SQF 2000 Code 6th edition issued August 2008 (and update July 2010) and the SQF 2000 Certification Trade Mark rules for use 7th edition amended November 2005. See also http://www.sqfi.com.

and ISO 9000,[124] SQF combines a management quality system, like ISO 9000 and a food safety system (HACCP) with requirements for tracking and tracing. Besides the Critical Control Points (CCP) for food safety, Critical Quality Points are also identified, which makes SQF an integrated system.

The SQF codes (in particular the 1000 and 2000 Code) provide the food sector (primary producers, food manufacturers, retailers, agents and exporters) a food safety and quality management certification program that is tailored to its requirements and enables suppliers to meet regulatory, food safety and commercial quality criteria in a cost effective manner. In 1994 the Code was developed and pilot programs implemented to ensure its applicability to the food sector. It was circulated in draft form for comment to experts in quality management, food safety, and food regulation, food processing, agriculture production systems, food retailing, food distribution and HACCP.

The Food Marketing Institute (FMI) acquired the rights to the SQF Program in August 2003 and has established the SQF Institute (SQFI) Division to manage the program. The SQF 2000 Code is recognised by the Global Food Safety Initiative[125] as a standard that meets its benchmark requirements.

The SQF 2000 Code can be used by all sectors of the food industry. The Code is a HACCP based quality management system that encapsulates NACMCF[126] and CODEX HACCP Principles and Guidelines, proven methods used by the food industry to reduce the incidence of unsafe food reaching the marketplace (Textbox 3.7). It is designed to support industry or company branded product and to offer benefits to suppliers at all links in the food supply chain.

The SQF 2000 Code enables a supplier to demonstrate that they can supply food that is safe and that meets the quality specified by a customer. Certified SQF 2000 suppliers receiving raw materials from suppliers who have implemented the SQF 1000 Code can ensure that, through these complimentary systems, product is traceable from the producer to the consumer.[127]

The SQF 2000 Code also provides a mechanism for the food sectors of developing countries seeking to effectively enter the global food market to implement a management system that addresses their needs and the needs of their customers.

[124] On quality management in general. See here after section 3.22 for ISO 22000 on food.
[125] Discussed in section 3.23.
[126] National Advisory Committee on Microbiological Criteria for Foods.
[127] Here we find an example where a private scheme goes beyond compliance. Unlike EU food law, traceability is not mandatory in US food law.

Textbox 3.7. SQF 2000 on HACCP (edition 6 August 2008, amended July 2010).

9. Principles and applications of HACCP

Table 1. A description of the 12 HACCP steps that comprise the HACCP method (Adapted from Codex Alimentarius Commisssion – recommended International Code of Practice Principles of Food Hygiene, CAC/RCP 1-1969, Rev. 4-2003)

Preliminary Steps	1. Assemble HACCP team with expertise in product and processes 2. Describe product 3. Identify intended use 4. Construct flow diagram 5. Confirm flow diagram against process in operation (or planned process)
HACCP Principle	**HACCP Application**
1. Conduct a hazard analysis.	6. List all potential hazards associated with each step and consider any measures to control identified hazards.
2. Determine Critical Control Points (CCPs).	7. Determine CCPs.
3. Establish critical limit(s).	8. Establish critical limits and tolerance levels. Determine at what point critical limit is exceeded based on known limits or risk assessment if unknown.
4. Establish system to monitor control of CCP(s).	9. Establish a monitoring system for CCP that is able to detect loss of control i.e. when critical limits are exceeded. Consider continuous monitoring and/or periodic audit.
5. Establish corrective action to be taken when monitoring indicates CCP(s) are not under control.	10. Establish corrective actions that are able to deal with loss of control when it occurs and is capable of determining when CCP has been brought under control.
6. Establish procedures for verification to confirm that the HACCP system is working effectively.	11. Establish procedures for verification or audit that include review of HACCP system and records, records of deviations and actions taken in order to confirm that CCPs are kept under control.
7. Establish documentation covering all procedures and records appropriate to these principles and their application.	12. Documentation and record keeping should be appropriate to the nature and scale of the operation.

3.22 FS22000[128]

The youngest of the main private food safety schemes is ISO 22000. When GFSI evaluated the ISO 22000 for approval, they determined that they wanted more requirements identified for Prerequisite Programs. To fill this gap, the British Standards Institution wrote a document called PAS 220. The combination of PAS 220 and ISO 22000 has been approved by GFSI for a registration scheme, and is called FS22000 (previously FSSC 22000). This scheme is run by the Foundation for Food Safety Certification (FSSC).

The International Organization of Standardization (ISO) has decades of experience developing standards for many different types of applications. One of the most popular and most recognised is the Quality Management System standard ISO 9001. This standard was developed to provide a uniform standard worldwide for quality management. A buyer in one part of the world would have a degree of confidence in the quality practices of a registered company in another part of the world. This standard was used as the basis for other more specific standards for quality management in the automotive industry, the medical device industry, and the aerospace industry.

Now this approach has been taken for food safety management. ISO and its member countries used the Quality Management System approach, and tailored it to apply to food safety, incorporating the HACCP principles into the quality management system. The resulting standard is ISO 22000. ISO 22000 requires that the business design and document a Food Safety Management System (FSMS). The standard contains the specific requirements to be addressed by the FSMS. Generally the standard addresses:

- Having an overall Food Safety Policy for the organisation, developed by top management.
- Setting objectives that will drive companies' efforts to comply with this policy.
- Planning and designing a management system and documenting the system.
- Maintaining records of the performance of the system.
- Establishing a group of qualified individuals to make up a Food Safety Team.
- Defining communication procedures to ensure effective communication with important contacts outside the company (regulatory, customers, suppliers and others) and for effective internal communication.
- Having an emergency plan.
- Holding management review meetings to evaluate the performance of the FSMS.
- Providing adequate resources for the effective operation of the FSMS including appropriately trained and qualified personnel, sufficient infrastructure and appropriate work environment to ensure food safety.
- Following HACCP principles.

[128] http://www.22000-tools.com.

- Establishing a traceability system for identification of product.
- Establishing a corrective action system and control of nonconforming product.
- Maintaining a documented procedure for handling withdrawal of product.
- Controlling monitoring and measuring devices.
- Establishing and maintaining and internal audit program.
- Continually updating and improving the FSFM.

ISO 22000 provides brand holders with a quality system equivalent to the other systems mentioned in this chapter that have primarily been developed with a view to retailers' brands.

3.23 GFSI[129]

The Global Food Safety Initiative (GFSI) co-ordinated by CIES (Comité International d'Entreprise à Succursales; The Food Business Forum), was launched in May 2000 as a reaction to the proliferation of private standards. Much of the advantages of private food law will be lost, if every important player in the marketplace defines a separate set of private standards.

GFSI started to benchmark private standards. This gives businesses the opportunity to send out the message in the market that for them it does not matter which particular standard is applied, as long as it is GFSI endorsed. In this way GFSI is developing into the standard of standards. The most powerful retailers in Europe, Asia and the USA have agreed to demand GFSI compliant certification for their own private label products. The GFSI mission is: continuous improvement in food safety management systems to ensure confidence in the delivery of safe food to consumers.

The GFSI objectives are:
1. Convergence between food safety standards through maintaining a benchmarking process for food safety management schemes.
2. To improve cost efficiency throughout the food supply chain through the common acceptance of GFSI recognised standards by retailers around the world.
3. To provide a unique international stakeholder platform for networking, knowledge exchange and sharing of best food safety practices and information.

In the light of the plethora of food safety standards, the GFSI Task Force decided not to write a new standard. Instead, they compiled a set of 'Key Elements' to serve as the requirements against which existing food safety standards will be benchmarked. The 'Key Elements' as defined by the Task Force are:

[129] The information for this section has been taken from the GFSI pages on http://www.ciesnet.com and http://www.mygfsi.com/.

1. Food Safety Management Systems.
2. Good Practices for Agriculture, Manufacturing and Distribution.
3. HACCP.

Under the umbrella of the Global Food Safety Initiative, seven major retailers have come to a common acceptance of initially four GFSI benchmarked food safety schemes. Retailers accept certificates based on standards in order to be able to make an assessment of their suppliers of private-label products and fresh products and meat, to ensure that production is carried out in a safe manner. There are many of these standards and suppliers with many customers may be audited many times per year, at a high cost and with little added benefit.

The GFSI Guidance Document Sixth Edition (released January 2011), contains commonly agreed criteria for food safety standards, against which any food or farm assurance standard can be benchmarked. In addition to the 'key elements' the Guidance Document holds 'requirements for the delivery of food safety management systems' regarding certification and accreditation. GFSI does not undertake any accreditation or certification activities.

The benchmarking work undertaken by the standard owners and other key stakeholders on four food safety schemes (initially BRC, IFS, Dutch HACCP and SQF) has reached a point of convergence. Each scheme aligned itself with common criteria defined by food safety experts from the food business, with the objective of making food manufacture as safe as possible. As a result, this will also drive cost efficiency in the supply chain and reduce the duplication of audits.

The GFSI vision of 'once certified, accepted everywhere' has become a reality in the sense that Carrefour, Metro, Migros, Ahold, Wal-Mart and Delhaize have agreed to reduce duplication in the supply chain through the common acceptance of any of the GFSI benchmark schemes. Tesco, however, has withdrawn. It was unwilling to accept other schemes as equivalent to its own 'Nature's choice'. The number of schemes recognised by GFSI has continuously grown. See Table 3.1 for an overview.

The GFSI Foundation Board, a retailer-driven group, with manufacturer advisory members, provides the strategic direction and oversees the daily management. Membership of the Board is by invitation only.

The GFSI Technical Committee was formed in September 2006 and is composed of retailers, manufacturers, standard owners, certification bodies, accreditation bodies, industry associations and other technical experts. It provides technical expertise and advice for the GFSI Board and replaces a previous GFSI retailer-only Task Force. Membership of the Technical Committee is by invitation only.

Table 3.1. 7 GFSI Recognised Schemes.[1]

Manufacturing Schemes:
- BRC Global Standard Version 5
- Dutch HACCP (Option B)
- FSSC 22000
- Global Aquaculture Alliance BAP Issue 2 (GAA Seafood Processing Standard)
- Global Red Meat Standard Version 3
- International Food Standard Version 5
- SQF 2000 Level 2
- Synergy 22000

Primary Production Schemes:
- CanadaGAP
- GlobalGAP IFA Scheme V3
- General Regulations: V3.1_Nov09 (all scopes)
- Fruit and Vegetables: 3.0-2_Sep07
- Livestock Base: 3.0-4_Mar10
- Aquaculture: V1.02_March10
- SQF 1000 Level 2

Primary and Manufacturing Scheme:
- PrimusGFS

[1]As published on http://www.mygfsi.com/about-gfsi/gfsi-recognised-schemes.html.

A crucial moment in its development was the entry of Wal-Mart into the scheme (February 2008). Noteworthy is Wal-Mart's comment: 'GFSI Standards provide real time details on where suppliers fall short in food safety on a plant-by-plant basis and go beyond the current FDA[130] or USDA[131] required audit process.' Wal-Mart is the first US-based grocery chain to require GFSI, the company claims. The company has published a schedule to suppliers requiring completion of initial certification between July and December 2008, with full certification required by July 2009.

Through GFSI benchmarking private food law seems to be achieving what public food law (through the Codex Alimentarius) never could: truly global harmonisation of food safety standards.

3.24 Public law on private food law

The European Commission has reacted to the emergence of private food law. A study conducted for DG Agri identified 441 different schemes.[132] The European

[130] Food and Drug Administration. The food regulatory agency in the USA.
[131] The US Department of Agriculture.
[132] See: http://ec.europa.eu/agriculture/quality/index_en.htm.

Commission decided that legislative action was not warranted to address the potential drawbacks in certification schemes at this stage. Instead, drawing on comments from stakeholders, the Commission undertook to develop guidelines for certification schemes for agricultural products and foodstuffs. So on 16 December 2010, it published 'Commission Communication – EU best practice guidelines for voluntary certification schemes for agricultural products and foodstuffs' (2010/OJ C 341/04).[133]

These guidelines are designed to describe the existing legal framework and to help improving the transparency, credibility and effectiveness of voluntary certification schemes and ensuring that they do not conflict with regulatory requirements. The guidelines warn the member states to respect the state aid rules when they give support to certain schemes. Businesses are reminded of the rules on competition. Certification schemes may not lead to anticompetitive behaviour.

From food law provisions are quoted that consumers may not be misled. I am mystified by the quote in full of Article 5(1) of Regulation 178/2002 'Food law shall pursue one or more of the general objectives of a high level of protection of human life and health and the protection of consumers' interests, including fair practices in food trade, taking account of, where appropriate, the protection of animal health and welfare, plant health and the environment'. Is the Commission implying that private certification schemes are within the scope of the concept of 'food law' and have to comply with the objectives of food law? It does not seem a very likely interpretation, but no other interpretation readily presents itself. In fact I would be inclined to believe that private regulation is not limited to serving the objectives listed in Article 5, but can also legitimately serve other interests even those of the business sector itself.[134]

The question about the public status of private food law also presents itself in the context of the WTO.

3.25 WTO

Private food law is rapidly replacing public law as the determining factor in international food chains. Within the WTO the question has been raised if private standards constitute a new generation of trade barriers that member states of the

[133] EU, 2010. Commission Communication – EU best practice guidelines for voluntary certification schemes for agricultural products and foodstuffs. Official Journal of the European Union C 341, 16/12/2010: 5-11. See also Appendix I to this book. Also noteworthy is the Communication from the Commission Contributing to Sustainable Development: The role of Fair Trade and non-governmental trade-related sustainability schemes, Brussels 5.5.2009 COM(2009) 215 final, Appendix II to this book.
[134] On Article 5 of Regulation 178/2002, see: Van der Meulen, B., 2010. The function of Food Law. On the objectives of food law, legitimate factors and interests taken into account. European Food and Feed Law Review 5(2): 83-90.

WTO should take on.[135] Two WTO agreements may be of relevance in this context; the TBT Agreement and the SPS Agreement.

3.25.1 TBT Agreement

In the Agreement on Technical Barriers to Trade, WTO members have set requirements for technical regulations and standards to ensure that they will not constitute unjustified barriers to international trade. WTO members are strongly encouraged to use international standards and to support international standardising bodies. In the agreement ISO/IEC is positioned as an overarching international standardising body. Other standardising bodies within the ambit of the TBT Agreement must be notified to ISO/IEC. This seems to imply that WTO members relying on ISO standards may believe to comply with TBT requirements.

3.25.2 SPS Agreement

The Agreement on the application of Sanitary and Phytosanitary measures (the SPS Agreement) deals with measures aiming to protect health of humans, animals and plants. Such measures are acceptable if the do not go beyond what is necessary and do not constitute disguised trade barriers. Measures to protect food safety are by definition SPS measures.

Some members of the WTO believe that private standards are within the ambit of the SPS Agreement and furthermore do not conform to the requirements. St. Vincent and the Grenadines, supported by Jamaica, Peru, Ecuador, and Argentina, complained that 'EurepGAP' SPS standards imposed by the Euro-Retailer Produce Working Group, composed primarily of food retailers, were more strict than EU governments' requirements. Referring to Article 13 of the SPS Agreement, which says that member governments 'shall take such reasonable measures as may be available to them to ensure that non-governmental entities within their territories (...) comply with the relevant provisions of this agreement,' these countries argue that only the public law EU rules should apply to the private sector (Textbox 3.8).[136]

So far it is believed that Article 13 of the SPS Agreement aims at entities that in reality are governmental in a private law guise. The text, however, leaves room for the interpretation that it also applies to 'real' private actors, in particular when they take on the role of regulator traditionally reserved for governments.[137] From a

[135] 29-30 meeting of the Committee on Sanitary and Phytosanitary (SPS) Measures.

[136] See ICTSD Bridges weekly news digest Volume 9 Number 24 6 July 2005 and Volume 7 Number 13, 6 July 2007.

[137] As we have seen above governments may participate in the formulation of private standards aiming to influence behaviour abroad. In such situation it is less evident that private standards originate from 'real' private actors and applicability of the SPS Agreement becomes likely, even in its more limited interpretation

Textbox 3.8. Article 13 SPS Agreement.

Article 13
Implementation
Members are fully responsible under this Agreement for the observance of all obligations set forth herein. Members shall formulate and implement positive measures and mechanisms in support of the observance of the provisions of this Agreement by other than central government bodies. Members shall take such reasonable measures as may be available to them to ensure that non-governmental entities within their territories, as well as regional bodies in which relevant entities within their territories are members, comply with the relevant provisions of this Agreement. In addition, Members shall not take measures which have the effect of, directly or indirectly, requiring or encouraging such regional or non-governmental entities, or local governmental bodies, to act in a manner inconsistent with the provisions of this Agreement. Members shall ensure that they rely on the services of non-governmental entities for implementing sanitary or phytosanitary measures only if these entities comply with the provisions of this Agreement.

legal point of view there is a fundamental difference between the situation where a product may not be brought to the market (because it does not comply with public law requirements) and the situation where a product legally brought to the market is not bought by its intended customers (because it does not comply with these customers' private law requirements). From an economic point of view and for all practical purposes these two situations amount to the same thing where the customers concerned dominate the market.

This discussion pinpoints a weak aspect of private food law. It seems at present underdeveloped in checks and balances.[138]

3.26 Conclusions

The content of private food safety schemes, just as public food safety law, is based on HACCP. The private sector is in the process of achieving what the public sector never could: world wide harmonisation of food safety standards.

The underlying legal structure is straight forward. Contractual requirements – flanked with instruments from (intellectual) property and business law – are used by dominant players in the food chain to impose 'voluntary' requirements on all players upstream, regardless in which country they are situated. Contractual requirements, audits and certification can be applied across national borders. In this sense, private food law is more *global* than international food law (such as the

[138] See Chapter 6 by Marinus Huigen for more detail on this topic.

SPS Agreement and the Codex Alimentarius). International (public) food law does not govern behaviour of stakeholders, but sets a meta-framework for (national) food law that in turn applies to stakeholders' behaviour.[139] Private food law does govern stakeholders' behaviour and in this sense private food law is more *law* than international food law.

References

De Bakker, E., Backus, G., Selnes, T., Meeusen, M., Ingenbleek, P. and Van Wagenberg C., 2010. Nieuwe rollen, nieuwe kansen? Een programmeringsstudie voor toezicht op controle in het agro-foodcomplex. Rapport 6.07.08. LEI, The Hague, the Netherlands. Available at: http://www.lei.wur.nl/NL/publicaties+en+producten/LEIpublicaties/?id=786.

Beekman, V., Kornelis, M., Pronk, B., Smelt, A. en Teeuw, J., 2006. Stimulering eigen verantwoordelijkheid. Zorgen dat producenten en consumenten zorgen voor voedselveiligheid, Rapport 5.06.05. LEI, The Hague, the Netherlands. Available at: http://www.lei.wur.nl/NL/publicaties+en+producten/LEIpublicaties/?id=704.

Bergsma, N., 2010. Voedselveiligheid: certificatie en overheidstoezicht, Praktijkgids Warenwet. Sdu, The Hague, the Netherlands.

Bondt, N., Deneux, S.D.C., Van Dijke, I., De Jong, O., Smelt, A., Splinter, G., Tromp, S.O. and De Vlieger, J.J., 2006. Voedselveiligheid, ketens en toezicht op controle. Rapport 5.06.01, LEI, The Hague, the Netherlands. Available at: http://www.lei.dlo.nl/publicaties/PDF/2006/5_xxx/5_06_01.pdf.

Bunte, F., Van Galen, M., De Winter, M., Dobson, P., Bergès-Sennou, F., Monier-Dilhan, S., Juhász, A., Moro, D., Sckokai, P., Soregaroli, C., Van der Meulen, B. and Szajkowska, A., 2011. The impact of private labels on the competitiveness of the European food supply chain. Publications Office of the European Union, Luxembourg, Luxembourg. Available at: http://ec.europa.eu/enterprise/sectors/food/files/study_privlab04042011_en.pdf.

Chia-Hui Lee, G., 2006. Private food standards and their impacts on developing countries. European Commission, DG Trade Unit G2. Available at: http://trade.ec.europa.eu/doclib/docs/2006/november/tradoc_127969.pdf.

Codex Alimentarius, 2009. Basis texts on food hygiene, 4th edition. FAO, Rome, Italy. Available at: ftp://ftp.fao.org/codex/Publications/Booklets/Hygiene/FoodHygiene_2009e.pdf.

EU, 1985. Council Directive 85/374/EEC of 25 July 1985 on the approximation of the laws, regulations and administrative provisions of the Member States concerning liability for defective products. Official Journal of the European Union L 210, 7/8/1985: 29-33.

EU, 2002. Regulation (EC) No 178/2002 of the European Parliament and of the Council of 28 January 2002 laying down the general principles and requirements of food law, establishing the European Food Safety Authority and laying down procedures in matters of food safety. Official Journal of the European Union L 31, 1/2/2002: 1-24.

EU, 2003. European Parliament, Council and Commission. Interinstitutional agreement on better law-making. Official Journal of the European Union C 321, 31/12/2003: 1-5.

[139] See: Van der Meulen, B., 2010. The global arena of food law: emerging contours of a meta-framework. Erasmus Law Review 3(4).

EU, 2004a. Regulation (EC) No 852/2004 of the European Parliament and of the Council of 29 April 2004 on the hygiene of foodstuffs. Official Journal of the European Union L 139, 30/4/2004: 1-54.

EU, 2004b. Regulation (EC) No 882/2004 of the European Parliament and of the Council of 29 April 2004 on official controls performed to ensure the verification of compliance with feed and food law, animal health and animal welfare rules. Official Journal of the European Union L 65, 30/4/2004: 1-141.

EU, 2005. Commission Regulation (EC) No 2073/2005 of 15 November 2005 on microbiological criteria for foodstuffs. Official Journal of the European Union L 338, 22/12/2005: 1-26.

EU, 2008. Regulation (EC) No 765/2008 of the European Parliament and of the Council of 9 July 2008 setting out the requirements for accreditation and market surveillance relating to the marketing of products and repealing Regulation (EEC) No 339/93. Official Journal of the European Union L 218, 13/8/2008: 30-47.

EU, 2010a. Commission communication – EU best practice guidelines for voluntary certification schemes for agricultural products and foodstuffs. Official Journal of the European Union C 341, 16/12/2010: 5-11.

EU, 2010b. European Parliament, Council and Commission. The interinstitutional agreement on better law-making. Official Journal of the European Union C321, 31/12/2003: 1-5.

EU, 2010c. Standing Committee on the Food Chain and Animal Health. Guidance on the Implementation of Articles 11, 12, 14, 17, 18, 19 and 20 of Regulation (EC) N° 178/2002 on General Food Law. Conclusions of the standing committee on the food chain and animal health. Available at: http://ec.europa.eu/food/food/foodlaw/guidance/guidance_rev_8_en.pdf.

FAO, 2003. Committee on agriculture, development of a framework for good agricultural practices, Seventeenth Session, Rome, 31 March-4 April 2003. Available at: http://www.fao.org/docrep/meeting/006/y8704e.htm.

Hospes, O. and Van der Meulen, B. (eds.) Fed up with the right to food? The Netherlands' policies and practices regarding the human right to adequate food. Wageningen Academic Publishers, Wageningen, the Netherlands.

Luning, P.A., Marcelis, W.J. and Jongen, W.M.F., 2009. Food Quality Management. A techno-managerial approach. Wageningen Academic Publishers, Wageningen, the Netherlands.

OECD, 2006. Working party on agricultural policies and markets, final report on private standards and the shaping of the agro-food system. Available at: http://ec.europa.eu/food/international/organisations/sps/docs/agr-cd-apm200621.pdf.

Van der Meulen, B., 2009a. Reconciling food law to competitiveness. Report on the regulatory environment of the European food and dairy sector. Wageningen Academic Publishers, Wageningen, the Netherlands.

Van der, Meulen, B.M.J., 2009b. The system of food law in the European Union. Deakin Law Review 14(2): 305-339.

Van der Meulen, B., 2010a. The function of Food Law. On the objectives of food law, legitimate factors and interests taken into account. European Food and Feed Law Review 5(2): 83-90.

Van der Meulen, B., 2010b. The global arena of food law: emerging contours of a meta-framework. Erasmus Law Review 3(4). Available at: http://www.erasmuslawreview.nl/files/the_global_arena_of_food_law.

Van der Meulen, B.M.J. and Freriks, A.A., 2006. Millefeuille. The emergence of a multi-layered controls system in the European food sector. Utrecht Law Review 2(1): 156-176.

Van der Meulen, B. and Van der Velde, M., 2008. European food law handbook. Wageningen Academic Publishers, Wageningen, the Netherlands.

Van der Meulen, B. and Van der Velde, M., 2010. The general food law and EU food legislation. In: Oskam, A., Meester, G. and Silvis, H. (eds.) EU Policy for agriculture, food and rural areas. Wageningen Academic Publishers, Wageningen, the Netherlands, pp. 211-224.

Van Plaggenhoef, W., Batterink, M. and Trienekens, J.H., 2003. International trade and food safety. Overview of legislation and standards. Wageningen University, Wageningen, the Netherlands.

Will, M. and Guenther, D., 2007. Food quality and safety standards, as required by EU law and the private industry with special reference to the MEDA countries' exports of fresh and processed fruits & vegetables, herbs & spices. A practitioners' reference book. 2nd edition GTZ 2007. Available at: http://www2.gtz.de/dokumente/bib/07-0800.pdf.

4. Inventory of private food law

Theo Appelhof and Ronald van den Heuvel[140]

4.1 Introduction

4.1.1 Requirements of the General food Law

The development of private standards is initiated by the food industry. This does not exclude the ambition of the authorities to influence the process. The authorities actively stimulate the food industry and cooperate in the development of e.g. hygiene guides.

In Regulation 178/2002 (known as the 'General Food Law')[141] the responsibilities on food safety of both authorities and food business are laid down. In Article 13, on international standards, the Union and Member States are stimulated among others to:

> '(a) contribute to the development of international technical standards for food and feed and sanitary and phytosanitary standards;
> (e) promote consistency between international technical standards and food law while ensuring that the high level of protection adopted in the Community is not reduced.'

Article 17, in section 1, lays down the responsibilities of food businesses on food safety as follows:

> 'Food and feed business operators at all stages of production, processing and distribution within the businesses under their control shall ensure that foods or feeds satisfy the requirements of food law which are relevant to their activities and shall verify that such requirements are met.'

In Article 18 is laid down that food businesses must be able to trace their products in all stages of the production chain so that in case of incidents unsafe products can be removed from the market (recall).

---

[140] This chapter is based on a contribution in Scholten-Verheijen, I., Appelhof, T. and Van der Meulen, B., 2011. Roadmap to EU food law. Eleven International Publishers, the Hague, the Netherlands. The Roadmap provides a graphic overview of food legislation at European, global and private level accompanied by explanatory text.

[141] EU, 2002. Regulation (EC) No 178/2002 of the European Parliament and of the Council of 28 January 2002 laying down the general principles and requirements of food law, establishing the European Food Safety Authority and laying down procedures in matters of food safety. Official Journal of the European Union L 31, 1/2/2002: 1-24.

All food businesses are required to execute a risk analysis on their production (methods), implement control measures and document the results. This is laid down in detail in Article 5 of Regulation 852/2004.[142]

4.1.2 Hygiene guides

Small and medium sized companies can choose to follow the rules laid down in an appropriate hygiene guide to comply with food law. However, the use of a hygiene guide is not mandatory and companies are allowed to develop their own food safety plan. Hygiene guides are usually developed by trade organisations, but need to receive final approval by the minister responsible for the food safety policy. A risk analysis on the standard activities within the branch is included in each hygiene guide.

A hygiene guide provides instructions on how a food business can comply with all relevant legislation regarding food production, storage, transport and distribution. Often, next to legal requirements, and/or some additional branch specific requirements are included. The addition of specific requirements aims at improving the quality and by this improving public opinion on companies that form a part of the branch. The implementation of hygiene guides has developed itself in quite different ways within the different Member States. The Netherlands played a pioneering role in the development of hygiene guides. The first guide was published in 1995, many years before the General Food Law laid down this option as an alternative for a self-generated food safety plan. In total more than 600 hygiene guides have been developed by the EU Member States.[143] Some countries like Spain and Italy have developed over 100 guides, while others like Greece and Ireland do not have more than six or seven. Furthermore it should be remarked that the scope differs from country to country. In some countries the guides describe the complete production processes, while in others, e.g. Spain, many guides are limited to e.g. the implementation of traceability. Table 4.1 shows an overview of the number of hygiene guides per country at the time of writing.

4.1.3 Major food safety management systems and standards

Multinational corporations often choose to develop their own food safety management system. These systems do not only aim at compliance with the (international) law but also cover the requirements and expectations of suppliers and users in the production chain. For the users in the chain it is impossible to check all different management systems of their suppliers. In the last few decades

[142] EU, 2004. Regulation (EC) No 852/2004 of the European Parliament and of the Council of 29 April 2004 on the hygiene of foodstuffs. Official Journal of the European Union L 139, 30/4/2004: 1-54.
[143] Available at: http://ec.europa.eu/food/food/biosafety/hygienelegislation/register_national_guides_en.pdf.

Table 4.1. The number of hygiene guides per country (2010).

Country	Quantity	Country	Quantity
Austria	13	Italy	104
Belgium	24	Lithuania	9
Switzerland	2	Latvia	20
Cyprus	6	Luxembourg	9
Czech Republic	27	The Netherlands	40
Germany	47	Poland	8
Denmark	24	Portugal	31
Estonia	1	Romania	17
Greece	6	Sweden	4
Spain	126	Slovenia	6
France	34	Slovakia	9
Hungary	21	United Kingdom	11
Ireland	7		

united retailers and buying associations of agricultural produce therefore have developed 'uniform' food safety management systems (also known as standards or schemes), laying down in detail their requirements for producers and service providers. Every supplier needs to be able to demonstrate compliance with the quality management system and also obtain certification. Suppliers are furthermore obliged to let independent audits be performed to verify compliance with the standard(s). Compliance with legal requirements is one of the pre-requisites of all standards. In 2001 the Codex Alimentarius has provided guidelines on the design and use of certificates.[144]

Because of the many different standards and requirements asked for by the customers, suppliers often have to obtain multiple certificates to be able to supply all their customers. This situation can be very burdensome to many suppliers as standards do differ on certain parts; being developed with the same objectives, the principles of the standards are the same and differences are mainly of a bureaucratic nature. The Global Food Safety Initiative (GFSI) aims at merging the different standards as much as possible by accepting only those standards that are of an adequate level.[145]

The GFSI is an initiative started in 2000 by international retailers as a benchmarking instrument for food safety standards. The GFSI is governed by the CIES (a worldwide

[144] Codex Alimentarius Commission, 2001. Guidelines for design, production, issuance and use of generic official certificates. CAC/GL 38-2001. FAO, Rome, Italy.
[145] http://www.ciesnet.com/.

food business forum that includes as members all major retailers such as Tesco, Marks & Spencers, Metro, Carrefour, Auchean, Casino and Royal Ahold) and plays an important role in the certification of food safety standards. All retailers together, that are members of the CIES, generate an annual turnover of more than € 2.1 trillion.

One of the main objectives of the GFSI is improving the efficiency of audits at the suppliers of the different standards. To achieve this, the GFSI has developed a model with which standards need to comply before they can receive approval by the CIES members. The GFSI thus focuses on harmonisation between countries and achieve efficiency for suppliers. Approval for certification schemes can be applied for at the GFSI. Approval by the GFSI acts as worldwide recognition and acceptance of the certification scheme ('certified once, accepted everywhere'). On the time of writing the following standards were approved by the GFSI:[146]

Manufacturing:
- BRC (British Retail Consortium) Global Standard Version 5
- Dutch HACCP (option B)
- FSSC 22000 (Food Safety System Certification)
- Global Aquaculture Alliance BAP, issue 2 (GAA Seafood Processing Standard)
- Global Red Meat Standard Version 3
- IFS (International Featured Standards) Version 5
- SQF (Safe Quality Food) 2000 Level 2
- Synergy 22000

Primary production:
- GlobalGAP IFA Scheme Version 3
- Canada Gap
- SQF (Safe Quality Food) 1000 Level 2

Manufacturing and primary production:
- PrimusGFS

As mentioned before, national authorities are interested in the role of private standards in ensuring safe food. For example, the Dutch competent authority (nVWA), has started a supervision policy where certification by the manufacturers against GFSI standards is taken into account. The same goes for some regional standards drawn up by large food businesses, (e.g. Vion Food Group and IKB-egg) which have been proven transparent and of high quality.

[146] See http://www.mygfsi.com/about-gfsi/gfsi-recognised-schemes.html.

4.1.4 Additional standards

Besides the major internationally accepted food safety management systems many more standards have been developed in the EU that are less known. These standards are more focussed on the quality than the safety of foodstuffs; objectives include among others care for the environment and sustainability. Some standards are based on existing standards, such as GlobalGAP, while others have been developed independently, such as 'Fruitnet' developed in Belgium. Often, but not always, these standards have a local function to protect specific quality aspects of certain products (e.g. 'Gepruefte Qualitaet Thüringen'). Products may carry a logo and/ or nomination, indicating that the product complies with the requirements of the standard. In recent years the number of this type of standards has exploded. Most standards are applied for 'business to consumer' (B2C) marketing and only a limited number find their way in 'business to business' (B2B) marketing.

In general the EU has a positive opinion on the development of certification schemes, but it seems that concern has been raised over the large number of schemes that have found their way to the market over the last few years in both the EU and the rest of the world. The EU has started a project with the objective to inventory all existing schemes for fruit and vegetables on the European market.[147] In 2010 a report[148] summarising the results for agricultural products and foodstuffs was published. From Figure 4.1 follows that the schemes and standards focus on many politically important subjects, e.g. it already includes a standard on climate change.

The development of these standards varies greatly over the different Member States (Bulgaria had only a single scheme where Germany had as many as 107 at the time of writing). See Figure 4.2 on the distribution of the number of schemes by country of origin, also some non-EU countries have been included for reference.

When reading the report it should be kept in mind that typical food safety management systems, such as hygiene guides, as well as specific quality management systems, such as Amboa Quality Label (for Belgian chocolate), have been included in the results. There is therefore a clear overlap with the information on hygiene guides in the European database (see section 4.1.2 for more information). On the other hand it is clear that not all quality marks and labels are included in the research.

---

[147] EU, Directorate L. Economic analysis, perspectives and evaluations: L.4. Evaluation of measures applicable to agriculture; studies Subject: Letter of Invitation to Tender – Contract Notice 2009/S 086-123210. Ref: Marketing standards in the fruit and vegetable sector, AGRI-2009-EVAL-07. Available at: http://ec.europa.eu/dgs/agriculture/tenderdocs/2009/72981/invite_en.pdf.

[148] Anonymous, 2010. Inventory of certification schemes for agricultural products and foodstuffs marketed in the EU Member States. Areté Research and Consulting in Economics, Bologna, Italy.

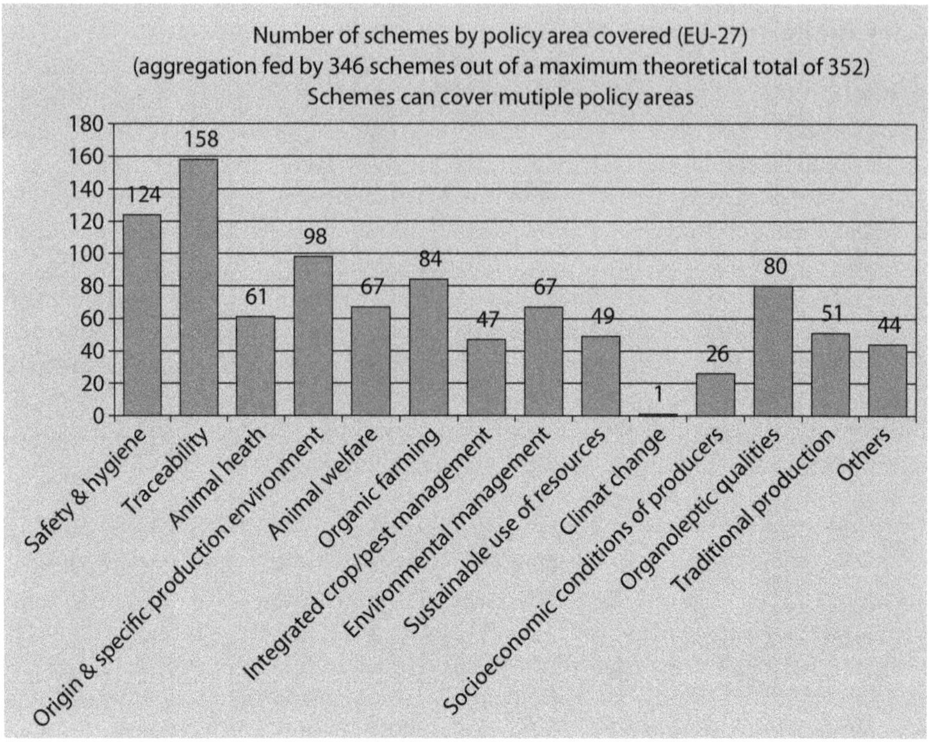

Figure 4.1. Number of certification schemes by policy area.

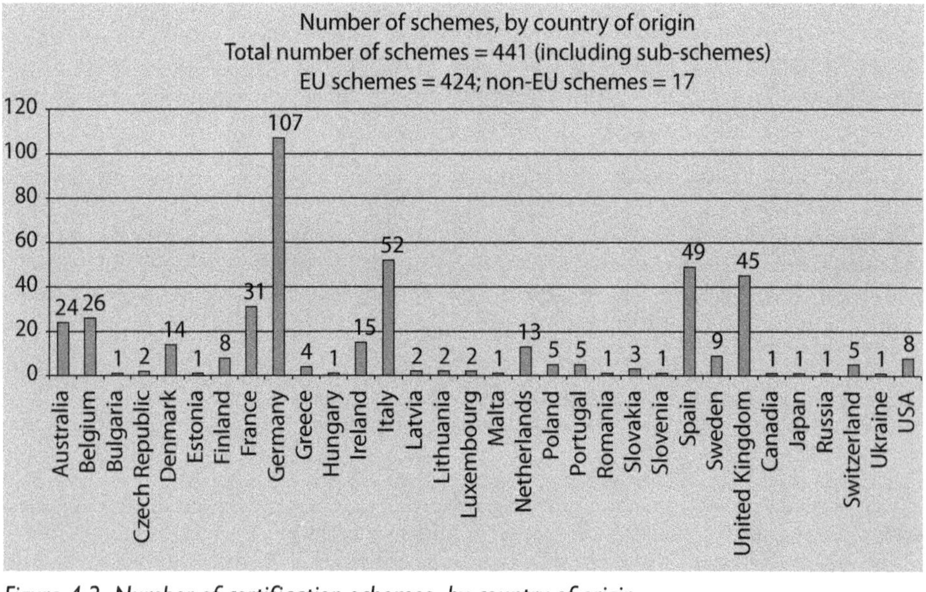

Figure 4.2. Number of certification schemes, by country of origin.

We can elaborate on this by studying the results from the report on the Dutch market. A total of 13 schemes are mentioned including 'IKB pork', this scheme is also included in the table of hygiene guides presented in section 4.1.2. The hygiene guides 'IKB egg' and 'IKB poultry' are not included in this database but do enjoy the status of hygiene guide in the Netherlands. Of the 13 schemes mentioned in the report, six are actually hygiene guides. The other schemes that are mentioned are quality marks or labels on e.g. sustainability, environment, organic, quality and animal welfare. Many other quality marks have not been included at all in the report, such as for products that are halal, vegetarian and gluten free. The report therefore provides a general impression of the market in the EU Member States, but is not always accurate.

After having carefully assessed the situation, the Commission developed guidelines showing best practice for the operation and implementation of such schemes[149]. These guidelines were drawn up in consultation with stakeholders.

4.2 Controlling food safety by quality management system/standard

4.2.1 Contents of a standard

In this section the contents of a food safety standard is discussed in more detail. The British Retail Consortium (BRC) standard, developed by the big British supermarket chains and used by many thousands of food businesses will be used as a reference.

The BRC global standards for Food Safety contain several different chapters and sections. Some of those are seen as fundamental in the preparation of safe food, while others are given less importance. The following ten requirements are indicated as being fundamental:

1. Management commitment and continuous improvement clause
2. Food safety plan – hazard analysis and critical control points
3.5. Internal audits
3.8. Corrective action and preventative action
3.9. Traceability
4.3.1. Layout, product flow and segregation
4.9. Housekeeping and hygiene clause
5.2. Handling requirements for specific materials – materials containing allergens and identity preserved materials
6.1. Control of operations
7.1. Training

[149] EU, 2010. Commission communication – EU best practice guidelines for voluntary certification schemes for agricultural products and foodstuffs. Official Journal of the European Union C 341, 16/12/2010: 5-11. See Appendix 1.

Other standards are structured differently but have similar or the same primary fundamental requirements. A good example is the International Featured Standard (IFS) for food developed by the German and French retailers. Both systems have been developed with the objective to be able to guarantee uniform and safe food products. BRC and IFS can be applied to different steps in the food production chain. Manufacturers, packaging or other types of businesses handling food can make use of these standards. In practice many companies will be certified against both standards to prevent trade restrictions. In both standards requirements are laid down in accordance to the European rules on production and packaging of foodstuffs. Securing food safety by implementation of HACCP ensures compliance with the requirements of the General Food Law. Furthermore the standards include rules that provide the specific requirements of the users on safety, quality and continuity.

4.2.2 Supervision

As mentioned before the practice of certification against good standards can be very interesting for national authorities. When good quality management systems have been implemented, authorities can reduce their supervision activities significantly. The condition for any reduction of supervision would be that authorities can trust the system to function in practice as is laid down in the standards. All organisations developing standards, apply a system of independent inspections (audits) to ensure the correct implementation of the standard by the food business operator. The audits are executed by certification bodies (Figure 4.3).

Certification bodies (CBs) receive their accreditation from Accreditation Bodies (ABs) when compliant with their requirements. Certification bodies exchange their experiences on the certification scheme with, for example, a Technical Committee or Board of Experts and the executive board of the standard.

The operational procedures of accreditation bodies in the EU are supervised by international Accreditation Bodies through peer assessment. ABs are accepted into Multilateral Agreements in Europe (EA-MLA) and outside of Europe (IAF-MLA and ILAC-MRA).[150] Regulation 765/2008 lays down the requirements on accreditation and market surveillance relating to the marketing of products[151]. In Article 4 of this Regulation the general principles for accreditation are laid down. Some examples are:

[150] http://www.european-accreditation.org/content/mla/what.htm.

[151] EU, 2008. Regulation (EC) No 765/2008 of the European Parliament and of the Council of 9 July 2008 setting out the requirements for accreditation and market surveillance relating to the marketing of products and repealing Regulation (EEC) No 339/93. Official Journal of the European Union L 218, 13/8/2008: 30-47.

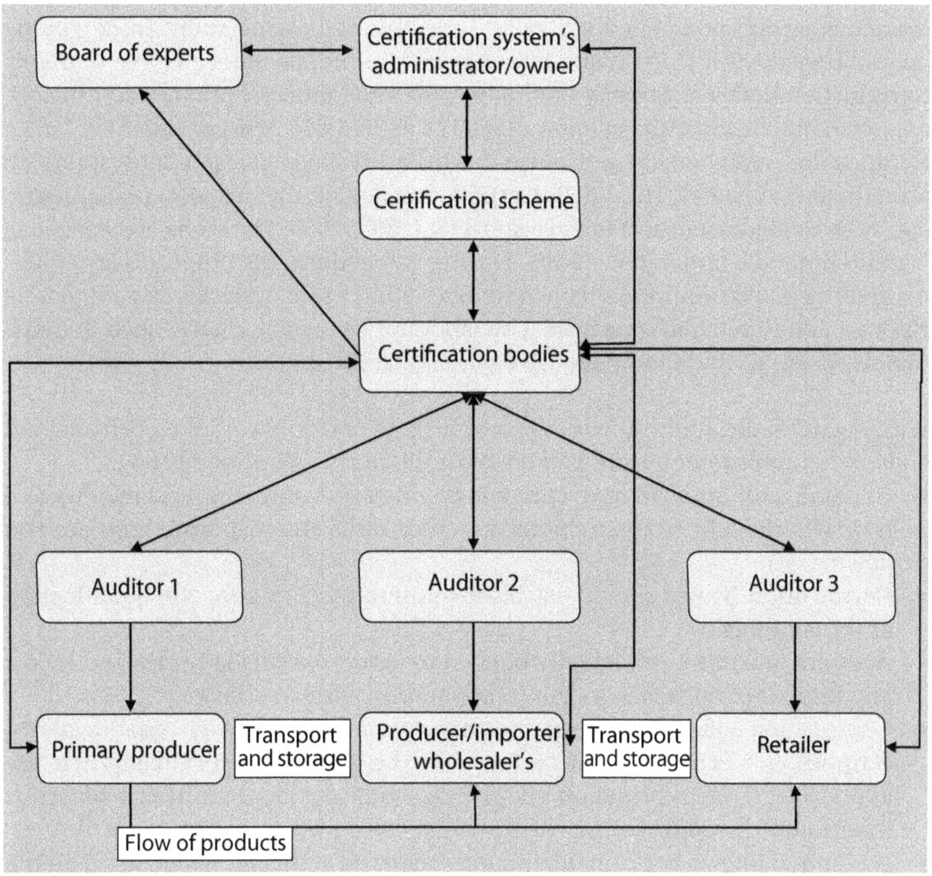

Figure 4.3. A schematic overview of the certification process.

1. Each Member State shall appoint a single national accreditation body.
2. Where a Member State considers that it is not economically meaningful or sustainable to have a national accreditation body or to provide certain accreditation services, it shall, as far as possible, have recourse to the national accreditation body of another Member State.
3. A Member State shall inform the Commission and the other Member States where, in accordance with paragraph 2, recourse is given to the national accreditation body of another Member State.
6. The responsibilities and tasks of the national accreditation body shall be clearly distinguished from those of other national authorities.
7. The national accreditation body shall operate on a not-for profit basis.
8. The national accreditation body shall not offer or provide any activities or services that conformity assessment bodies provide, nor shall it provide consultancy services, own shares in or otherwise have a financial or managerial interest in a conformity assessment body.

An accreditation body shall act as an independent organisation. In deploying an auditing team it gives much attention to the independence of the auditors. Certification bodies that carry out audits and issue process/product certificates, are accredited against the standard ISO/IEC Guide 65 (1996) or ISO/IEC 17021. Certification bodies which carry out audits to certify food management systems are accredited against ISO/IEC 17021. Certification bodies, that inspect companies on e.g. hygiene guides, are accredited against ISO/IEC 17020. The scope of inspections is more limited compared to audits. During inspections, the emphasis lays more on meeting fixed requirements, while in an audit more attention is paid to how risks are identified and managed. The laboratories, which analyse food and raw materials, are accredited by the AB against ISO/IEC 17025.

With regard to the auditors, with the authority to perform work for a certification body, strict requirements are laid down in the norms mentioned above:

- General: auditors shall have competence in performing technical reviews and have clearly defined instructions, in which tasks and responsibilities are laid down.
- Certification bodies should establish minimum criteria for the qualification of the Auditors.
- Auditors must be contractually obliged to follow established rules and report any kind of previous or on-going cooperation with an auditee.
- Certification bodies must keep track of information on the right qualifications, training and experience of auditors on the different specialisms that are applicable. This information should also indicate the date from which this information is valid.
- Auditors shall not perform other audits than the audits for which they, on the basis of their training and experience, have authority.

The certification protocol ISO/TS 22003, laying down requirements on among others the auditor and duration of the audit, is available since February 2007.

4.2.3 Audit frequency

The major international systems have set different rules regarding auditing frequencies. BRC and IFS know an annual frequency while Dutch HACCP almost always keeps a bi-annual audit frequency. In other systems an audit is performed two or three times per year at the start. If a company has correctly implemented and controlled the system, the frequency may drop to a single audit per year. When non-conformities are found, additional audits are performed.

The results of an audit are reviewed by an expert panel at the certification body, before the certificate is issued. The auditor, on performing the audit, does therefore not determine the end result.

4.2.4 The scope of a certificate

The scope (the scale of the activities concerning the control of food safety in the company that is being certified) can generally be divided into 3 parts:

- Pre-requisite programmes of Good Manufacturing Practices (GMP). These programs describe basic requirements for hygiene, establishment and such. It may also include requirements that are specific to the organisation that developed the standard.
- HACCP, in which the requirements are set for the risk analysis, as required by the General Food Law. Usually the model published by the Codex Alimentarius is applied.
- System management, where the requirements are laid down on a coherent quality system of the entire organisation, allowing better management of operational processes.

The standards that are used globally, benchmarked by the Global Food Safety Initiative, are based on these 3 elements. Less extensive systems mainly use the pre-requisite programmes combined with a risk analysis performed for the sector which is based on HACCP.

4.2.5 Sanctions policy of certification bodies

For certification systems that do not meet the requirements generally the following measures apply:

- Non-conformities are discussed with the auditee.
- A date is agreed upon which corrective measures shall have been introduced.
- Possibly a re-audit needs to be performed to establish the effectiveness of taken measures.
- If this proves to be insufficient, the causes will be discussed again.
- Disciplinary measures may be applied.

If non-conformity is not resolved within the agreed upon period, suspension of a certificate is usually the first disciplinarian measure, which is followed by revocation of the certificate. Depending on the severity of the non-conformity CBs may also decide on immediate revocation of the certificate.

Where non-conformity in the management system is related to only a specific part of the organisation, CBs can also consider a limitation of the scope of certification as an alternative to suspension. For example, an auditee may no longer be allowed to produce, but is allowed to sell. CBs can also choose to only give a warning to the food business operator before suspension is considered.

4.3 Description of commonly used Standards

Table 4.2 shows where some commonly applied food safety standards are used in the food production chain. The different food safety standards are very diverse

Table 4.2. Commonly used private food and feed systems and their place in the food production chain.

Scheme	Stage of application in the food and feed chain														Content
	Primary production					Processing									
	AGF	Meat and fish	Dairy	Juices	Feed	General	Bakery	Meat	Juices	Feed	Packaging	Storage and distribution	Retail	Catering	
BRC						X					X	X	X		safety
Carbon Trust						X							X		sustainability
Dutch HACCP						X					X	X	X		safety
Fair Trade	X					X							X		social responsibility
Fami QS					X										safety
FSSC 22000 / ISO 22000						X									safety
Gluten free							X				X		X		safety
GlobalGAP	X	X			X										safety
GMP+					X										safety
Halal	X	X				X		X			X	X	X	X	religious
Hygiene guide	X	X	X	X	X	X	X	X	X	X	X	X	X	X	safety
IFS						X					X	X	X		safety
ISO 26000	X	X	X	X	X	X	X	X	X	X	X	X	X	X	social responsibility
Kosher		X	X			X	X	X					X	X	religious
MSC		X											X		sustainability
Organic	X	X	X			X	X	X	X				X		sustainability
QS	X	X													safety
Rainforest alliance	X														sustainability
SGF/IQCS									X			X			safety
SGF/IRMA				X											safety/social responsibility
SQF 1000	X	X	X												safety
SQF 2000						X									safety
TrusQ					X										safety
UTZ cert.	X														sustainability
Vegetarian						X					X		X		animal welfare

in nature. Brief descriptions of the major international standards, applied by thousands of businesses worldwide, are given. These standards are often applicable to all sectors of the food chain and are often developed by groups of stakeholders in the food production and trade. Examples are BRC and IFS, developed by the united retailers from respectively the United Kingdom, France and Germany.

4.3.1 Primary production of food

GlobalGAP

GlobalGAP[152] stands for Global Good Agricultural Practice. This standard lays down the worldwide requirements for farmers and horticulturists on food safety, sustainability and quality. A group of twenty-six European supermarket organisations has taken the initiative in 1997 to harmonise the requirements for their suppliers of fresh produce. The final goal is that GlobalGAP shall become an 'umbrella' standard for food safety in the primary production sector. A large number of quality systems from all parts of the world is being or expecting to be benchmarked against the GlobalGAP standard.

Food safety is the most important part of GlobalGAP. In addition, GlobalGAP also includes requirements for animal welfare, environment and working conditions. The ultimate goal is a single worldwide-accepted set of quality requirements, clarifying the 'forest' of quality systems currently used. Harmonisation should therefore take place first. GlobalGAP is the quality standard that can help. GlobalGAP is meant to become the quality system for all farmers in the world. Also transporters of animals shall have to comply with the requirements set by the GlobalGAP standard for animal transport.

The GlobalGAP standard is divided into several modules, where each module covers different areas or activity levels on a production site. There are scopes covering general subjects and sub-scopes covering more production specific details, such as bulk coffee, flowers and dairy cattle. GlobalGAP lays down requirements only to the primary production of food. It is a business to business quality mark, which should not appear on consumer packaging.

QS

QS[153] stands for 'Qualität und Sicherheit für Lebensmittel vom Erzeuger bis zum Verbraucher'. The administration of the scheme is held by the 'Central Marketing Gesellschaft der Deutschen Agrarwirtschaft'. The QS-system was developed after the BSE scandal by the German food industry as a counterpart of EurepGAP (now

152 http://www.globalgap.org.

153 http://www.q-s.de/.

GlobalGAP). Initially it was developed as a certification system for meat and meat products, set up to provide security to the consumers on the origin of meat. A group of German business chains has enlarged the application of the standard to fruit and vegetables.

The QS-system requires quality controls for the meat sector over the entire production chain, from birth to slaughter and processing. Traceability of the raw materials and transparency of production are the key building blocks of the standard. Part of the standard forms the protection of animals. The rules apply for German products and for products which are imported by Germany.

For the fruit and vegetables sector the QS-system can be comparable to GlobalGAP. However, QS is a consumer label, with more concrete requirements than GlobalGAP, QS requirements for plant protection products and fertilisers fit the Western Europe situation better. The fruit and vegetables sector has more than 20,000 QS certified companies.

QS knows no borders. In an attempt to achieve a uniform level of food safety at the European level as well, QS puts its faith in international cooperation and integration. The aim is to avoid double auditing of the economic participants and to enable the flow of goods between the various quality assurance systems. For this reason, QS has already come to agreements with various standards in neighbouring European countries:

Austria:	Pastus +
Belgium:	Ovocom/Bemefa and Certus
Denmark:	Global Red Meat Standard
The Netherlands:	GMP+ and IKB+

SGF/IRMA and SGF/RQCS

SGF International eV, formerly known as the 'Schutzgemeinschaft der Fruchtsaftindustrie' is a model for industrial self-regulation in the fruit juice industry.[154] It can also be applied in other sectors of the food industry. SGF stands for Sure Global Fair, which also describes the 'Programm des Branchenverbandes'. SGF is a registered association with headquarters in Frankfurt am Main (Germany). SGF has grown to some 600 affiliated businesses in approximately 50 countries. SGF acts as a partner for business on all issues of safety and quality of fruit juices.

GSF/IRMA (International Raw Material Assurance) supervises the production of raw materials and fruit growers, mixing stations, distributors, (cold) storage and transportation companies, who participate in the standard on a voluntary basis. The standard uses a hygiene checklist, which among others is based on Regulation

[154] http://www.sgf.org/.

852/2004[155] and the AIJN-Guide of Good Hygiene Practice (Association of the Industry of Juices and Nectars from Fruits and Vegetables of the EU). The checklist is specifically designed to supervise the monitoring of all relevant aspects in the production of semi-finished products.

SGF has developed a Code of Conduct to support raw material suppliers, participating in the voluntary control system of SGF. This Code of Conduct supports the increasing awareness that generally accepted standards for ethical behaviour should be taken into account.

SQF 1000

SQF stands for Safe Quality Food.[156] It is a management system developed by the West Australian Department of agriculture, it was introduced in 1994. Since 2004 SQF is managed by the SQF Institute (SQFI), a division of the Food & Marketing Institute (FMI) in Washington. SQF is the only standard, which has its head office outside of Europe, which is accepted by the Global Food Safety Initiative.

SQF is a scheme that has many provisions on the quality aspects of products, on which it distinguishes itself from other standards. According to SQF a single standard for all businesses of the food production chain is not possible. The major international standards fit industrial producers, but for the primary sector is far too complicated and extensive. SQF has worked around this by using two customised standards:
- SQF 1000 is intended for the primary agricultural sector and small-scale processors and service providers. These players are often 'chained' in a product-market organisation. The risks are mostly limited. The code is based on HACCP, but uses a simplified method.
- SQF 2000 is intended for the 'larger' supplying/processing industry, where the risks are greater. In this code HACCP is fully integrated.

SQF combines a quality management system (ISO 9001) with a food safety system (HACCP) and adds requirements on identification and traceability (tracking and tracing). This is secured using both the well-known CCPs (Critical Control Points) and CQP's (Critical Quality Points). Using this approach, an integrated system was developed, that complies with the requirements of the Global Food Safety Initiative (GFSI). SQF 2000 can, on a voluntary basis, be supplemented with modules on environmental protection and corporate social responsibility. SQF is intended to be used by all types of food businesses. SQF plays an important role in the North American food industry. In addition, suppliers and retailers in

[155] EU, 2004. Regulation (EC) No 852/2004 of the European Parliament and of the Council of 29 April 2004 on the hygiene of foodstuffs. Official Journal of the European Union L 139, 30/4/2004: 1-54.
[156] http://www.sqfi.com/.

Australia, Asia, Africa, the Middle East and South America also use the standard. In Europe SQF still holds little foothold. Altogether more than 10,000 certificates have been issued to various food businesses.

4.3.2 Feed production

FAMI-QS Code of Practice for Feed Additive and Premixture Operators

FAMI-QS[157] was officially and successfully assessed by the EU authorities in January 2007. They considered this Guide as practicable throughout the European Union and suitable for compliance with the requirements of Regulation 183/2005 on Feed Hygiene[158].

This official adoption established the formal FAMI-QS link with the Feed Hygiene Regulation, meaning that a feed additive or premixture operator who puts in place the FAMI-QS principles is meeting the requirements of this Regulation. FAMI-QS also expects that this will facilitate the work of the control authorities at national level. Positive signals are already received in this sense and it is expected that it will minimise the need for controls of FAMI-QS certified operators by national authorities and customers. A sense of complementarity with officials control is installing itself in an increasing number of EU countries.

The FAMI-QS Code of Practice is a public document and its content can be freely followed by any feed additive or premixture operator. The text of the Code is designed to set out general requirements. Questionnaires are included in order to further detail the requirements. In addition it can be used as a tool for the operators and in auditing operators. A guidance document is provided as annex to the Code. While the requirements of the Code are mandatory for every operator desiring certification, the guidance is not mandatory but provides information on how to deal with specific issues in a more detailed and practical way. The guidance documents provide support to the operator when implementing the Code.

The number of certified sites has considerably increased. Since December 2009, it exceeds 500 sites, while in December 2008 this numbered 298 sites, an important increase of 72.15%. Besides the unconditional support from the European Industry to the FAMI-QS Code, the international presence of FAMI-QS Code is shown as the number of certificates in third countries is actually higher than in the EU. Especially in China there are many participants (172).

---

157 http://www.fami-qs.org/documents/FAMI-QS%20Report%202009.pdf.

158 EU, 2005. Regulation (EC) No 183/2005 of the European Parliament and of the Council of 12 January 2005 laying down requirements for feed hygiene. Official Journal of the European Union L 35, 8/2/2005: 1-22.

GMP+

GMP+[159] standard (Good Manufacturing Practice) has been drawn up by the Dutch Animal Feed Sector Central College of Experts (CCvDD), who were responsible for the standard as owners. This College of Experts has representatives from the entire feed production chain. Product Board Animal Feed supports this college since 1993. The latest version of this standard is GMP+ 2006. In 2010 the organisation has been transformed to a private foundation.

The basis of the GMP+ standard is formed by a combination of the standards EN-ISO 9001: 2000, ISO 22000 and the HACCP principles, as laid down in the documents of the Codex Alimentarius and elaborated upon in the requirements for a HACCP based food safety system, supplemented with general GMP+ requirements.

The standard was initially intended for use in the Netherlands. Today, more than 11,000 companies in the feed sector from over 65 countries worldwide let their products and services be certified according to the GMP+ 2006 standard. Characteristic for the current scheme is that it provides an integral assurance system for all products and stages of the production in the animal feed chain. These stages are: farming, trade, production/processing, transport, laboratory research and feeding.

TrusQ

TrusQ[160] started in 2003. It is an important collaboration in the area of food safety between Dutch and Belgian feed suppliers. The TrusQ partners deliver feed of guaranteed quality to cattle farmers. TrusQ is based on GMP+, but considers this insufficient to be able to guarantee the quality of all raw materials for feed used by the participants of the scheme. Also by-products, used in the production of feed, are within the TrusQ scope.

In particular with suppliers from abroad guaranteed quality regularly forms a problem. TrusQ therefore analyses many raw materials to reduce the risks of contamination for farmers. This leads to a high concentration of knowledge at TrusQ. As an additional security measure TrusQ uses a traffic light system to indicate the reliability of suppliers and products. The following indications are used:

- Green: product is in order; the supplier may supply.
- Orange: measures have been agreed upon (e.g. more controls) to ensure a product is safe.
- Red: business with the supplier is not allowed and its raw materials are not used.

[159] http://www.gmpplus.org/en/.

[160] http://www.trusq.nl/pages.php?cID=2&pID=&ln=2.

In practice it may be so that a supplier cannot supply at all, but it is also possible that the supplier is not allowed to supply a certain raw material, but is allowed to supply another.

TrusQ is used by the participating feed production companies (the nine TrusQ companies) to ensure their requirements towards suppliers of raw materials and additives are being complied with, both at home and abroad. Cattle farmers participating in the scheme can therefore trust that the supplied feed is safe to use. Each of the nine TrusQ-companies supplies feed to thousands of cattle farmers in primarily the Netherlands, Belgium and Germany.

4.3.3 Manufacturing sector

Dutch HACCP

Dutch HACCP[161] is a certification scheme developed in 1996 by the National Board of Experts HACCP (NBE). Since its start, the standard is regularly adapted to new developments. On the initiative of the Central Body of Experts food safety and its affiliated certification bodies in 2004 founded the 'Stichting Certificatie Voedselveiligheid' (SCV). Dutch HACCP is the only standard developed by certifying bodies themselves. The Foundation was established to provide a legal entity to the NBE. SCV facilitates the board and is the legal owner of the standard 'requirements for a HACCP based food safety system' (Dutch HACCP) and manages the copyright on this standard through user agreements.

The standards has 2 versions: management/system certification (option A) and process/product certification (option B). A distinctive difference between the two versions is how the pre-requisites programme is implemented. There is also a difference in the auditing and reporting method by the certification body. Dutch HACCP option B is approved by the Global Food Safety Initiative (GFSI).

Dutch HACCP can be applied in the entire food production chain: primary production, processing and manufacturing industry, transport, storage, distribution and trade. The standard does not apply to supplying and service providers, such as suppliers of machinery, packing materials and cleaning companies. At the time of writing approximately 2,000 certificates were issued by the 12 certifying bodies.

BRC Global Standards

In the UK in the 1990s a number of supermarket chains (including Tesco, Sainsbury, Somerfield, and Safeway) united themselves on the area of quality and founded

[161] http://www.foodsafetymanagement.info.

the British Retail Consortium (BRC).[162] They developed a standard (the BRC Global Standard for Food Safety) and then made compliance with this standard a requirement for all suppliers (food businesses). The BRC Global Standard requires that a quality system is used, that HACCP is applied and that the establishment, the product, process and personnel are included into this system.

The BRC-scheme (see also section 4.2) consists of an inspection protocol and a technical standard. The inspection protocol was developed for inspecting bodies. The technical standard, an extensive checklist, is relevant to the suppliers of food. The technical standard was set up in 1998 and celebrated its 5th version in 2008. The BRC code was corrected and approved in 2008 by the GFSI. The 6th version is expected at the of 2011.

With a BRC-certificate a producer complies in principle at once with all the requirements of the British (and also other) international supermarkets. Because this is cost-saving for both the producers and users the BRC certificate is widely appreciated. Most British and many other European large supermarkets and brand owners only do business with suppliers certified for to BRC Global Standard for Food Safety.

IFS

In Germany and France a number of supermarket chains (among others Aldi and Metro) have developed a food safety standard and made it a requirement to their suppliers. This standard was called the International Food Standard (IFS), now International Featured Standards.[163] In many ways, it is similar to the BRC Standard: there is an overlap of 90 to 95%. The first version of the IFS standard dates back to 2002 and was developed by the German retailers. The fourth version, launched in 2007, was published together with the French companies. The administration of IFS is held by the German retailers united in 'Hauptverband des Deutschen Einzelhandels' (HDE) and the French retailers united in the 'Fédération des entreprises du Commerce et la Distribution' (FCD). Just like the BRC standard, IFS requires a quality system to be present, in which HACCP is applied and that requirements on the establishment, product, process and personnel are included in the system.

IFS is applied by food businesses that supply mainly to retailers in Germany and France. The IFS standard can be applied in all sectors; there are no specific requirements for sectors of the chain or for specific product groups. Many retailers accept BRC and IFS as equivalent.

---

[162] http://www.brc.org.uk.

[163] http://www.ifs-online.eu.

FSSC 22000, ISO 22000 and PAS 220

With the appearance of many large and small standards to ensure the safety of food over the last decades, the desire to stop the appearance of new standards and to develop a single uniformly accepted standard was expressed by the international food business community. Not less than thirty countries contributed to the development of the ISO 22000 standard.[164] This has resulted in this standard being recognised by the 'big brand' holders and enjoys worldwide support and recognition. ISO 22000 therefore is an international, chain oriented standard for food safety published in September 2005 by 'The International Organization for Standardization (ISO). The extent to which retailers will accept this standard is not yet clear. This is one of the reasons why at the end of 2008 the industry developed an additional module: PAS 220 (Publicly Available Specification). Early February 2009 the new standard FSSC (Food Safety System Certification) 22000 was developed by The Foundation for Food Safety Certification, the standard integrates ISO 22000 and PAS 220. In May 2009 the GFSI accepted this new FSSC 22000.

ISO 22000 follows the structure and approach of ISO 9001 and integrates these with food safety assurance based on the HACCP principles. ISO 22000 does not replace standards like ISO 9001. ISO 22000 includes only requirements on food safety while ISO 9001 also includes quality aspects. Since it is an ISO standard, requirements are less clearly described and defined as with BRC and IFS. The BRC and IFS standards include more requirements on GMP.

The additional PAS 220 specifies requirements on the establishment, implementation and maintenance of preconditions for food safety programs that food business operators have to comply with to keep food safety risks under control. PAS 220 is not designed or intended for use in other parts of the food supply chain.

ISO 22000 is applicable to all organisations that are directly or indirectly part of the food chain, regardless of the size or complexity of the organisation. The standard allows also for smaller and/or less-developed organisations (such as a small farm, a small packaging and distribution company or a small food shop) to implement a food safety management system. The standard allows small companies to implement an externally developed combination of management measures. This approach is similar to that used in hygiene guides, which also use generically developed measures to comply with the HACCP requirements. ISO 22000 is also suitable for organisations that are involved indirectly in the food production chain, such as suppliers of machinery and tools, cleaning and disinfection products and packaging materials. So far the standard is little used. There are also still relatively few qualified auditors.

---

[164] http://www.FSSC22000.com.

SQF 2000

The Safe Quality Food (SQF) 2000[165] Code provides for the food supplier a food safety and quality management certification program. In 1994 the Code was developed and pilot programs implemented to ensure its applicability to the food industry. The Food Marketing Institute (FMI) acquired the rights to the SQF Program in August 2003 and has established the SQF Institute (SQFI) Division to manage the Program. The SQF 2000 Code is recognised by the Global Food Safety Initiative as a standard that meets its benchmark requirements. The SQFI Technical Advisory Council reviews and makes recommendations on changes to the Code in line with the current requirements and expectations of the global food sector and other comments received from stakeholders.

4.3.4 Packaging sector

As it is possible for foodstuffs to get contaminated with microbiological or chemical hazards when being packaged, authorities have put relevant legislation with appropriate norms into force. In the end quality management systems should protect all steps in the food production chain against unknown risks. From some food safety and quality management standards, such as Dutch HACCP, it is claimed that they are applicable to all steps of the production chain. Other international standards have developed special modules to fit this need.

BRC-IOP

For the specific details on BRC, see section 4.3.3. The BRC has developed the 'Global Standard for Packaging and Packaging Materials' specifically for the food (packaging) industry. This standard is known as the BRC-IOP [166] and is approved as well by the GFSI.

4.3.5 Transport sector

Transporting foodstuffs also involve risks regarding safety and quality. This is especially true for fresh and deep frozen products. A well-known example is the famous nitrite incident of 1980. When a leaking pipe of the cooling unit caused contamination of a shipment of spinach with extremely high concentrations of nitrite, one of the consumers that ate from the spinach did not survive the incident.

Standards covering the whole chain can be applied to this sector, but sector specific standards also exist.

---

[165] http://www.sqfi.com/wp-content/uploads/SQF-2000-Code.pdf.
[166] http://www.brcglobalstandards.com/standards/packaging-and-packaging-materials/.

BRC for Storage and Distribution

For the specific details on BRC, see section 4.3.3. The Global Standard for Storage and Distribution [167] has been developed specifically for the storage and distribution sector in 2006.

IFS/ILS

For the specific details on IFS, see section 4.3.3. Since May 2006 transporting companies can achieve certification against the IFS-transport standard, also known as the International Logistic Standard (ILS)[168].

4.3.6 Other standards

Besides the previously discussed quality standards, with the main focus on the production, from raw materials to end product, of safe foods, some other standards and schemes exist that focus on quite other aspects in which consumers are highly interested.

ISO 26000

ISO 26000 is Guidance on Social Responsibility[169]. Companies that implement ISO 26000 have, next to conventional business objectives, a set of specific company objectives on:
- environment;
- human rights;
- labour practices;
- organisational governance;
- fair operating practices;
- consumer issues;
- community involvement/society development.

ISO 26000 provides guidance to corporate social responsibility. It does not lay down specific requirements and it is therefore not possible, contrary to other ISO management standards, to achieve certification. The impossibility of obtaining a certificate is one of the principles of this standard.

The scope of ISO 26000 is to make social responsibility operational. ISO 26000 provides guidance for all types of organisation, regardless of their size or location, on:

---

[167] http://www.brcglobalstandards.com/standards/storage-and-distribution/.
[168] http://www.gs1.org/transportlogistics/forum/work_groups/ll/.
[169] http://www.iso.org/iso/iso_catalogue/management_and_leadership_standards/social_responsibility/sr_discovering_iso26000.htm#std-1.

1. Concepts, terms and definitions related to social responsibility;
2. Background, trends and characteristics of social responsibility;
3. Principles and practices relating to social responsibility;
4. Core subjects and issues of social responsibility;
5. Integrating, implementing and promoting socially responsible behaviour throughout the organisation and, through its policies and practices, within its sphere of influence;
6. Identifying and engaging with stakeholders;
7. Communicating commitments, performance and other information related to social responsibility.

ISO 26000 is intended to assist organisations in contributing to sustainable development. It is intended to encourage them to go beyond legal compliance, recognising that compliance with law is a fundamental duty of any organisation and an essential part of their social responsibility. It is intended to promote common understanding in the field of social responsibility, and to complement other instruments and initiatives for social responsibility, not to replace them.

ISO 26000 was approved on 12 September 2010 in Oslo with 93% of all votes and has now been published. Only 5 countries voted against.

Fair trade

Fair trade encourages sustainable development in international trade,[170] most importantly the export from poor countries to richer Western countries. Fair trade means that e.g. coffee bean growers, cacao bean growers and banana growers in Latin America, Africa and Southeast Asia receive an honest price for their export products. This is a price that is based on production costs rather than a price that is subject to the situation on the international commodities market. Fair trade products also need to comply with very strict environmental requirements.

Already in the 40's and 50's of last century some religious and non-profit organisations were actively promoting products from third world countries in the Western world. The Fair Trade organisation took its current shape in the sixties. In those years, fair trade practices were often seen as a political statement against neo-imperialism. Students opposed the multinational corporations and trade practices with the indigenous population. In that period the slogan 'trade not aid' was invented. Organisations such as UNCTAD and the British NGO Oxfam were involved in the foundation. In 1969 the first Dutch 'Worldshop' was opened, quickly followed by many others in the Benelux, Germany and many other Western European countries. At first, the products sold under the label were

[170] http://en.wikipedia.org/wiki/Fair_trade.

mainly traditionally handcrafted; over time more and more foodstuffs were also included in the product range.

The credibility of Fair Trade is dependent on strict criteria and a permanent and good supervision of those criteria. Currently products sold under the Fair Trade label (Figure 4.4) are produced in 23 countries that are interconnected through the autonomous umbrella organisation: Fair Trade Labelling Organizations International eingetrager Verein (FLO-eV). This organisation determines the criteria and assists producers to comply with the requirements. The inspection body of the FLO-eV operates totally independent. Fair Trade certification can be achieved after evaluation of a set of 250 criteria related to working conditions and investment in environmental friendly and economic development.

Figure 4.4. The International Fair Trade certification mark.

The following organisations are involved in the Fair trade movement:
- The World Fair Trade Organization (formerly the International Fair Trade Association) is a global association created in 1989 of fair trade producer cooperatives and associations, export marketing companies, importers, retailers, national, and regional fair trade networks and fair trade support organisations.
- The Network of European Worldshops, created in 1994, is the umbrella network of 15 national Worldshop associations in 13 different countries all over Europe.
- The European Fair Trade Association (EFTA), created in 1990, is a network of European alternative trading organisations which import products from some 400 economically disadvantaged producer groups in Africa, Asia, and Latin America.
- FINE, created in 1998, an informal association[171] whose goal is to harmonise fair trade standards and guidelines, increase the quality and efficiency of fair trade monitoring systems, and advocate fair trade politically,

[171] FINE stands for (F) Fairtrade Labelling Organizations International (FLO) (I) International Fair Trade Association, now the World Fair Trade Organization (WFTO) (N) Network of European Worldshops (NEWS!) and (E) European Fair Trade Association (EFTA).

- The Fair Trade Federation (FTF), created in 1994, is an association of Canadian and American fair trade wholesalers, importers, and retailers. The organisation links its members to fair trade producer groups while acting as a clearinghouse for information on fair trade and providing resources and networking opportunities to its members.
- The Fair Trade Action Network, created in 2007, is an international fair trade volunteer web-based network. The association links volunteers from a dozen of European and North American countries, actively supports Fair Trade Towns initiatives and encourages grassroots networking at the international level.

The Carbon Trust Standard

The Carbon Trust Standard[172] awards organisations for real carbon reduction. Carbon Trust certifies organisations that have measured, managed and genuinely reduced their carbon footprint and committed to making further reductions year on year. In March 2011 London Metropolitan University has officially been rewarded the Carbon Trust Standard, after reducing carbon emissions by nearly 12%.

The Carbon Trust Standard is the world's first certification scheme designed to allow companies to measure the carbon footprint of their operations and facilitate an independent, specialist review of energy management practices. The use of the Carbon Trust Standard logo (Figure 4.5) allows companies to then communicate their commitment to combating climate change. The award requires organisations to provide evidence on real reductions in their own CO_2 emissions rather than paying third parties to reduce emissions via off-setting, like planting trees or green tariffs.

The Carbon Trust is set up by government in response to the threat of climate change, to accelerate the move to a low carbon economy by working with

Figure 4.5. The Carbon Trust Standard logo.

[172] http://www.carbontrust.co.uk/cut-carbon-reduce-costs/promote/carbon-trust-standard/Pages/carbon-trust-standard.aspx.

organisations to reduce carbon emissions and develop commercial low carbon technologies. The Carbon Trust works with the UK business and public sector through its work in five complementary areas: insights, solutions, innovations, enterprises and investments. Together these help to explain, deliver, develop, create and finance low carbon enterprise. The Carbon Trust is funded by the Department for Environment, Food and Rural Affairs (Defra), the Department for Business, Enterprise and Regulatory Reform (BERR), the Scottish Government, the Welsh Assembly Government and Invest Northern Ireland.

Over 350 organisations have achieved the Standard with a total carbon footprint of nearly 35 million tonnes of CO_2. Organisations awarded the Standard include well-known names such as First Direct, Tesco's, O2 and public sector organisations such as HM Treasury, London Fire Brigade, DSM and Manchester University. To achieve certification against the Standard, an organisation will need to meet the requirements in three areas:

- measure the carbon footprint over 2-3 years;
- demonstrate a reduction in carbon emissions;
- provide evidence of good carbon management.

Rainforest Alliance

Rainforest Alliance[173] is an environmental organisation based in the United States. Their main objective is to protect ecosystems and the people and animals that are depending on it. The Chiquita company collaborates with the Rainforest Alliance since 1992 and was rewarded with certification in 2005. This is shown by the little green frog next to famous blue sticker on Chiquita bananas. By obtaining the Rainforest Alliance certificate, Chiquita assures that its bananas are produced using sustainable methods on certified plantations. The quality mark can also be found on coffee and tea (Figure 4.6).

Figure 4.6. The Rainforest Alliance quality mark.

[173] http://www.rainforest-alliance.org/.

Inspections take place on farms or plantations. The inspections are performed by independent and accredited organisations, for this reason the Rainforest Alliance quality mark is considered reliable. To be allowed to use the Rainforest Alliance mark, farmers have to comply with 200 strict requirements. The requirements are divers: there are requirements on nature preservation, water preservation and forest management. The workforce on the plantations should receive the minimum wage and good secondary working conditions, including a safe environment to live. The Rainforest Alliance does not guarantee prices to farmers.

Marine Stewardship Council

The Marine Stewardship Council[174] is one of the most important organisations promoting a sustainable fishing industry. Some other organisations are: Friends of the Sea (FOS), Marine Ecolabel Japan (MEL-Japan), Global Aquaculture Alliance (GAA), GlobalGAP, Naturland, DEWHA Environment Protection and Biodiversity Conservation Act and Thai Quality Shrimp (TQS). These are all organisations that issues certificates. There are also other organisations providing fish recommendations, producing 'Sustainable Seafood Guides' or providing information and recommendations to businesses and consumers on sustainable fisheries and aquaculture. Most of the schemes are improving their conformance with the FAO guidelines. Key attributes for these guidelines are Scope; Accuracy; Independence; Precision; Transparency; Standardisation; and Cost effectiveness.

Of the certification schemes, the MSC makes the most comprehensive, robust, and transparent assessment of performance.[175] MSC is the only scheme that specifically requires the data and information to be sufficient for achieving the other objectives (stock status and ecosystem impacts). MSC uses the most recently available stock-specific assessment results directly from fishery managers and stock assessment scientists. MSC criteria require that the target population(s) and associated ecological community are maintained at high productivity relative to their potential productivity. The assessment of this considers outcome indicators (stock status, reference points and stock rebuilding) and harvest strategy indicators (the harvest strategy, control rules, monitoring and stock assessment procedures). The consideration of stock status includes a peer review of the stock assessment information. Certification by the MSC's program is rather expensive (on average 30,000 euro). For small businesses these costs usually are too high.

MSC evaluation is a cycle where one or more ships are audited by an independent commission (therefore not by the MSC itself) against the MSC standard. An

[174] http://www.msc.org/home-page?set_language=en.

[175] Parkes, G., Walmsley, S., Cambridge, T., Trumble, R., Clarke, S., Lamberts, D., Souter, D. and White C., 2009. Review of Fish Sustainability Information Schemes - Final Report, October 2009. MRAG, London, UK. Available at: http://cels.uri.edu/urissi/docs/FSIG_Report.pdf.

evaluation cycle normally takes around 12-15 months, but longer periods are possible. The whole process is open to the public and any stakeholder (including fishing industry, scientists and environmental organisations) can participate in the evaluation cycle. There are possibilities for participation oriented approaches and grievance procedures. A certificate is issued for a period of five years with a minimum number of inspections of one per year.

The MSC's fishery certification program and seafood ecolabel recognise and reward sustainable fishing (Figure 4.7). The Marine Stewardship Council is a global non-profit organisation working with fisheries, seafood companies, scientists, conservation groups and the public to promote the best environmental choice in seafood. Consumers can recognise MSC-certified products by the blue MSC-logo. MSC was founded 1997 by Unilever and World Wildlife Fund as a result of concerns over the state of fisheries as expressed by a wide range of stakeholders. The total number of available MSC-certified products and countries involved is growing. Anno 2009 approximately 8% of the total world offshore fishing used for human consumption took place under the MSC-program.

Figure 4.7. The Marine Stewardships Council certificate trade mark.

The head office of the Marine Stewardship Council is located in London and has branches in Seattle and Sydney. Furthermore there are offices in Scotland (opened in 2008), Germany (2008), the Netherlands (2007), South Africa (2008), Japan (2007), France (2009), Sweden (2010) and Spain (2011) Worldwide more than 10,000 products certified under MSC are available in 74 countries. In total, over 240 fisheries are engaged in the MSC programme with 105 certified and over 140 under full assessment.

The three main principles of the MSC standard for sustainable fishing are:
1. the stock status for fish or shellfish shall be and stay healthy;
2. the impact of the fishing industry shall be and stay small;
3. the supervision of the fishing sector shall be organised correctly, compliance shall be controlled and verified.

Irrespective of the size, range, the fished species and catch area, a fishery business can request to be audited by an independent certification body to the MSC environmental standard. If the business meets the requirements, it can achieve certification. Businesses that want to carry the MSC label on their products also have to obtain a certificate for chain control to ensure the traceability of the certified fish products.

UTZ certified

Started in 2002, UTZ Certified[176] is dedicated to creating an open and transparent Marketplace for agricultural products. It offers coffee, tea and cocoa certification programs and manages traceability for RSPO certified palm oil. UTZ Certified's vision is to achieve sustainable agricultural supply chains where farmers are professionals implementing good practices which lead to better business, where the food industry take responsibility by demanding and rewarding sustainable grown products, and where consumers buy products which meet their standard for social and environmental responsibility (Figure 4.8).

Figure 4.8. The UTZ Certified trade mark.

Until 2007 UTZ was known as UTZ Kapeh, which means 'good coffee' in Maya language from Guatemala. In just over five years UTZ Certified has grown to be one of the leading coffee certification programs worldwide, and is now expanding to become a multi-commodity program. UTZ Certified's vision is to achieve sustainable agricultural supply chains, that meet the growing needs and expectations of farmers, the food industry and consumers alike.

In response to the urgent and pressing global call for sustainably produced palm oil, the Roundtable on Sustainable Palm Oil (RSPO) was formed in 2004 with the objective promoting the growth and use of sustainable oil palm products through credible global standards and engagement of stakeholders. Its members accepted

[176] http://www.utzcertified.org/.

its principles and criteria for sustainable production in November 2007, the time it contracted UTZ Certified as its provider for traceability services.

By the end of 2008 the first plantations were officially certified and the first sustainably produced palm oil was traced by UTZ. Since then the volume has been growing at an ever increasing pace. In 2009 the total volume of physically traced RSPO-certified palm oil was 100,000 metric tons. In 2010 it nearly quadrupled to almost 400,000 metric tons.

European logo for organic production

If a business operator wants to place organic products on the market he or she must comply with all relevant legislation. One of the requirements, to enable consumers to easily recognise the products as for what they are, is the use of the European logo for organic products (Figure 4.9). The logo communicates that the foodstuff is produced according to all requirements relevant to the production of organic production, processing and trade. Organic products are produced under the following conditions:

- The seeds are of organic origin.
- Fertilisers are of organic origin.
- The use of plant protection products is restricted and limited.
- Compound products are produced using (almost) uniquely raw materials of organic origins. An exception can be made for those ingredients that are temporarily unavailable.
- Transportation and trade took place under controlled conditions.

Figure 4.9. The European logo for Organic Farming.

Logo for gluten free

Patients suffering from coeliac disease (food intolerance) cannot tolerate food containing gluten. Foodstuffs containing gluten cause damaging of the mucous membrane of the small intestine, this in turn causes the intestines to stop functioning correctly. Healthy small intestines have a large number of intestinal villi on the inside of the intestine. Altogether this results in a large surface area for the absorption of nutrients. The villi of coeliac patients do not tolerate gluten, resulting in a bad absorption of nutrients. The human body needs these nutrients to function correctly and in children they are also essential in growth.

Patients suffering from gluten intolerance require foodstuffs that do not contain gluten and are therefore identifiable as such. The authorities have laid down rules that aim to enable patients to make a well informed choice. Regulation 41/2009[177] states that a product is only seen as 'gluten free' when the concentration of gluten is not higher than 20 mg/kg in the foodstuff. If this criterion is met a logo (Figure 4.10), indicating the foodstuff is gluten free, can be used with the product. When the concentration is higher than 20 mg/kg, but not higher than 100 mg/kg the term 'very low gluten' can be used. Products that contain gluten fall under the legislation concerning allergen labelling.

Figure 4.10. The logo indicating a product is 'gluten free'.

Logo for vegetarian products

Vegetarians do not consume meat, poultry and fish products. Most vegetarians do consume product of animal origin that are obtained without killing the animal, such as milk (products), cheese and eggs. Some vegetarians do consume fish. A healthy diet is possible without the consumption of meat. A vegetarian diet composed of sufficient vegetables, fruit, bread, potatoes wheat products, pulse crops, dairy products, eggs and meat replacers provide all the nutrients a human body requires.[178] Vegans do not eat any products of animal origin, so also no dairy products or eggs.

[177] EU, 2009. Commission Regulation (EC) No 41/2009 of 20 January 2009 concerning the composition and labelling of foodstuffs suitable for people intolerant to gluten. Official Journal of the European Union L 16, 21/01/2009: 3-5.

[178] http://findarticles.com/p/articles/mi_m0FDE/is_3_25/ai_n26957000/.

The European V-label (Figure 4.11) is used as a logo indicating the product is 'suitable for vegetarians', without limiting the product to vegetarians only. It is registered in 1985 by the European Vegetarian Union (EVU) and is recognised and used in more than 16 European countries. In all these countries the same logo is used. The logo is available in two versions: one for vegetarian products (that can still contain ingredients of animal origin such as dairy and eggs) and one for vegan products not containing any ingredient of animal origins.

Figure 4.11. The European V-label.

Currently, the EVU label is displayed mostly on food products and in restaurants in Europe, although some products displaying the label may be exported to the United States. A website (V-label) containing information on vegetarian products, producers and restaurants that have accepted the criteria of the V-label is available.[179]

Halal food

In Arabic 'halal' stands for foodstuffs that are 'clean', 'allowed' or 'permitted'. The term indicating the opposite is haram (in Arabic: 'unclean', 'not allowed'), some examples are alcohol and pork. The definitions apply to both activities and foodstuffs. Halal products are prepared according to religious requirements set by Islam. Halal refers to food law laying down rules for Muslims.[180] Halal food is prepared in compliance with the carefully laid down Islamic food laws. As food plays an important role in the daily life, Islamic food law plays a very important role in the life of a Muslim. Meat can only be Halal if the animal was slaughtered according to religious rituals and has lived a dignified life with as little stress as possible.

Food businesses producing Halal products can achieve certification by the Halal International Control.[181] HIC is an international organisation based in Cairo, Egypt, providing services to clients in the Middle-East and Europe.

[179] http://www.v-label.info/en/home/start.html.
[180] http://en.wikipedia.org/wiki/Halal. For more details see Chapter 12 by Tetty Havinga.
[181] http://www.hic-quality.com/.

HIC ensures that raw materials and ingredients used in the production of Halal foodstuffs come from companies that perform ritual slaughtering and respect the general Halal rules. Before any meat is purchased, the supplier is evaluated and it is assessed if the criteria set by HIC are being met. Also herbs and spices require being Halal. HIC controls the herbs and spices for the presence of animal meat extracts and for E-numbers that are not Halal.

HIC can issue Halal certificates for the following processes:
- ritual slaughtering of poultry or other animals;
- manufacturing, from raw materials to final or intermediate products;
- processing of intermediate products;
- preparation of readymade foods (including catering and other large kitchens);
- retail and out of home market (restaurants, fast food chains, take aways, canteens);
- export (as might be requested by customers from e.g. Saudi Arabia or Malaysia);
- storage (distribution centres, warehouses, harbours);
- transport (e.g. fluids such as liquids, such as medicines).

Similar certification bodies exist in other parts of the world such as Asia (Malaysia), United States, Canada:
- INFANCA, Illinois, USA;
- ISNA, Ontario, Canada;
- Malaysian Halal Certification (Figure 4.12).

In 2010 the Malaysian Halal Certification has prepared an overview of 49 different certification bodies in many countries from all over the world.[182] Harmonisation is attempted by the International Halal Integrity Alliance (IHIA).

Figure 4.12. The Malaysian Halal certification stamp.

[182] http://www.halal.gov.my/v2/cms/content/editor/files/File/Badan%20Islam/badan_islam_update_may2010%5B1%5D.pdf.

Kosher food

The basic for 'kashrut', the food law of Jewish religion,[183] is written in the Torah, to be more specific in the books known in the Christian Bible as Leviticus en Deuteronomium 14:3-21. In Deuteronomium 14:6 is written: 'You may eat any animal that has completely split hooves and chews the cud'. Fish must have fins and scales and not all birds are allowed to be eaten. Furthermore, it is forbidden to eat carrion and insects with wings.

It is therefore allowed to eat beef, sheep, goat and deer. It is not allowed to eat camel, pork, horse or rabbit. Finally, it is also not allowed to eat eel, shrimp and lobster. Also in the application of additives there are specific conditions. There are few conditions in the food law regarding the usage of colours. A problem may exist with the colour red made from scale insects (cochenille red). According to the Jewish law it is not allowed to eat insects. Food that is not in accordance with the kashrut is called 'trefah'.

According to the rules of the Kashrut, approved animals shall be slaughtered using unstunned and ritual methods. This is called sjechita. Knives should be razor sharp and straight to instantly kill the animal. After the blood has been drained from the carotid artery, the remaining blood is removed by heavily salting the meat. Besides rules on the slaughtering of animals, other rules on the consumption of kosher foods exist. It is not allowed to use dairy and meat products together in a product or meal. Fish and meat can be used within a single meal but not within the same dish. These rules have implications also to the machinery used to produce or store the foodstuffs.

Hechher (sometimes Anglicized as hechsher) is the Hebrew word for a declaration of kashrut issued by a rabbi (kosher certification)[184]. All kosher shops (except for supermarkets) should have a Teudat Hechsher (Kosher certificate). A rabbi or rabbinical court/organisation can issue certificates that are valid for a limited period of time. Businesses that require a Teudat Hechsher include: (fast food) restaurants and bakeries. Especially in Israel, but increasingly in the United States, a small symbol, usually the logo of the rabbi or rabbinate is placed on prepacked products. Just placing the word kosher alone is not sufficient as multiple levels in kosher certification exist. Most orthodox Jews only eat food labelled with the hechsher of a selected group of hechsherim; the most widely accepted hechsherim is that of Edah HaChareidis. Some other well-known hechsherim from Israel are, among others, those of Rechovot; the rabbinical court of Chassam Sofer from Bnei Brak; the rabbinical court of rabbi Ovadia Yosef; Machzikei HaDas of the chassidish movement Belz, Shearis Yisroel of the Litvaks and the sephardic rabbi Shlomo

---

183 http://www.oukosher.org/index.php/basics/what.

184 http://en.wikipedia.org/wiki/Hechsher.

Machpoud. More than 800 hechher images have been gathered in a database[185] According to the Talmud the practice of certification has been applied since the second century before the existence of a calendar in the time of Chanoeka in the year 1964 before the current era.

In the United States the best known organisations are the Orthodox Union (OU), Organized Kashrus (OK) and the Central Rabbinical Congress (CRC). Over the past decades, the demand for kosher food products in the United States and around the world has greatly impacted the food industry. Established in 1935, the OK Kosher Certification is one of the world's most respected symbols of kosher approval (Figure 4.13). The OK provides certification for food giants and products such as IFF, Kraft, ConAgra and Tropicana. Operating on six continents and supported by more than 350 of the world's leading kosher experts, the OK certifies more than 500,000 products, produced by over 2,500 companies.

Figure 4.13. Logo of Organized Kashrus (OK) certification.

4.4 To conclude

At this moment in time, proliferation of private schemes in the food sector is such that an all-encompassing inventory is no longer possible. However, what this chapter did achieve is to provide an overview of the systems that are currently most important and to show the richness and diversity of private regulation of the food sector.

References

Anonymous, 2010. Inventory of certification schemes for agricultural products and foodstuffs marketed in the EU Member States. Areté Research and Consulting in Economics, Bologna, Italy.

Codex Alimentarius Commission, 2001. Guidelines for design, production, issuance and use of generic official certificates. CAC/GL 38-2001. FAO, Rome, Italy.

[185] http://www.hechshers.info/.

EU, 2002. Regulation (EC) No 178/2002 of the European Parliament and of the Council of 28 January 2002 laying down the general principles and requirements of food law, establishing the European Food Safety Authority and laying down procedures in matters of food safety. Official Journal of the European Union L 31, 1/2/2002: 1-24.

EU, 2004. Regulation (EC) No 852/2004 of the European Parliament and of the Council of 29 April 2004 on the hygiene of foodstuffs. Official Journal of the European Union L 139, 30/4/2004: 1-54.

EU, 2005. Regulation (EC) No 183/2005 of the European Parliament and of the Council of 12 January 2005 laying down requirements for feed hygiene. Official Journal of the European Union L 35, 8/2/2005: 1-22.

EU, 2008. Regulation (EC) No 765/2008 of the European Parliament and of the Council of 9 July 2008 setting out the requirements for accreditation and market surveillance relating to the marketing of products and repealing Regulation (EEC) No 339/93. Official Journal of the European Union L 218, 13/8/2008: 30-47.

EU, 2010. Commission communication – EU best practice guidelines for voluntary certification schemes for agricultural products and foodstuffs. Official Journal of the European Union C 341, 16/12/2010: 5-11.

Parkes, G., Walmsley, S., Cambridge, T., Trumble, R., Clarke, S., Lamberts, D., Souter, D. and White C., 2009. Review of Fish Sustainability Information Schemes - Final Report, October 2009. MRAG, London, UK. Available at: http://cels.uri.edu/urissi/docs/FSIG_Report.pdf.

Scholten-Verheijen, I., Appelhof, T. and Van der Meulen, B., 2011. Roadmap to EU food law. Eleven International Publishers, the Hague, the Netherlands.

5. Codex Alimentarius and private standards

Spencer Henson and John Humphrey

5.1 Background

A key trend in the governance of global agri-food value chains in the last 10 to 15 years is the increasing prevalence of private standards.[186] Private firms and non-governmental organisations (NGOs) have progressively laid down standards for food safety, food quality and environmental and social aspects of agri-food production, which are generally linked in turn to processes of second or third party certification.[187] While not subject to the same legal processes of enforcement as public regulations, it is argued that market forces can make compliance with private standards mandatory in practice.[188] Thus, in the sphere of food safety on which we focus here, many global agri-food value chains are governed by an array of public and private standards, which are variously interconnected and play a leadership role in driving the implementation of food safety controls.[189]

The evolution of private food safety standards has raised profound questions about the role of public and private institutions in governing food safety. Embedded in this dialogue are concerns about the legitimacy of private modes of governance where public regulation has been the dominant institution[190] with accusations that they are not risk-based and fail to adhere to basic democratic norms.[191] More generally, there are concerns that private standards could undermine processes of public policy-making in the area of food safety, both within nation states and trans-nationally.

---

[186] Jaffee, S. and Henson, S.J., 2004. Standards and agri-food exports from developing countries: rebalancing the debate. World Bank Policy Research Working Paper 3348, The World Bank, Washington, DC, USA; OECD, 2004. Private standards and the shaping of the agri-food system. OECD, Paris, France.
[187] Busch, L., Thiagarajan, D., Hatanaka, M., Bain, C., Flores, L. and Frahm, M., 2005. The relationship of third-party certification (TPC) to sanitary/phytosanitary (SPS) measures and the international agri-food trade: final report. RAISE SPS Global Analytical Report 9, USAID, Washington, DC, USA.
[188] Henson, S.J., 2007. The role of public and private standards in regulating international food markets. Journal of International Agricultural Trade and Development 4(1): 52-66.
[189] Henson, S.J. and Humphrey, J., 2009. The impacts of private food safety standards on the food chain and on the public standards-setting process. Paper prepared for FAO/WHO, ALINORM 09/32/9D-Part II, Rome, Codex Alimentarius Commission; Henson, S.J. and Humphrey, J., 2010. Understanding the complexities of private standards in global agri-food chains as they impact developing countries. Journal of Development Studies 46(9): 1628-1646.
[190] Henson, S.J. and Caswell, J.A., 1999. Food safety regulation: an overview of contemporary issues. Food Policy 24(6): 589-603.
[191] See for example, Fuchs, D. and Kalfagianni, A., 2010. The democratic legitimacy of private authority in the food chain. In: Porter, T. and Ronit, K. (eds.) The challenges of global business authority: democratic renewal, stalemate or decay? State University of New York Press, Albany, NY, USA, pp. 65-88.

In the global context, the rise of private standards has been seen as a challenge to the position of established international institutions that lay down rules for the promulgation of public food safety regulations, notably the World Trade Organization (WTO) and Codex Alimentarius Commission (CAC).[192] While the trade effects of private standards have been raised within the WTO, there is considerable uncertainty as to whether it has any legal jurisdiction over private standardisation activities.[193] Within Codex, significant anxiety has been expressed that the rapid pervasion of private food safety standards is serving to undermine the Commission's role in establishing science-based standards, guidelines and recommendations that guide national rule-making and provide the legal reference point for the SPS Agreement.[194] A number of developing country members of the WTO and/or Codex, have been the predominant 'voice' behind these concerns. Thus, private standards have been discussed at each of the last three sessions of the CAC[195], for example, although no clear conclusions have been reached and action around this issue appears to have stalled.

Private standards are remarkably varied with respect to who they are developed by, who adopts them, the parameters of agri-food systems they address, etc. Reflecting this diversity, there has been a lack of clarity about which standards count as 'private', the functions they perform and the potential impacts that they have.[196] There is often also a failure to appreciate the distinctions and inter-relationships between public regulations and private standards. This lack of clarity has served to cloud debates about the impacts of private standards and the trajectory we might expect in their future evolution, and has tended to throw all private standards into the same (often negative) basket. This chapter attempts to add some coherence to this debate, by providing a reasoned analysis of how and why private standards have evolved and, in particular, reflecting on the implications for the role of the international standards promulgated by organisations such as Codex.[197] The particular focus is on private standards relating to food safety, whilst recognising that private standards also govern other attributes of agri-food products.

[192] Henson, S.J., 2007. The role of public and private standards in regulating international food markets. Journal of International Agricultural Trade and Development 4(1): 52-66.
[193] Roberts, M.T., 2009. Private standards and multilateral trade rules. Paper prepared for FAO, Rome, Italy.
[194] See for example Roberts, D. and Unnevehr, L.J., 2005. Resolving trade disputes arising from trends in food safety regulation: the role of the multilateral governance framework. World Trade Review 4(3): 469-497.
[195] CAC, 2008. Report of the 31st Session of the Codex Alimentarius Commission, ALINORM 08/31/REP. Codex Alimentarius Commission, Rome, Italy; CAC, 2009. Report of the 32nd Session of the Codex Alimentarius Commission, ALINORM 09/32/REP. Codex Alimentarius Commission, Rome, Italy; CAC, 2010. Report of the 33rd Session of the Codex Alimentarius Commission, ALINORM 10/33/REP. Codex Alimentarius Commission, Rome, Italy.
[196] Henson, S.J. and Humphrey, J., 2010. Understanding the complexities of private standards in global agri-food chains as they impact developing countries. Journal of Development Studies 46(9): 1628-1646.
[197] In so doing, it draws heavily on a report by Henson, S.J. and Humphrey, J., 2009. The impacts of private food safety standards on the food chain and on the public standards-setting process. Paper prepared for FAO/WHO, ALINORM 09/32/9D-Part II, Codex Alimentarius Commission, Rome, Italy.

5.2 Nature of private food safety standards

Before reflecting on what private standards mean for Codex, it is important to recognise the institutional forms they take and how and why they have evolved over time. Private standards have emerged as an important mode of market governance in many industrialised countries.[198] This is particularly true of the agri-food sector, where they have increasingly pervaded both domestic business and international trade. These standards may relate to food safety and the integrity of food safety systems, but can also refer to other food attributes, such as provenance, environmental impact, animal welfare, etc.

One of the defining characteristics of these private standards, particularly as they relate to food safety, is an increasing focus on the processes by which food is produced. Such 'process' standards necessarily involve the following:

- They provide a basis for making claims about processes and practices relating to how food is produced, transported or processed.
- They necessarily involve some form of monitoring and enforcement, through second or (increasingly) third party certification.
- They are codified into a written statement that sets out rules and procedures and provides clear instructions as to how rules are to be implemented, monitored and enforced.
- They include some form of traceability to link particular food products at some point downstream in the value chain to the point of at which the standard specifies and controls processes.

Critically, private standards involve not only a specification of what outcomes are to be achieved, but also sets of rules to show how this should be accomplished, a governance structure of certification and enforcement, as well as systems to generate and approve changes to each of these elements as the standards evolve over time. It is for this reason that some bodies involved in setting and/or administering private standards, such as the Global Food Safety Initiative (GFSI), refer to 'schemes' rather than standards. This has important implications for Codex and, more generally, the relationship between public regulations and private standards.

[198] See for example: Henson, S.J. and Humphrey, J., 2010. Understanding the complexities of private standards in global agri-food chains as they impact developing countries. Journal of Development Studies 46(9): 1628-1646; Humphrey, J., 2008. Private standards, small farmers and donor policy: EUREPGAP in Kenya. IDS Working Paper 308, Institute of Development Studies, Brighton, UK; Jaffee, S. and Henson, S.J., 2004. Standards and agri-food exports from developing countries: rebalancing the debate. Policy Research Working Paper 3348, The World Bank, Washington, DC, USA; OECD, 2004. Private standards and the shaping of the agri-food system, Paris, France; World Bank, 2005. Food safety and agricultural health standards: challenges and opportunities for developing country exports. Report 31207, The World Bank, Poverty Reduction and Economic Management Trade Unit, Washington, DC, USA.

In the literature, the terms 'private standards' and 'voluntary standards' are frequently used interchangeably. Indeed, private standards developed collectively by private sector actors are frequently referred to as 'private voluntary standards'.[199] Implicitly this equates the actions of public authorities with rules backed by legal sanctions,[200] leaving the territory of voluntary standards to non-governmental entities.[201] In practice, this distinction is not so stark. Governments may promulgate standards with which compliance is voluntary, or conversely, they may require compliance with private standards. Indeed, there is a decided 'blurring' of traditional governance roles in the agri-food system, suggesting the emergence of a continuum between public and private modes of regulation.[202]

To provide clarification, Table 5.1 distinguishes between mandatory and voluntary standards, and between standards set by public and private entities.[203] Here, private standards are represented by the right-hand column; they are standards that are set by commercial or non-commercial private entities, including firms, industry organisations and NGOs. In turn, the extent to which private standards are truly voluntary depends on the form and level of power wielded by the entities adopting those standards; that is the nature of the entities requiring that the standard be implemented by other value chain actors.[204] Private standards can be adopted by non-state (private) actors; even if they become *de facto* mandatory in a commercial sense through adoption by dominant market actors, there is no legal penalty from

Table 5.1. Forms of standards.

	Public	**Private**
Mandatory	regulations	legally-mandated private standards
Voluntary	public voluntary standards	private voluntary standards

[199] See for example OECD, 2004. Private standards and the shaping of the agri-food system, Paris, France.
[200] Black, J., 2002. Critical reflections on regulation, Centre for Analysis of Risk and Regulation, London School of Economics and Political Science, London, UK; Havinga, T., 2006. Private regulation of food safety by supermarkets. Law and Policy 28(4): 515-533.
[201] The distinction between single-company private standards and private standards developed by coalitions of private actors will be discussed below.
[202] Havinga, T., 2006. Private regulation of food safety by supermarkets. Law and Policy 28(4): 515-533; Havinga, T., 2008. Actors in private food regulation: taking responsibility or passing the buck to someone else? Paper presented at the symposium Private Governance in the Global Agro-Food System, Munster, Germany, 23-25 April 2008.
[203] Henson, S.J. and Humphrey, J., 2008. Understanding the complexities of private standards in global agri-food chains. Paper presented at the workshop Globalization, Global Governance and Private Standards, University of Leuven, Belgium, November 2008; Henson, S.J. and Humphrey, J., 2009. The impacts of private food safety standards on the food chain and on the public standards-setting process. Paper prepared for FAO/WHO, ALINORM 09/32/9D-Part II, Codex Alimentarius Commission, Rome, Italy.
[204] Brunsson, N. and Jacobsson, B., 2000. The contemporary expansion of standardization. In: Brunsson, N. and Jacobsson, B. (eds.) A world of standards. Oxford University Press, Oxford, UK, pp. 1-17.

non-compliance. However, private standards can also be adopted by state actors and invested with statutory power. In this case, compliance is mandatory. We refer to these as legally-mandated private standards.[205]

With respect to the middle column in Table 5.1, public standards, the most familiar form is the regulations promulgated by governments that are mandatory within the sphere of competence of the government. However, governments also promote standards that are voluntary, which have elsewhere been termed 'optional laws'.[206] The Food Safety Enhancement Programme (FSEP) in Canada is one example.[207] While governments may put in place compelling incentives for compliance with such public voluntary standards, including the threat of regulation should an industry not 'regulate itself' through their implementation, in a legal sense there is no compulsion.[208]

Following the classification of the WTO with respect to private (voluntary) standards, three forms of private agri-food standard are distinguished in Table 5.2 according to the institutional and geographical characteristics of the entities that generate them.[209] This table provides a non-exhaustive list of private standards in four European countries as illustration:

- *Individual company standards.* These are set by individual firms, predominantly large food retailers, and adopted across their global supply chains. They are frequently communicated to consumers as sub-brands on own label products. Examples include Tesco's Nature's Choice and Carrefour's Filières Qualité. These standards frequently include food safety elements, but when presented to consumers they tend to emphasise non-safety aspects, such as environmental impact.
- *Collective national standards.* These standards are set by collective organisations that operate within the boundaries of individual countries. These organisations can represent the interests of commercial entities (for example food retailers, processors or producers) or be NGOs. These and other entities are then free to adopt the established standards if they so wish. It is important to note that

[205] This process is seen, for example, with the referencing of ISO 9000 in EU directives covering CE marking for telecommunications and electronic products.
[206] Brunsson, N. and Jacobsson, B., 2000. The contemporary expansion of standardization. In: Brunsson, N. and Jacobsson, B. (eds.) A world of standards. Oxford University Press, Oxford, UK, pp. 1-17.
[207] Martinez, M.G., Fearne, A., Caswell, J.A. and Henson, S.J., 2007. Co-regulation as a possible model for food safety governance: opportunities for public-private partnerships. Food Policy 32(2): 299-314.
[208] The position of a particular standard within the grid in Table 4.1 may change over time. It is not uncommon for standards to migrate between cells. For example, the Safe Quality Food (SQF) series of standards was originally developed by the Government of Western Australia, which we would categorise as a public voluntary standard, but these were subsequently acquired by the Food Marketing Institute, an industry organisation representing the US food retail and wholesale sectors, implying reclassification as a private voluntary standard.
[209] This typology aims to present the dominant forms of private standards, but given the various forms taken by private standards, is necessarily incomplete.

some of these standards are inherently national, while others have international reach. Collective national standards may be specifically designed to establish claims about food from particular countries or regions. For example, the UK Farm Assured British Beef and Lamb scheme sustains claims about the superior attributes (safety, quality, environmental impact, etc.) of conforming products as a means of differentiation against imports. As a result, they are usually 'visible' to the consumer though labels and trademarks. Other standards are national in character only because they have been developed by national agencies, but may have international reach through the global supply chains of their adopters. An example is the British Retail Consortium (BRC) Global Standard for Food Safety.

- *Collective international standards:* This category of private standard is often defined by its reach; that it is intentionally designed to be adopted by firms or other entities in different countries. This frequently means that the organisation that sets the standard has international membership. For example, GlobalGAP (formerly EurepGAP) was initially created by an international coalition of

Table 5.2. Examples of private standards in agri-food chains.[1]

Individual firm standards	Collective national standards	Collective international standards
• Nature's Choice (Tesco) • Filières Qualité (Carrefour) – version applied in multiple countries • Field-to-Fork (Marks & Spencer) • Filière Controlleé (Auchan) – version applied in multiple countries • P.Q.C. (Percorso Qualità Conad) • Albert Heijn BV: AH Excellent	• Assured Food Standards (UK) • British Retail Consortium Global Standard • Freedom Food (UK) • Qualität Sicherheit (QS) • Assured Combinable Crops Scheme (UK) • Farm Assured British Beef and Lamb • Sachsens Ahrenwort • Sachsen Qualitätslammfleisch • QC Emilia Romagna • Stichting Streekproduction Vlaams Brabant	• GlobalGAP • International Food Standard • Safe Quality Food (SQF) 1000/ 2000 • Marine Stewardship Council (MSC) • Forest Stewardship Council (FSC)

[1] Based on WTO, 2007. Private standards and the SPS agreement, note by the Secretariat, G/SPS/GEN/746, WTO, Committee on Sanitary and Phytosanitary Measures, Geneva, Switzerland, with additional examples from Aragrande, M., Segre, A., Gentile, E., Malorgio, G., Giraud Heraud, E., Robles, R., Halicka, E., Loi, A. and Bruni, M., 2005. Food supply chains dynamics and quality certification, final report. EU DG Joint Research Centre, Brussels, Belgium.

European retailers, whilst the International Food Standard (IFS) is *set* by German and French retailers. Such standard-setting organisations may have non-business actors, as with the Forest Stewardship Council.

Critically, the private standards landscape is highly dynamic, with new forms of standard emerging, which in turn induce changes in the relative importance of particular forms of standard. For example, a number of large UK food retailers established their own private standards in the early 1990s and employed second or third party audits of their suppliers in order to assess compliance.[210] Later, many of these retailers participated in the promulgation of a collective national standard, the BRC Global Standard for Food Safety. More recently, the scope of collective private standards has tended to become international rather than national, as is seen with GlobalGAP and the International Food Standard (IFS), and at the same time national firm or collective standards are increasingly benchmarked through the GFSI. While these processes have driven broad trends of collective action and the internationalisation of private standards, at the same time individual firm standards have emerged in new spaces of product and process attribute standardisation.

While it is important to recognise the institutional nature of the entities involved in establishing private standards in the agri-food sector, comprehending the different functions involved in making a standard operational is perhaps more critical. In this regard, we can define five specific functions:

- *Standard-setting.* The introduction and operationalisation of a standard through the formulation of written rules and procedures.
- *Adoption.* A decision by an entity to adopt the standard. This can take various forms. A private company can adopt a standard by requiring implementation by its suppliers. This could be a standard developed by the company itself, or one it helped to develop, for example as part of a standards-setting coalition, or a standard created by another body. Equally, groups of producers can develop a standard which they themselves adopt. The decision to adopt is an important driver of the spread and influence of private standards. This stage of standards development is sometimes under-emphasised in the categorisation of standards. For example, recent typologies of standards that identify the actors that define and implement standards[211], but not actors that adopt them, do not sufficiently recognise the way in which standards are integrally related to increasing globalised agribusiness value chains.
- *Implementation.* The implementation of the rule is carried out by the organisation that is conforming to the standard. This will not be the standards-setter. In the

[210] Henson, S.J. and Northen, J.R., 1998. Economic determinants of food safety controls in the supply of retailer own-branded products in the UK. Agribusiness 14(2): 113-126.

[211] See for example WTO, 2007. Considerations relevant to private standards in the field of animal health, food safety and animal welfare. Submission by the World Organisation for Animal Health, G/SPS/GEN/822, WTO, Committee on Sanitary and Phytosanitary Measures, Geneva, Switzerland.

case of a standard like the BRC Global Standard for Food Safety, the implementer is the company that applies the standard in its own operations.
- *Conformity assessment*. This involves the procedures employed to verify that those claiming to comply with the standard and provide documented evidence to show that this is the case. There are various means of assessing conformity, including self-declaration by the implementer of the standard, inspection by the standards adopter (so-called second-party certification) and inspection by a third party (so-called third party certification). Third-party certification carried out by independent certification bodies has become the norm for many private food safety standards. We term these certification-based private standards. Standards schemes include processes for recognising the certification bodies that are allowed to verify compliance.
- *Enforcement*. This refers to the approaches employed in response to non-compliance and sanctions if corrective action is not taken. The standard setter has to have some procedure for responding to the results of the conformity assessment, either by invoking corrective action or withdrawing the recognition of the organisation as conforming to the standard.

These functions may be carried out by public or private entities according to the nature of the standard (Table 5.3). In the case of regulations, all of the functions (aside from implementation) are typically undertaken by the public sector. With voluntary public standards and mandatory private standards, these functions can be divided between the public and private sectors. Such divisions, however, are

Table 5.3. Functions associated with standards schemes.

Function	Regulations	Public voluntary standards	Legally-mandated private standards	Private voluntary standards
Standard-setting	legislature and/or public regulator	legislature and/or public regulator	commercial or non-commercial private body	commercial or non-commercial private body
Adoption	legislature and/or public regulator	private firms or organisations	legislature and/or public regulator	private firms or organisations
Implementation	private firms and public bodies	private firms	private firms	private firms
Conformity assessment	official inspectorate	public/private auditor	public/private auditor	private auditor
Enforcement	criminal or administrative courts	public/private certification body	criminal or administrative courts	private certification body

not hard and fast. Thus, new conceptualisations of regulation are ceding a role for the private sector, for example through private firms undertaking conformity assessment on their own compliance with public regulations.[212] Even private standards, for which these functions are predominantly undertaken by non-state entities, may build on the public standards infrastructure through their use of accreditation bodies to govern certifiers and the specification of public laboratories to be used in product testing.[213]

The distinction between standard setting and adoption also clarifies the issue of compulsion and obligation. First, it is possible for private standards to be made mandatory by public bodies. This is the situation with legally-mandated private standards. An example is obligations placed on companies to have relevant production processes certified to ISO 9000 before products can be imported into the European Union (EU). Second, there are situations where companies freely adopt private standards, either because they see this as a signal to potential buyers or because it promotes firm efficiency. Third, while private standards not adopted by public bodies remain voluntary in that there is no legal compulsion to comply, they can become quasi-mandatory if powerful companies, alone or *en masse*, make the standard a condition of entry to their supply chains. Concentration in global food retailing and processing may increase this tendency. It is this type of relationship along the value chain that drives much of the development of private standards.

5.3 Trends in the development and functions of private food safety standards

Having recognised the diversity of private standards and the often blurred distinctions between regulations and private standards, we now turn to the questions of why private standards have emerged as an increasingly dominant mode of governance of attributes such as food safety and what has driven this process. There appear to be four key factors here:[214]

- Business responses to increasing consumer and government concerns about food safety, notably in the wake of a history of food safety scares that have undermined public confidence in established controls in many industrialised countries.[215]

[212] Havinga, T., 2006. Private regulation of food safety by supermarkets. Law and Policy 28(4): 515-533.
[213] Further, standards such as GlobalGAP involve selected national certification bodies for processes involved in benchmarking of national standards to the global standard (Sheehan, K., 2007. Benchmarking of Gap schemes, EUREPGAP Asia conference, Bangkok, 6-7 September 2007. Available at: http://www.globalgap.org/cms/upload/Resources/Presentations/Bangkok/3_K_Sheehan.pdf).
[214] Henson, S.J. and Humphrey, J., 2010. Understanding the complexities of private standards in global agri-food chains as they impact developing countries. Journal of Development Studies 46(9): 1628-1646.
[215] Jaffee, S. and Henson, S.J., 2004. Standards and agri-food exports from developing countries: rebalancing the debate. Policy Research Working Paper 3348, The World Bank, Washington, DC, USA.

- Progressive changes in the expectations and demands of consumers with respect to the safety and quality attributes of food, reflecting broader demographic and social trends.[216] These attributes include the manner in which products are produced and the existence of substances in food that are perceived to be risky, including those purposefully used in food production (for example pesticides) and contaminants (for example dioxins). Thus, food safety is no longer defined simply as 'fit for human consumption', but rather in terms of a wider array of safety and quality attributes that range from search, through experience, to credence attributes.[217]
- The globalisation of agri-food chains whereby supply chains for agricultural and food products increasingly extend beyond national boundaries, in part facilitated by new food, communications and transportation technologies and a more liberal trade environment.[218] Global sourcing creates new sources of risk as food is subject to greater transformation and transportation, while supply chains are increasingly fragmented across enterprises, production systems, environments and regulatory frameworks.
- The shift in responsibility for food safety from the public to the private sector, which has both driven and ceded a space for private 'regulation'.[219] This reflects a broad political trend towards more liberalised markets in many industrialised countries, and also a change in the philosophy of food regulation.[220]

These four trends have combined to create an environment in which businesses are under increasing pressure to deliver food safety and to maintain the integrity of their brands. They need to do this in the face of increasingly globalised and

[216] Buzby, J., Frenzen, P.D. and Rasco, B., 2001. Product liability and microbial food-borne illness. Agricultural Economic Report 828, United States Department of Agriculture, Economics Research Service, Washington, DC, USA; Jaffee, S. and Henson, S.J., 2004. Standards and agri-food exports from developing countries: rebalancing the debate. Policy Research Working Paper 3348, The World Bank, Washington, DC, USA.

[217] Reardon, T., Codron, J.-M., Busch, L., Bingen, J. and Harris, C., 2001. Global change in agrifood grades and standards: agribusiness strategic responses in developing countries. International Food and Agribusiness Management Review 2(3): 421-435.

[218] Nadvi, K. and Waltring, F., 2003. Making sense of global standards. In: Schmitz, H. (ed.), Local enterprises in the global economy: issues of governance and upgrading. Elgar, Cheltenham, UK; OECD, 2004. Private standards and the shaping of the agri-food system. OECD, Paris, France; Henson, S.J. and Reardon, T., 2005. Private agri-food standards: implications for food policy and the agri-food system. Food Policy 30(3): 241-253; Fulponi, L., 2005. Private voluntary standards in the food system: the perspective of major food retailers in OECD countries. Food Policy 30(2): 115-128.

[219] EU, 2002. Regulation (EC) No 178/2002 Laying Down the General Principles and Requirements of Food Law, Establishing the European Food Safety Authority and Laying Down Procedures in Matters of Food Safety. Official Journal of the European Communities, 1 February 2002.

[220] For example, the preamble to the European Union's General Food Law states that: 'a food business operator is best placed to devise a safe system for supplying food and ensuring that the food it supplies is safe; thus, it should have primary legal responsibility for ensuring food safety.' (EU, 2002. Regulation (EC) No 178/2002 Laying Down the General Principles and Requirements of Food Law, Establishing the European Food Safety Authority and Laying Down Procedures in Matters of Food Safety. Official Journal of the European Communities, 1 February 2002).

complex food supply chains. Private standards are part of the response to this challenge. The key role of standards, whether public or private, mandatory or voluntary, is to facilitate the coordination of agri-food value chains across space and between producers/firms and, in so doing, to transmit credible information on the nature of products and the conditions under which they are produced, processed and transported.[221] In other words, one of the primary functions of private standards relating to food safety is risk management. This means providing a level of assurance that a product is in compliance with defined minimum product and/or process requirements.

Many of the assurances over food safety that firms strive to deliver are defined by regulations. This is particularly the case with levels of contaminants such as pesticides, where critical limits are defined by legal Maximum Residue Levels (MRLs), but also more general aspects of hygiene where legal norms allocate liability in the event of food safety failures. Further, private standards can only operate in an environment of credible infrastructure (for example laboratories) and recognised processes (for example HACCP) that are built around public standards at the national and/or international levels. Which begs the question, why are private standards needed at all? And perhaps more critically, why have profit-oriented firms invested resources in their design, adoption, implementation and enforcement?

One explanation for this effort, which ironically is also why national regulators and international standards-setting bodies such as Codex are concerned about their rise, is that private standards lay down requirements that are in some way in excess of legal requirements. Thus, it is important to understand precisely in which respects this is the case in practice. We would suggest that there are three dimensions to this:

- The standard sets stricter requirements with respect to particular food product attributes, whether by extending these requirements and/or by setting lower thresholds. For example, the Field to Fork standard of the UK retailer Marks and Spencer specifically prohibits residues of around 70 pesticides in fruit and vegetables that are sold fresh or used as ingredients in the manufacture of prepared foods to be sold under the firm's own label.[222] This is the dimension on which critics of private standards tend to focus.

[221] Humphrey, J. and Schmitz, H., 2000. Governance and upgrading: linking industrial cluster and global value chain research. IDS Working Paper 120, Institute of Development Studies, Brighton, UK; Humphrey, J. and Schmitz, H., 2001. Governance in global value chains. IDS Bulletin 32(3): 19-29; Humphrey, J., 2008. Private standards, small farmers and donor policy: EUREPGAP in Kenya, IDS Working Paper 308, Institute of Development Studies, Brighton, UK; Henson, S.J. and Jaffee, S., 2008 Understanding developing country strategic responses to the enhancement of food safety standards. The World Economy 31(1): 1-15.

[222] Henson, S.J. and Humphrey, J., 2009. The impacts of private food safety standards on the food chain and on the public standards-setting process. Paper prepared for FAO/WHO, ALINORM 09/32/9D-Part II, Codex Alimentarius Commission, Rome, Italy.

- The standard lays down more specific instructions as to how to achieve desired end-product attributes and/or how to operationalise particular process parameters. We would argue that this is the most important function of private standards in the area of food safety. In many cases regulations and/or international standards lay down the basic parameters of a food safety control system, while private standards elaborate on what this system should 'look like' in order to be effective. For example, the Codes of Practice and Guidelines of Codex specify what controls need to be in place in food processing operations, but do not provide specific instructions on what these might look like in practice and/or how they might be monitored so as to ensure effective enforcement. This 'gap' is filled by the BRC Global Standard for Food Safety, IFS and SQF 2000 (for example) that lay down very detailed and auditable instructions on the specific controls to be in place. Here, the predominant aim of private standards is to provide for a level of protection against food safety failures beyond that inherent in regulations and associated systems of enforcement, and in a way that is consistent across supply chains that are increasingly global and so traverse regulatory jurisdictions.
- The standard extends the range of process controls beyond that required by regulations, vertically and/or horizontally. Increased vertical coverage, for example, might mean extending traceability requirements beyond the 'one up, one down' requirement of legislation. Alternatively, standards can extend requirements horizontally. Thus, GlobalGAP not only lays down requirements for food safety, but also sustainability and worker rights, which lay outside of the purview of current regulatory requirements.

Implementing a system of conformity assessment that provides a greater level of oversight than is afforded by prevailing systems of regulatory enforcement involves two elements. First, the predominant use of third party certification that takes both the standards adopter and the standards implementer out of the conformity assessment process.[223] This allows for an independent system of conformity assessment against an agreed and objective protocol. Second, the application of a governance structure and support system that ensures this system of third party certification works effectively. Examples include processes for certifier approval, complaint handling, compliance monitoring, etc. The parameters for this governance structure are largely laid down by the international standards developed by ISO.

Thus, we see that private standards in the realm of risk management are multi-layered 'schemes' consisting of the standard *per se*, systems of certification and a standards and conformity assessment governance structure. While private

[223] There are some exceptions here. In the case of private standards organisations and companies (see below), conformity assessment tends to be undertaken through second party certification, which is using the certifier's own auditing staff.

standards, notably for food safety, do lay down requirements that are 'additional' to legal requirements, the predominant focus of these is regulatory compliance. At the same time, standards directed at risk management are progressively encompassing a broader range of attributes that may lie outside of the realm of regulatory requirements.

These concerns with regulatory compliance and the mechanisms to achieve them are not, in fact, restricted to the private sector. The EU establishes specific conditions for importing food from Third Countries that require the competent authority in these countries to demonstrate that their food safety systems offer equivalent levels of safety to those provided by EU legislation[224]. In so doing, the EU is also going beyond Codex norms. Inspections and subsequent recommendations by the EU Food and Veterinary Office provide for monitoring of the efficacy of the enforcement system in Third Countries and impose penalties for non-compliance. Here, as with many private standards, the issue is not the process standards themselves, but how effectively compliance is monitored and enforced.

However, private standards can address issues that are not covered at all by public regulations. One motive for doing this is product differentiation. Standards can be adopted to support claims to consumers that products have certain desirable characteristics. Generally speaking, claims about credence characteristics[225] are backed up by standards which provide a credible basis for these claims. Examples include fair trade, animal welfare, etc. There is relatively little evidence, however, that private standards aim to achieve product differentiation on the basis of food safety, except perhaps as part of a blend of product and process attributes cutting across environmental protection, ethical and social issues and food safety. Most of the major European food retailers, for example, recognise that market competition on the basis of food safety is likely to erode consumer confidence. More often, food safety claims are bundled with other claims. For example, Tesco's Nature's Choice standard is being used to support a broader branding strategy 'Nurture' that differentiates fresh fruit and vegetables from those sold by other UK food retailers, predominantly on the basis of environmental protection.[226] At the same time, private standards can be employed to present additional guarantees to consumers about the safety of the food at the sectoral rather than the firm level. Thus, the origin of produce-origin standards, such as the UK's 'Red Tractor' label, lies in the damage to consumer confidence caused by previous food scares.

[224] See for example, EU, 2004. Corrigendum to Regulation (EC) No 882/2004 of the European Parliament and of the Council of 29 April 2004 on Official Controls Performed to Ensure the Verification of Compliance with Feed and Food Law, Animal Health and Animal Welfare Rules. Official Journal of the European Union, 28 May 2004, Article 48.
[225] Attributes of a product that cannot be verified through direct examination pre or post-consumption.
[226] See http://www.tesco.com/nurture/?page = nurturevalues.

The foregoing discussion suggests that the predominant focus of private standards in the realm of food safety is risk management, predominantly motivated by the need of dominant players in the value chain to achieve a higher level of assurance with respect to regulatory compliance. This generally involves the development and/or adoption of private standards that drive the implementation of more rigorous process controls, either reinforcing regulatory requirements at a particular level of the value chain or extending process controls along the value chain. The main adopters of these private standards are dominant buyers, predominantly large food retailers and food service companies. Where attempts to differentiate on the basis of food safety are observed, private standards are generally developed and implemented further up the value chain, notably in production, as a means to communicate to consumers that food of a particular origin or from a particular system of production is safe. It is not unsurprising, therefore, that private food safety standards have come to be developed collectively.

5.4 Role of Codex in the context of private standards

Having set out the ways in which private standards have evolved and the roles these perform in the governance of food safety, we now turn to the functions of Codex in general and specifically in the context of private standards. It is only through an understanding of the work that Codex does that we can begin to assess whether private standards do indeed undermine established international standards, as some fear.

While the work of Codex is generally described in terms of standards-setting, it is more useful to think about its activities as defining a set of rules within which national governments establish their own regulatory requirements.[227] It is possible to discern three distinct types of rule in this regard (Table 5.4). Thus, Codex standards, guidelines and recommendations both provide guidance to governments and also act as the reference point for compliance with obligations under the WTO. ISO standards play a similar and often complementary role. At the same time, Codex principles provide guidance, and set rules, for the development and implementation of private standards. Indeed, many private food safety standards make explicit reference to Codex Standards, guidelines and recommendations (for example SQF 2000).

The first group in Table 5.4 refers to rules about products. For example, Codex has established a rule on veterinary drugs in meat that provides recommendations on maximum residue levels for veterinary drugs.[228] This product standard can

[227] Humphrey, J., 2008. Private standards, small farmers and donor policy: EUREPGAP in Kenya, IDS Working Paper 308, Institute of Development Studies, Brighton, UK.
[228] CAC, 2006. Maximum residue limits for veterinary drugs in foods: updated as at the 29th session of the Codex Alimentarius Commission. Codex Alimentarius Commission, Rome, Italy.

Table 5.4. Three types of rules promulgated by Codex Alimentarius.[1]

Codex standards
- Referring to specific commodities – standards for specific products
- Referring to ranges of commodities – standards for ranges of products
- Methods of sampling and analysis

Codex codes of practice for production, processing, manufacturing, transport and storage
- For individual foods
- For groups of foods
- General principles for all products, such as the Codex General Principles of Food Hygiene

Codex guidelines
- Principles that set out policy in key areas
- Guidelines for the interpretation of these principles or for the interpretation of other Codex standards
- Interpretative Codex guidelines for labelling and claims about food
- Guidelines for interpreting Codex principles for food import and export inspection and certification, etc.

[1] CAC, 1997. Understanding the Codex Alimentarius. Codex Alimentarius Commission, Rome, Italy.

also be thought of as an outcome standard; the outputs of a food safety system should result in a residue of this particular veterinary drug being no greater than the recommended limit. Note also that Codex defines rules (or recommendations) about methods of analysis and sampling for veterinary drugs in food. In other words, as well as defining rules about product characteristics, it also suggests ways in which these rules should be implemented through testing procedures. These rules have no direct legal force. They are recommendations aimed primarily at governments to guide their own rule-making. The legal implications of these recommendations lie in the SPS Agreement, which leaves regulations not based on these recommendations open to challenge within the WTO if they cannot be demonstrated to be justified by science-based risk assessment.

Within this first group of rules, Codex also has set product standards that are more concerned with establishing common reference points.[229] The issue here is not whether one reference point is better than another, but that everyone uses the same reference point in order to facilitate transactions, interfaces between products, etc. These can include definitions of products and terminology, for example.

[229] David, P.A., 1995. standardization policies for network technologies: the flux between freedom and order revisited. In: Hawkins, R., Mansell, R. and Skea, J. (eds.), Standards, innovation, and competitiveness: the politics and economics of standards in national and technical environments. Edward Elgar, Aldershot, UK, pp. 15-35.

The second group in Table 5.4 refers to Codex codes of practice for production, processing, manufacturing, transport and storage. These are the meta-standards that are incorporated into specific standards that relate to processes; the means by which products are produced, handled and processed along the value chain. Process controls have three main objectives. First, they provide a means of controlling quality and safety in a way that is more efficacious and cost-efficient than testing.[230] Second, process standards are a means of controlling for food safety hazards that are either impossible or very difficult to detect, such that the most effective approach is to implement food safety and hygiene controls at source to reduce the risk of contamination. Third, they facilitate the monitoring and control of characteristics that are extrinsic to the product; which have no physical presence in the product and so are not revealed by inspection.

The Codex codes of practice for production, processing, manufacturing, transport and storage referred to in Table 5.4 are frequently expressed in guidelines that have been drawn from best practice on food safety, codified by Codex and incorporated into numerous specific standards. These meta-standards include Good Agricultural Practice (GAP) and Good Manufacturing Practice (GMP), which are then adopted by both private standards setters and governments.[231] For example, the Recommended International Code of Practice – General Principles of Food Hygiene has evidently formed the basis of many private food safety standards for food processing[232], including the BRC Global Standard for Food Safety, IFS and SQF 2000, and also the GFSI Guidance Document for the benchmarking of such standards. Likewise, ISO 22000 substantively defines a HACCP-based food safety management system in accordance with Codex guidelines.[233]

The third group of Codex guidelines listed in Table4.4 are more general, setting out principles and providing guidelines for interpreting principles. In effect, these are rules that specify the ways in which food safety rules are formulated and implemented; for example, inspection and controls on imports and/or exports. They are addressed to governments, but many private food safety standards are constructed around these same principles. There are at least three reasons for this. First, these guidelines represent best practice, and private firms often participate in

[230] Unnevehr, L., 2000. Food safety issues and fresh food product exports from LDCs. Agricultural Economics 23(3): 231-240.

[231] Busch, L., Thiagarajan, D., Hatanaka, M., Bain, C., Flores, L. and Frahm, M., 2005. The relationship of third-party certification (TPC) to sanitary/phytosanitary (SPS) measures and the international agri-food trade: Final Report, RAISE SPS Global Analytical Report 9, USAID, Washington, DC, USA; Henson, S.J., 2007. The role of public and private standards in regulating international food markets. Journal of International Agricultural Trade and Development 4(1): 52-66.

[232] WTO, 2007. Private Standards and the SPS Agreement, note by the Secretariat, G/SPS/GEN/746. WTO, Committee on Sanitary and Phytosanitary Measures, Geneva, Switzerland.

[233] WTO, 2007. Submission by the International Organisation for Standardisation (ISO) to the SPS Committee Meeting 28 February, 1 March 2007, G/SPS/GEN/750, WTO, Committee on Sanitary and Phytosanitary Measures, Geneva, Switzerland.

their formulation through their membership of bodies such as the ISO or through their participation in national Codex committees and/or International NGOs recognised by Codex that participate in Commission meetings. Second, private voluntary standards for food safety are often responses to government regulations and are aimed at the same outcome. Third, by building on the framework of regulations rather than 'reinventing the wheel', the cost of formulating and enforcing private standards is reduced. Thus, private standards can use the facilities provided by public regulatory and standards infrastructures, for example recognition of laboratories or rules regulating certification bodies.

5.5 Do private standards jeopardise the work of Codex?

The rapid rise of private food safety standards has apparently sent 'shock waves' though public policy-makers, and especially those engaged in establishing international standards through organisations such as Codex[234] and in the WTO. As noted above, private standards have been discussed extensively within Codex and the WTO's SPS Committee. Further, Codex has commissioned two papers on the implications of private standards, with a particular focus on developing countries, and their compatibility with international standards.[235]

The concern is that private food safety standards are acting to supplant or weaken Codex's role in establishing standards for food safety, and in turn the functioning of the SPS Agreement within the WTO. It is important here to recognise the wider context, with wider debates about the legitimacy of Codex and the extent to which its current governance structures are compatible with defining legal benchmarks for the purposes of the WTO, facilitating inclusiveness of decision-making processes and elaborating standards in a timely manner.[236] However, while it is understandable that public regulators may feel some discomfort at

[234] See for example: CAC, 2008. Report of the 31st Session of the Codex Alimentarius Commission, ALINORM 08/31/REP, Codex Alimentarius Commission, Rome, Italy; CAC, 2009. Report of the 32nd Session of the Codex Alimentarius Commission, ALINORM 09/32/REP, Codex Alimentarius Commission, Rome, Italy; CAC, 2010. Report of the 33rd Session of the Codex Alimentarius Commission, ALINORM 10/33/REP, Codex Alimentarius Commission, Rome, Italy; WTO, 2008. Private Standards and the SPS Agreement, note by the Secretariat, G/SPS/GEN/746, WTO, Committee on Sanitary and Phytosanitary Measures, Geneva, Switzerland; WTO, 2008. Report of the STDF Information Session on Private Standards, G/SPS/R/50, WTO, Committee on Sanitary and Phytosanitary Measures, Geneva, Switzerland.
[235] Henson, S.J. and Humphrey, J., 2009. The impacts of private food safety standards on the food chain and on the public standards-setting process. Paper prepared for FAO/WHO, ALINORM 09/32/9D-Part II, Codex Alimentarius Commission, Rome, Italy; CAC, 2010. Consideration of the impact of private standards, CX/CAC 10/33/13. Codex Alimentarius Commission, Rome, Italy.
[236] Henson, S.J., Preibisch, K.L. and Masakure, O., 2001. Enhancing developing country participation in international standards-setting organizations. Department for International Development London, UK; Rosman, L., 1993. Public participation in international pesticide regulation: when the Codex Commission decides, who will listen? Virginia Environmental Law Journal 12: 329-365; Livermore, M.A., 2006. Authority and legitimacy in global governance: deliberation, institutional differentiation and the Codex Alimentarius. New York University Law Review 81: 766-801.

seeing their traditional monopoly over food safety governance being challenged, is there any real evidence that private standards are undermining international standards? We would argue that such concerns have their basis in false premises regarding why private standards have emerged as mechanisms of food safety governance, and the roles that national regulations and international standards play in this context.

Seeing Codex as an organisation that defines rules for the elaboration of public and private standards by other entities – member governments, firms and NGOs– suggests that Codex has, perhaps ironically, had a role in guiding the development of private standards. It has set out both a framework and common vocabulary that enables the developers and adopters of private standards across the globe to communicate with one another and to agree on what these standards should strive to achieve. More generally, Codex standards reflect current international consensus on food safety issues. In the same way that national regulations are formulated to build on and elaborate Codex guidelines, through turning rules into standards schemes, private standards setters interpret and elaborate Codex standards, guidelines and recommendations. Thus, Codex arguably serves to promote the legitimacy of private food safety standards and at the same time reduces the costs of standards development.

Private standards setters can thus be seen as translating the rules of Codex into standards that provide sufficient guidance for implementers to know what they are required to do in order to comply and also for conformity assessors to undertake an objective assessment of when compliance has been achieved. Indeed, this process of translation is necessary in order that such standards can be audited in a manner that is compatible with ISO guidelines.[237] For example, Codex's Recommended International Code of Practice – General Principles of Food Hygiene stipulates that a food safety system should enable traceability, while private standards such as the BRC Global Standard for Food Safety and IFS specify the substantive elements this system should contain, how this system should perform and how the effectiveness of this system should be monitored. It is perhaps not surprising, therefore, that the report on private standards prepared for Codex by FAO and which was discussed at the Commission meeting in 2010 concluded that collective private food safety standards were largely consistent with Codex.[238,239]

It is important to recognise that the scope of many private food safety standards extends beyond single Codex standards, guidelines and recommendations, at

[237] For example, Guide 65 on General Requirements for Bodies Operating Product Certification Systems.
[238] CAC, 2010. Consideration of the impact of private standards, CX/CAC 10/33/13, Codex Alimentarius Commission, Rome, Italy.
[239] This report did, however, conclude that there was a tendency for private company standards to be more stringent than Codex standards, specifically with respect to numerical limits, for example on pesticide residues.

times making it difficult to discern where and the extent to which there is a disconnect between the two. Thus, it is more accurate to see private food safety standard as substantively packaging multiple Codex standards, guidelines and recommendations, along with national legislation that will variously be based on these Codex documents. For example, the GFSI Guidance Document contains substantive elements of all of the following:[240, 241]

- Recommended International Code of Practice-General Principles of Food Hygiene 1969 Rev 4 2003.
- Principles for Food Import and Export Inspection and Certification, 1969.
- Guidelines for the Validation of Food Safety Control Measures, 2008.
- Principles for Traceability/Product Tracing as a Tool within a Food Inspection and Certification System, 2006.

Private food safety standards can be seen as defining a system around these core principles in terms of their substantive elements and how these are managed, and related systems of conformity assessment. The international standards of ISO provide many of the key principles (or rules) underlying these systems.

Of course private food safety standards do not confine themselves to areas where Codex has defined international standards, guidelines and recommendations. Here, private standards can be seen as filling a 'void' in international rules. This is seen, for example, with the GlobalGAP standard that defines requirements for GAP in primary production where international and national regulatory standards are scarce. It should be recognised, however, that a major driver behind pre-farm-gate standards is regulatory requirements, for example with respect to MRLs for pesticides in fresh produce. Collective private standards generally do not define such parameters. Rather, target levels for pesticides residues in the end product tend to be stipulated by national governments, which may or may not be based on Codex MRLs.[242] To the extent that national governments do or do not base their legal requirements on Codex MRLs, private standards will or will not be directed at complying with Codex MRLs. Similarly, private food safety standards for food

[240] Swoffer, K., 2009. GFSI and the relationship with Codex. Presentation to CIES International Food Safety Conference, Paris, France.

[241] The GFSI clearly recognises the importance of Codex as a global reference point and is anxious to demonstrate where its Guidance Document and Codex standards, guidelines and recommendations coalesce. Thus, it is currently cross-referencing the Guidance Document and the four recognised post-farm-gate standards with Codex standards.

[242] A number of individual company standards do specify MRLs for pesticides in fresh fruit and vegetables, often at levels that are stricter than Codex standards and/or national legislation. Thus, limits may be set at the Limit of Detection (LOD) or as a proportion of the regulatory MRL. Likewise, individual company standards may set limits for microbial pathogens or veterinary drugs that are stricter than Codex standards and/or national legislation or where no such limit has otherwise been established. This is the key area where private standards set stricter requirements than Codex standards according to the recent report on private standards prepared by FAO for Codex (CAC, 2010. Consideration of the impact of private standards, CX/CAC 10/33/13, Codex Alimentarius Commission, Rome, Italy).

processing, such as the BRC Global Standard for Food Safety and IFS, incorporate requirements that are not integral to the Recommended International Code of Practice-General Principles of Food Hygiene, for example on product analysis, internal audit, purchasing procedures, etc.

Finally, it is important to recognise that private food safety standards are far from universal. There are many areas where Codex standards, guidelines and recommendations, and national legislation, has been laid down and private standards are less important, or indeed do not exist at all. Thus, there are significant differences in the importance of private food safety standards across sectors (for example fresh fruit and vegetables versus dairy products), between levels of the value chains (for example food processing versus production) and geographically (for example Northern Europe versus the US or Japan). At the same time, it must not be forgotten that private standards are only relevant to the extent that they are adopted in agri-food value chains. While there is an evident increase in the use of private standards, this is far from universal. Despite the great attention given to these standards, at the current time more global markets make no reference to private standards than require strict compliance.

5.6 Challenges and opportunities for Codex

While there appears to be little compelling evidence that private food safety standards are appreciably undermining the role of Codex, their emergence as an increasingly dominant mechanism of governance in global agri-food value chains does raise certain challenges and opportunities. These relate predominantly to the speed and inclusiveness of the standards-setting process; an ongoing issue that predates the emergence of private standards and on which Codex is rather sensitive. Private food safety standards illustrate the ability and willingness of private sector stakeholders to bring about new governance institutions where existing arrangements are not deemed to provide the required level of protection, both against non-compliance with legal food safety requirements and against losses to market share and brand capital. While private standards operate within the framework of rules defined by Codex[243], they are also able to step outside of this framework when it is perceived that this is required. The challenge for Codex, thus, is to continue to elaborate standards, guidelines and recommendations that are relevant to adopters, both in the public and private sectors.

The speed and complexity of the standards-setting process within Codex has long been a cause of concern,[244] including by the official evaluation of Codex concluded

[243] As well as (for example) ISO and the International Organisation for Animal Health (OIE).

[244] Henson, S.J., Preibisch, K.L. and Masakure, O., 2001. Enhancing developing country participation in international standards-setting organizations. Department for International Development, London, UK.

in 2002.[245] The concern here is that Codex is not able to elaborate new or revised standards at the rate that adopters require them. This is in contrast to the relatively rapid development of private standards,[246] reflecting the limited membership, narrower focus and more common interests of the firms and organisations involved. For example, the Recommended International Code of Practice – General Principles of Food Hygiene has been revised four times since its original adoption in 1969, while the BRC Global Standard for Food Safety has been revised five times since its initial implementation in 1998. Many Codex standards take appreciably longer than this to be established and/or revised. While the emergence of private food safety standards arguably provides scope for Codex's influence within the global food safety system to be enhanced (rather than diminished as has been implied by some), this will be dependent on its ability to elaborate standards, guidelines and recommendations at a faster rate as new issues emerge, established approaches and practices change, etc.

The rise of private food safety standards also implies that the clientele of Codex is changing, or perhaps more accurately is expanding. Traditionally, the role of Codex has been to establish rules for the implementation of official food control systems, suggesting that the main beneficiaries are public regulators. Private standards have added an additional layer to food safety governance and Codex needs to take account of this in directing its work programme and in elaborating standards. It must be remembered that Codex's influence and relevance is dependent on the adoption of its standards, guidelines and recommendations, both by governments and private standards setters. This latter group are not bound by the rules of the WTO; they will base their standards on Codex to the extent that it reflects recognised good practice, but will look elsewhere if not. Private standards provide considerable scope for Codex to have more influence, provided it meets the needs of the full range of adopters. Just as an increasing number of regulatory authorities in member countries are embracing private food safety standards as a means towards achieving higher levels of compliance and/or reducing costs,[247] Codex needs to see the adopters and setters of these standards as 'legitimate' clients.

While Codex remains the main international body for the elaboration of standards, guidelines and recommendations related to food safety, the emergence of organisations such as GlobalGAP and GFSI is serving to put a spotlight on Codex

[245] CAC, 2002. Report of the evaluation of the Codex Alimentarius commission and other FAO and WHO food Standards work, Codex Alimentarius Commission, Rome, Italy.

[246] Henson S.J. and Humphrey, J., 2009. The impacts of private food safety standards on the food chain and on the public standards-setting process. Paper prepared for FAO/WHO, ALINORM 09/32/9D-Part II, Codex Alimentarius Commission, Rome, Italy.

[247] For example, the UK Food Standards Agency tiers inspection of primary production facilities according to certification against approved private farm assurance standards (Henson, S.J. and Humphrey, J., 2008. Understanding the complexities of private standards in global agri-food chains. Paper presented at the workshop: globalization, global governance and private standards, University of Leuven, Belgium, November 2008).

in terms of the representation of stakeholders, globally and especially from developing countries.[248] The decision-making process of Codex is essentially driven by governments, which variously take account of national stakeholder interests. International NGOs can be recognised as observers at Codex, but have no decision-making power. The 'voice' of developing countries at the Codex table is also appreciated to be limited.[249] While the range of interests that typically feed into the elaboration of private food safety standards is much narrower than with Codex, the organisations involved have become significantly more open over time, and begun to incorporate a wider range of stakeholders. This is seen with the membership of GlobalGAP and of the GFSI, both of which have moved appreciably away from their original core of major European food retailers.[250] Certainly there are concerns about the legitimacy of decision-making processes underlying private standards.[251] However somewhat paradoxically, the private sector may have considerable interest in opening up the standards-setting process to an increasingly wide range of stakeholders, aimed at deflecting criticism and building the legitimacy of their standards.

5.7 Conclusions

A key trend in global agri-food systems in recent years has been the emergence of private standards. This trend has sparked a vigorous debate on the role of private governance of food safety and the degree to which this might act to undermine established (public) regulatory systems and transnational norms, the latter predominantly established through the work of Codex and the WTO. Critics argue that private standards need to be 'reined in' whilst at the same time fearing that the WTO has little or no jurisdiction such that these standards will multiply and evolve unchecked. More generally, public regulators are uncomfortable at seeing their traditional monopoly in the governance of food safety being challenged.

There is little doubt that private food safety standards do raise questions about the role of governmental and intergovernmental institutions in the regulation of food safety, and in an international context specifically the position of Codex. However, much of the debate about private food safety standards has been fuelled by misunderstandings of why such standards have evolved and the functions they perform. Key here is a failure to recognise that private food safety standards are

[248] Henson, S.J. and Humphrey, J., 2009. The impacts of private food safety standards on the food chain and on the public standards-setting process. Paper prepared for FAO/WHO, ALINORM 09/32/9D-Part II, Codex Alimentarius Commission, Rome, Italy.
[249] Henson, S.J., 2002. The current status and future directions of Codex Alimentarius. World Health Organisation, Geneva, Switzerland.
[250] It could be argued, perhaps provocatively, that the interests of developing country producers are heard more loudly in GlobalGAP than in Codex.
[251] Fuchs, D. and Kalfagianni, A., 2010. The democratic legitimacy of private authority in the food chain. In: Porter, T. and Ronit, K. (eds.) The challenges of global business authority: democratic renewal, stalemate or decay? State University of New York Press, Albany, NY, USA, pp. 65-88.

quite closely attuned to regulatory requirements; at times private food safety standards do extend beyond the requirements of public mandatory standards, but in many cases their key functions is to provide assurances to buyers in global agri-food value chains that regulatory requirements have been satisfied. Further, the great diversity of private food safety standards, in their institutional form, scope and prevalence across value chains, belies attempts to draw general conclusions.

The increasing adoption of private food safety standards in global agri-food value chains clearly raises important questions about the role played by Codex, both broadly and within the context of the SPS Agreement. There has been an undue tendency, however, to see private food safety standards as threatening the status of Codex standards, guidelines and recommendations, and undermining the Commission's mandate to promote consumer protection and fair agri-food trade. However, there is limited evidence to support this contention. Where private food safety standards exist, they predominantly appear to take Codex standards, guidelines and recommendations, alongside national regulatory requirements, as their starting point and build a system of process requirements and conformity assessment around these. There are also many commodities and markets where private food safety standards have not been elaborated and Codex remains the key driver of international food safety standards.

Evidently, Codex needs to respond to the challenges and opportunities presented by private food safety standards. There is certainly a need for an informed debate within Codex about the implications for its mandate and work programme. Admittedly this is difficult; the records of discussions at recent meetings of the Commission suggest established positions and misinformation persist amongst Codex members.[252] In the meantime, ways need to be found for Codex to engage more effectively with the organisations involved in setting and/or adopting private food safety standards in order to build trust and mutual understanding. A natural starting point here seems to be the GFSI. [253] There would appear to be more to gain from a cooperative relationship between international standards organisations such as Codex and private standards organisations than conflict.

References

Aragrande, M., Segre, A., Gentile, E., Malorgio, G., Giraud Heraud, E., Robles, R., Halicka, E., Loi, A. and Bruni, M., 2005. Food supply chains dynamics and quality certification, final report. EU DG Joint Research Centre, Brussels, Belgium.

[252] See in particular CAC, 2009. Report of the 32nd Session of the Codex Alimentarius Commission, ALINORM 09/32/REP, Codex Alimentarius Commission, Rome, Italy; CAC, 2010. Report of the 33rd Session of the Codex Alimentarius Commission, ALINORM 10/33/REP, Codex Alimentarius Commission, Rome, Italy.

[253] There are evidently the beginnings of constructive dialogue between Codex and the GFSI, which is to be welcomed.

Black, J., 2002. Critical reflections on regulation, Centre for Analysis of Risk and Regulation, London School of Economics and Political Science, London, UK.

Brunsson, N. and Jacobsson, B., 2000. The contemporary expansion of standardization. In: Brunsson, N. and Jacobsson, B. (eds.) A world of standards. Oxford University Press, Oxford, UK, pp. 1-17.

Busch, L., Thiagarajan, D., Hatanaka, M., Bain, C., Flores, L. and Frahm, M., 2005. The relationship of third-party certification (TPC) to sanitary/phytosanitary (SPS) measures and the international agri-food trade: final report. RAISE SPS Global Analytical Report 9, USAID, Washington, DC, USA.

Buzby, J., Frenzen, P.D. and Rasco, B., 2001. Product liability and microbial food-borne illness. Agricultural Economic Report 828, United States Department of Agriculture, Economics Research Service, Washington, DC, USA.

CAC, 1997. Understanding the Codex Alimentarius. Codex Alimentarius Commission, Rome, Italy.

CAC, 2002. Report of the evaluation of the Codex Alimentarius commission and other FAO and WHO food Standards work, Codex Alimentarius Commission, Rome, Italy.

CAC, 2006. Maximum residue limits for veterinary drugs in foods: updated as at the 29th session of the Codex Alimentarius Commission. Codex Alimentarius Commission, Rome, Italy.

CAC, 2008. Report of the 31st Session of the Codex Alimentarius Commission, ALINORM 08/31/REP. Codex Alimentarius Commission, Rome, Italy.

CAC, 2009. Report of the 32nd Session of the Codex Alimentarius Commission, ALINORM 09/32/REP. Codex Alimentarius Commission, Rome, Italy.

CAC, 2010a. Consideration of the impact of private standards, CX/CAC 10/33/13. Codex Alimentarius Commission, Rome, Italy.

CAC, 2010b. Report of the 33rd Session of the Codex Alimentarius Commission, ALINORM 10/33/REP. Codex Alimentarius Commission, Rome, Italy.

David, P.A., 1995. standardization policies for network technologies: the flux between freedom and order revisited. In: Hawkins, R., Mansell, R. and Skea, J. (eds.), Standards, innovation, and competitiveness: the politics and economics of standards in national and technical environments. Edward Elgar, Aldershot, UK, pp. 15-35.

EU, 2002. Regulation (EC) No 178/2002 Laying Down the General Principles and Requirements of Food Law, Establishing the European Food Safety Authority and Laying Down Procedures in Matters of Food Safety. Official Journal of the European Communities, 1 February 2002.

EU, 2004. Corrigendum to Regulation (EC) No 882/2004 of the European Parliament and of the Council of 29 April 2004 on Official Controls Performed to Ensure the Verification of Compliance with Feed and Food Law, Animal Health and Animal Welfare Rules. Official Journal of the European Union, 28 May 2004, Article 48.

Fuchs, D. and Kalfagianni, A., 2010. The democratic legitimacy of private authority in the food chain. In: Porter, T. and Ronit, K. (eds.) The challenges of global business authority: democratic renewal, stalemate or decay? State University of New York Press, Albany, NY, USA, pp. 65-88.

Fulponi, L., 2005. Private voluntary standards in the food system: the perspective of major food retailers in OECD countries. Food Policy 30(2): 115-128.

Havinga, T., 2006. Private regulation of food safety by supermarkets. Law and Policy 28(4): 515-533.

Havinga, T., 2008. Actors in private food regulation: taking responsibility or passing the buck to someone else? Paper presented at the symposium Private Governance in the Global Agro-Food System, Munster, Germany, 23-25 April 2008.

Henson, S.J., 2002. The current status and future directions of Codex Alimentarius. World Health Organisation, Geneva, Switzerland.

Henson, S.J., 2007. The role of public and private standards in regulating international food markets. Journal of International Agricultural Trade and Development 4(1): 52-66.

Henson, S.J. and Caswell, J.A., 1999. Food safety regulation: an overview of contemporary issues. Food Policy 24(6): 589-603.

Henson, S.J. and Humphrey, J., 2008. Understanding the complexities of private standards in global agri-food chains. Paper presented at the workshop: globalization, global governance and private standards, University of Leuven, Belgium, November 2008.

Henson, S.J. and Humphrey, J., 2009. The impacts of private food safety standards on the food chain and on the public standards-setting process. Paper prepared for FAO/WHO, ALINORM 09/32/9D-Part II, Codex Alimentarius Commission, Rome, Italy.

Henson, S.J. and Humphrey, J., 2010. Understanding the complexities of private standards in global agri-food chains as they impact developing countries. Journal of Development Studies 46(9): 1628-1646.

Henson, S.J. and Jaffee, S., 2008 Understanding developing country strategic responses to the enhancement of food safety standards. The World Economy 31(1): 1-15.

Henson, S.J. and Northen, J.R., 1998. Economic determinants of food safety controls in the supply of retailer own-branded products in the UK. Agribusiness 14(2): 113-126.

Henson, S.J. and Reardon, T., 2005. Private agri-food standards: implications for food policy and the agri-food system. Food Policy 30(3): 241-253.

Henson, S.J., Preibisch, K.L. and Masakure, O., 2001. Enhancing developing country participation in international standards-setting organizations. Department for International Development London, UK.

Humphrey, J., 2008. Private standards, small farmers and donor policy: EUREPGAP in Kenya. IDS Working Paper 308, Institute of Development Studies, Brighton, UK.

Humphrey, J. and Schmitz, H., 2000. Governance and upgrading: linking industrial cluster and global value chain research. IDS Working Paper 120, Institute of Development Studies, Brighton, UK.

Humphrey, J. and Schmitz, H., 2001. Governance in global value chains. IDS Bulletin 32(3): 19-29.

Jaffee, S. and Henson, S.J., 2004. Standards and agri-food exports from developing countries: rebalancing the debate. Policy Research Working Paper 3348, The World Bank, Washington, DC, USA.

Livermore, M.A., 2006. Authority and legitimacy in global governance: deliberation, institutional differentiation and the Codex Alimentarius. New York University Law Review 81: 766-801.

Martinez, M.G., Fearne, A., Caswell, J.A. and Henson, S.J., 2007. Co-regulation as a possible model for food safety governance: opportunities for public-private partnerships. Food Policy 32(2): 299-314.

Nadvi, K. and Waltring, F., 2003. Making sense of global standards. In: Schmitz, H. (ed.), Local enterprises in the global economy: issues of governance and upgrading. Elgar, Cheltenham, UK; OECD, 2004. Private standards and the shaping of the agri-food system. OECD, Paris, France.

OECD, 2004. Private standards and the shaping of the agri-food system. OECD, Paris, France.

Reardon, T., Codron, J.-M., Busch, L., Bingen, J. and Harris, C., 2001. Global change in agrifood grades and standards: agribusiness strategic responses in developing countries. International Food and Agribusiness Management Review 2(3): 421-435.

Roberts, D. and Unnevehr, L.J., 2005. Resolving trade disputes arising from trends in food safety regulation: the role of the multilateral governance framework. World Trade Review 4(3): 469-497.

Roberts, M.T., 2009. Private standards and multilateral trade rules. Paper prepared for FAO, Rome, Italy.

Rosman, L., 1993. Public participation in international pesticide regulation: when the Codex Commission decides, who will listen? Virginia Environmental Law Journal 12: 329-365.

Sheehan, K., 2007. Benchmarking of Gap schemes, EUREPGAP Asia conference, Bangkok, 6-7 September 2007. Available at: http://www.globalgap.org/cms/upload/Resources/Presentations/Bangkok/3_K_Sheehan.pdf.

Swoffer, K., 2009. GFSI and the relationship with Codex. Presentation to CIES International Food Safety Conference, Paris, France.

Unnevehr, L., 2000. Food safety issues and fresh food product exports from LDCs. Agricultural Economics 23(3): 231-240.

World Bank, 2005. Food safety and agricultural health standards: challenges and opportunities for developing country exports. Report 31207, The World Bank, Poverty Reduction and Economic Management Trade Unit, Washington, DC, USA.

WTO, 2007a. Considerations relevant to private standards in the field of animal health, food safety and animal welfare. Submission by the World Organisation for Animal Health, G/SPS/GEN/822, WTO, Committee on Sanitary and Phytosanitary Measures, Geneva, Switzerland.

WTO, 2007b. Private standards and the SPS agreement, note by the Secretariat, G/SPS/GEN/746, WTO, Committee on Sanitary and Phytosanitary Measures, Geneva, Switzerland.

WTO, 2007c. Submission by the International Organisation for Standardisation (ISO) to the SPS Committee Meeting 28 February, 1 March 2007, G/SPS/GEN/750, WTO, Committee on Sanitary and Phytosanitary Measures, Geneva, Switzerland.

WTO, 2008. Report of the STDF Information Session on Private Standards, G/SPS/R/50, WTO, Committee on Sanitary and Phytosanitary Measures, Geneva, Switzerland.

6. Private retail standards and the law of the World Trade Organisation

Marinus Huige

6.1 Introduction

Standards are meant to facilitate trade. However listening to discussions in international forums about standards, and more specifically private standards, one might get the idea the opposite is true. With the growing use of private standards by retailers, debate in international trade and standard setting bodies has emerged. In these discussions on private standards the trade facilitating aspect of the standards is particularly disputed.

At the international public level, harmonised food safety, animal health and plant health standards are established by organisations like the Codex Alimentarius Commission,[254] the International Organisation for Animal Health (OIE)[255] and the International Plant Protection Convention (IPPC)[256]. The use of international standards is encouraged by the World Trade Organisation (WTO) based on the logic that trade will flow if everybody adheres to uniform measures.[257]

The most relevant WTO agreement for agri-food standards is the Agreement on Sanitary and Phytosanitary Measures (SPS)[258] which particularly deals with animal health, plant health and food safety. It refers explicitly to the standards of Codex Alimentarius, OIE and IPPC. The SPS Agreement disciplines the use of requirements and standards in international trade in order to ensure that food safety and quality requirements are not misused as protectionist measures.

Apart from public standards, the so-called private standards have grown considerably in importance and are now a significant factor to take into account when dealing with market access issues. Market access is about trade and trade is governed and regulated by the WTO. WTO rules however, in principle only apply in government

[254] The Codex Alimentarius Commission is a United Nations organization that develops international standards for food safety.
[255] The International Organisation for Animal Health is known by its French acronym OIE (Office International des Epizooties) and develops international standards for animal health and welfare.
[256] The International Plant Protection Convention is a United Nations organisation that develops international standards for plant health.
[257] WTO, 1995. Agreement on the Application of Sanitary and Phytosanitary Measures. WTO, Geneva, Switzerland, Article 3.
[258] WTO, 1995. Agreement on the Application of Sanitary and Phytosanitary Measures. WTO, Geneva, Switzerland.

to government relations,[259] where the private standard is typically an issue between private parties. Because of their increased importance in international trade and their possible trade impact, private standards are subject to debate in the above mentioned organisations. Since 2005, the debate is particularly animated in the WTO SPS Committee. The main question in this Committee is whether and how amongst others, food safety requirements demanded by the private sector should be dealt with at the multilateral level.[260]

This chapter will focus on the debate in the WTO SPS Committee and the legal aspects of private standards from a WTO perspective.

6.2 What are private standards?

Private standards are sets of rules on how to produce, developed by private companies. There are different categories of standards, varying from so-called 'individual firm schemes' such as Tesco's Nature's Choice, to 'collective national and international schemes' such as GlobalGAP. The latter is a pan-European organisation of which most of the large European retailers are members. They may also be categorised according to pre-farm (i.e. GlobalGAP) versus post-farm gate standards (i.e. British Retail Consortium).[261] Coverage normally extends beyond food safety issues to include quality, environment, animal welfare and other societal issues. The SPS Committee has no mandate to deal with issues other than food safety.[262]

In the case of an individual firm scheme it is moreover questionable whether one can speak of a standard in a sense that it is used by a significant number of participants in the industry.

Finally, private schemes are mainly based on process control using the HACCP[263] system and rely on third party auditing by accredited certifying bodies for their implementation.

[259] See: Van der Meulen, B., 2010. The global arena of food law. Emerging contours of a meta-framework. Erasmus Law Review 3(4).

[260] WTO, 2007. Private Standards and the SPS Agreement, Note by the Secretariat, G/SPS/GEN/822. WTO, Committee on Sanitary and Phytosanitary Measures, Geneva, Switzerland.

[261] See Chapter 3 Codex Alimentarius and Private Standards, by Henson and Humphrey for examples of different types of private standards.

[262] And of course human, animal and plant health.

[263] HACCP: Hazard Analysis of Critical Control Points (HACCP) is a systematic preventive approach to food safety and pharmaceutical safety that addresses physical, chemical, and biological hazards as a means of prevention rather than finished product inspection. HACCP is used in the food industry to identify potential food safety hazards, so that key actions, known as Critical Control Points (CCPs) can be taken to reduce or eliminate the risk of the hazards being realised. The system is used at all stages of food production and preparation

6.3 Private standards, what drives them?

Since the nineties, the concentration of retail chains has resulted in a small number of large supermarket chains with enormous market and purchasing power with more control over the entire supply chain. These concentrations tend to by-pass wholesalers to source directly from domestic and foreign suppliers.

Apart from this consolidation of power in the supply chain, consumer expectations have changed. The more wealthy the consumers, the more importance they attach to social, ethical and environmental values. Increased consumer awareness is also a direct consequence of food scares such as BSE in Europe and the US, dioxins in Belgium and melamine in dairy products in China.[264] Consequently, governments have shifted food safety controls from the testing on final retail products to the production and distribution process. This approach is also known as the 'farm-to-fork' or 'stable-to-table' approach. An approach like this requires the food industry to put in place risk management systems that ensure food safety at all stages of the chain – from the farm level to final retail sale.

The introduction of the General Food Law[265] in the EU in 2002 has also been important for the development of private standards. The Regulation establishes the basic principle that the primary responsibility for ensuring compliance with food law, and in particular food safety, rests with the food business (companies, supermarkets). To complement and support this principle, there must be adequate and effective controls organised by the competent authorities of the Member States and operators must amongst others: work transparently, use a traceability system to identify suppliers and have an emergency system to be able to immediately withdraw food from the market. This Regulation requires all players along the supply chain to be able to prove that they have undertaken all possible steps to ensure that their product is safe and will not cause harm. To be able to provide proof of this so-called 'due diligence', the food industry developed systems of self-regulation. These started with codes of practice, such as Good Agricultural Practices (GAP) schemes, and a protocol of good hygiene practices (later the British Retailer Consortium Global Standard), and with that the phenomenon of private standards was born.

Technically these standards are voluntary as they are not required by law. However, in the case when there would be no choice but to sell your produce via a private

[264] Henson, S.J. and Humphrey, J, 2009. The impacts of private food safety standards on the food chain and on public standard-setting processes. Paper prepared for FAO/WHO, ALINORM 09/32/9D-Part II, Codex Alimentarius Commission, Rome, Italy, pp. 9-11.

[265] EU, 2002. Regulation (EC) No 178/2002 of the European Parliament and of the Council of 28 January 2002 laying down the general principles and requirements of food law, establishing the European Food Safety Authority and laying down procedures in matters of food safety. Official Journal of the European Union L 31, 1/2/2002: 1-24.

standard, these can effectively set the conditions for the import or sale of products, which makes them more mandatory in practice.

6.4 Private standards and the WTO SPS Committee

The SPS Agreement strikes a balance between the right of Members to protect human, animal and plant life or health and trade rules they have committed themselves to. Each WTO Member country is entitled to maintain a level of protection it considers appropriate to protect life or health within its territory. But when SPS measures directly or indirectly affect trade, members have the obligation to minimise negative impacts of such measures on international trade. This means that SPS measures must:

- be applied only to the extent necessary to protect life or health and not be more trade restrictive than required;
- be based on scientific principles and not be maintained without sufficient scientific evidence; and
- not constitute arbitrary or unjustifiable treatment or a disguised restriction on trade.

The preferred way of meeting the core principle of *scientific justification* is through the use of internationally developed food safety, plant and animal health protection standards - that is, those adopted by the Codex, IPPC and the OIE. The harmonisation of national requirements, on the basis of these international standards, facilitates trade by reducing the proliferation of distinct national requirements. The SPS Agreement allows only for national standards or measures beyond the above mentioned international standards if they can be justified on the basis of an appropriate risk assessment and if they meet the criterion of least trade restrictiveness to achieve the desired level of health protection.

The SPS Agreement also requires in Annex C that there be no unjustified costs in testing, certification or approval procedures, to ensure that these do not function as barriers to trade. Furthermore SPS measures or standards of individual WTO members need to be notified via the WTO secretariat to other members. Finally, the WTO agreement ensures that SPS requirements can be challenged by other trading partners, through the use of the WTO's dispute settlement mechanism.

While public SPS measures and standards need to fit the above mentioned criteria, private standards which address a mix of SPS and other objectives - including social and environmental concerns that are not related to food safety or plant/animal health protection - may have no scientific justification and may not be notified in a timely fashion. Together with a proliferation of the schemes without much harmonisation between them, it is understandable, that governments and international organisations get curious about the relationship between private standards and the standards set by the so called three sisters (OIE, CODEX, and IPPC).

The Committee on Sanitary and Phytosanitary Measures (SPS Committee) monitors the way Members implement their SPS-measures. If this is not done in conformity with the rules of the SPS Agreement and the measures affect trade, other Members can bring this up for discussion in the Committee (under the agenda item: specific trade concerns), which meets three times a year at WTO headquarters in Geneva.

The issue of private standards was introduced at the WTO as a specific trade concern by St. Vincent and the Grenadines complaining at a meeting of the June 2005 SPS Committee about the negative impact on its banana exports of EurepGAP (now GlobalGap) standards for pesticides. The European Commission rejected the complaint by stating that it did not concern any official requirement of the EU. After that however the issue of private standards has been on the agenda of the Committee as a general issue, not pointing to one specific WTO-member.[266]

The discussions so far have focused on three main concerns:[267]

1. Market access. It is acknowledged that private standards can help producers and traders by providing step-by-step guidelines showing what needs to be done to meet (government) regulations and market conditions. Several studies have shown that following the risk management approach defined by some of the private standards schemes results in better overall farm and business management as well as increased efficiency and profitability. However it is also recognised that these standards can have negative effects. Producers must be certified as meeting the private standards, and becoming certified is a very expensive business. Also it is argued that private standards can be both more restrictive (e.g. requiring lower levels of pesticide residues) and more prescriptive (accepting only one way of achieving a desired food safety outcome) than official import requirements, thus acting as additional barriers to market access.
2. Development. The costs of certification and compliance with private standards, can make the development of export-oriented schemes virtually impossible for small-scale producers in developing countries.
3. WTO law. While some are of the view that setting standards for the products they purchase is a legitimate private sector activity with which governments should not interfere, others are of the view that the SPS Agreement makes governments in importing countries responsible for the standards set by their private sectors. The latter maintain that the private standards do not meet WTO requirements such as transparency and scientific justification of food safety measures and are more trade-restrictive than necessary to protect health.

[266] Stanton, G.H. and Wolff, C., 2008. Private voluntary standards and the World Trade Organisation (WTO) Committee on Sanitary and Phytosanitary Measures. Fresh Perspectives 2008.
[267] WTO, 2007. Private standards and the SPS Agreement, note by the Secretariat of 24 January 2007. Committee on Sanitary and Phytosanitary Measures, WTO Doc. No. G/SPS/GEN/746.

WTO Members have had numerous debates on what problems private standards can potentially cause (some members identified concrete examples) and whether and how the SPS committee should deal with this phenomenon. The tendency is to focus on discussing the scope of the SPS Agreement (under point 3 above). A number of Members (mostly developing countries) follow the interpretation that the SPS Agreement is applicable to private standards while other Members (mostly developed countries) deny the applicability beyond public standards or government regulations. A discussion essentially concerned with the interpretation of the scope of Article 13 of the SPS Agreement. In June 2010 the SPS committee identified possible actions regarding SPS related private standards.[268]

Article 13 of the SPS Agreement indicates that:

> Members are fully responsible under this Agreement for the observance of all obligations set forth herein. Members shall formulate and implement positive measures and mechanisms in support of the observance of the provisions of this Agreement by other than central government bodies. Members shall take such reasonable measures as may be available to them to ensure that non-governmental entities within their territories, as well as regional bodies in which relevant entities within their territories are members, comply with the relevant provisions of this Agreement. In addition, Members shall not take measures which have the effect of, directly or indirectly, requiring or encouraging such regional or non-governmental entities, or local governmental bodies, to act in a manner inconsistent with the provisions of this Agreement. Members shall ensure that they rely on the services of non-governmental entities for implementing sanitary or phytosanitary measures only if these entities comply with the provisions of this Agreement.

The perception of most WTO Members in the SPS committee is that private standards go beyond the standards set up by international standard setting bodies that are referenced in the SPS Agreement. The question is then: what is meant by 'going beyond'? The private standards extend to ethics, environment, animal welfare and social accountability and they are much more specific about how to structure the production process.[269] Looking purely at the SPS aspects however there is little proof of many private standards going beyond public standards in terms of setting, for example, higher maximum residue levels.[270] Strictly speaking from an SPS point of view in most cases there would be no interference with the

[268] WTO, 2010. Possible actions for the SPS Committee regarding SPS-related private standards. Committee on Sanitary and Phytosanitary Measures, G/SPS/W/247/Rev.2.

[269] Henson, S.J. and Humphrey, J., 2009. The impacts of private food safety standards on the food chain and on public standard-setting processes. Paper prepared for FAO/WHO, ALINORM 09/32/9D-Part II, Codex Alimentarius Commission, Rome, Italy, p. 12.

[270] Rau, M.L., 2009. Public and private standards and certification in agri-food trade. Small assignment report prepared for the Ministry of Agriculture, Nature and Food Quality, The Hague, the Netherlands.

international standards as it seems that in most cases private standards take the official national requirements or standards, or the international standards as a basis to build upon.[271] If you look at for example the Codex Alimentarius Hygiene Code, which is very broad, it is only logical that private entities implement this Code with specific regulations for their industry.[272]

While the debate in the SPS committee focuses mainly on the question whether the scope of private standards go beyond the scope of public international standards, the discussion should better be driven by concerns about 'real' issues such as the costs of compliance, proliferation without enough harmonisation between the standards, transparency and the lack of stakeholder involvement.

Some of the private standard setting bodies have themselves recognised the problem of proliferation and efforts are now underway to 'benchmark' or accept other private standard schemes as equivalent. The Global Food Safety Initiative (GFSI)[273] is playing an active role in this. GlobalGAP is a good example of a private standard setting body that has developed initiatives to ensure stakeholder inclusiveness, and has taken into account the specific needs of smallholders in defining their own standards. For example it has appointed an 'Ambassador' for Africa. They also work with committees with representatives from different sectors.[274]

Obviously private standards give rise to concern, but the question is whether the SPS committee is addressing these concerns in the right way. By focussing too much on applicability of the SPS Agreement, WTO members risk wasting their time without progressing in any way towards solving real questions.

6.5 The current discussion on applicability of the SPS Agreement

National civil law in principle contains nothing that prevents buyers from demanding the supplier to comply with specific technical requirements to the

[271] Henson, S.J. and Humphrey, J., 2009. The impacts of private food safety standards on the food chain and on public standard-setting processes. Paper prepared for FAO/WHO, ALINORM 09/32/9D-Part II, Codex Alimentarius Commission, Rome, Italy, par 3 p. 12 and p. 39. See also Smith, G., 2009. Interaction of public and private standards in the food chain. OECD Food, Agriculture and Fisheries Working Papers, No. 15. OECD Publishing, Paris, France, par 51.

[272] Van Exel, L.D., 2009. Public vs. Private Standards. Does GlobalGAP exceed the international standard requirements, and if so, to what extent? Report for the Permanent Mission of the Netherlands, Geneva, Switzerland.

[273] The Global Food Safety Initiative (GFSI) is a collaboration between some of the world's leading food safety experts from retailer, manufacturer and food service companies, as well as service providers associated with the food supply chain.

[274] Henson, S.J. and Humphrey, J., 2009. The impacts of private food safety standards on the food chain and on public standard-setting processes. Paper prepared for FAO/WHO, ALINORM 09/32/9D-Part II, Codex Alimentarius Commission, Rome, Italy.

products they want to purchase. Especially for retailers who have to manage complex business risks, there are many good reasons to justify such requirements. If retailers use private standards as an instrument to abuse a dominant position, then domestic competition regulations may be applicable. [275]

Desirable or not, for the moment most of the attention is focused on the SPS Agreement as far as food safety aspects of private standards are concerned. The legal question which then rises: do private standards fall within the scope of the SPS Agreement?

The SPS Agreement is applicable to 'all sanitary and phytosanitary measures which may, directly or indirectly, affect international trade'. The definition of a 'sanitary or phytosanitary measure'[276] therefore does not (seem to) exclude the types of measures imposed by private standards. But the obligations under the SPS Agreement are addressed to WTO Members only (which are national states). As we have seen above, Article 13 of the SPS Agreement states that Members shall take such reasonable measures as may be available to them to ensure that non-governmental entities within their territories comply with the relevant provisions of this Agreement. But the term 'non-governmental entities' is not defined anywhere. While some argue that it covers private standard setting bodies, others are of the view that only those entities mandated by a government to carry out certain tasks are covered under this provision. Moreover, it is important to realise that some private standard setting bodies are 'multinational'. The question then arises which government can be held accountable if Article 13 would be applicable.

Who is right? Does Article 13 include actions taken by the private sector, or only those taken by governments (at national or sub-national levels)? Or more specificly can national governments, the Contracting Parties in WTO terms, be held accountable for actions of private parties?

One of the few legal analyses existing today on the relationship between Article 13 and private standards has been performed by Digby Cascoine.[277] According to Cascoine, very limited WTO case law exists in relation to Article 13 of the SPS Agreement and there is no case law in relation to non-governmental entities.[278] Some indications however can be drawn from the panel report in the case Australia

---

[275] On this issue, see Chapter 17 by Fabian Stancke.

[276] SPS measures are all measures whose purpose is to protect: human or animal health from food borne risks; human health from, animal or plant carried diseases; animals and plants from pests or diseases.

[277] Gascoine, D. and O'Connor and Company, 2006. Private voluntary standards within the wto multilateral framework. Report prepared for the Department for International Development, London, UK. Available at: http://www.handelenduurzaamheid.nl/web/index.php?page=agenda

[278] Gascoine, D. and O'Connor and Company, 2006. Private voluntary standards within the wto multilateral framework. Report prepared for the Department for International Development, London, UK, p. 10.

– Measures affecting the importation of salmon[279] from which he concludes that, when asked to analyse a possible violation of the SPS Agreement by a non-governmental body, a Panel would:

1. Look at Article 13 of the SPS Agreement to determine whether there is 'responsibility' of a WTO Member;
2. Establish in light of Article 1.1 of the SPS Agreement whether the measure at stake is an SPS measure;
3. Decide whether there is a violation of the SPS Agreement.

According to the Panel in Japan – measures affecting consumer photographic film,[280] there is need for a degree of government involvement to put a measure under the scrutiny of the WTO Agreements.[281] Furthermore Cascoine states: '... case law[282] on the entrustment and direction provisions of another WTO Agreement (Agreement on Subsidies and Countervailing Measures) concluded that the ordinary meaning of the words 'entrusts' and 'directs' requires an 'explicit and affirmative action of delegation of command'. This does not seem to be the case with private standards. Also, the negotiation history of the SPS Agreement in the WTO does not provide for any interpretative guidance other than that at the time of the negotiations there were no propositions made by individual member countries that the SPS Agreement would/should have an application to the activities of the private sector.[283]

Amendment of, or even reaching agreement by WTO Members on the interpretation of Article 13 of the SPS Agreement to get clarity on the applicability to private standards, is impossible with 153 Members clearly divided in 2 opposite camps.

[279] Panel Report, Australia – Measures affecting importation of salmon – recourse to Article 21.5 by Canada, WT/DS18/RW of 18 February 2000.

[280] Panel Report, Japan – Measures Affecting Consumer Photographic Film and Paper, WT/DS44/R of 31 March 1998, par 10.52 and Panel Report, EEC – Restrictions on Imports of Dessert Apples, WT/DS44/R, Par. 10.56 of 22 June 1989.

[281] Gascoine, D. and O'Connor and Company, 2006. Private voluntary standards within the wto multilateral framework. Report prepared for the Department for International Development, London, UK, p. 10, par 16(ii) and p. 80.

[282] Panel Report, European Communities – countervailing Measures on Dynamic Random Access Memory Chips from Korea, WT/DS299/R of 3 August 2005, par 7.18-146. and Appellate Body Report, United States Countervailing Duty Investigation on Dynamic Random Access Memory Semiconductors (DRAMS) from Korea, WT/DS296/ AB/R, par 102-197 and Panel Report, United States – Countervailing Duty Investigation on Dynamic Random Access Memory Semiconductors (DRAMS) from Korea, WT/DS296/R of 20 July 2005, par 7.10-178 and Panel Report, Korea – Measures Affecting Trade in Commercial Vessels, WT/DS273/R of 7 March 2005, par 7.350-426 and Panel Report, United –States – Measures Treating Export Restraints as Subsidies, WT/DS194/R of 29 June, par 8.15-75.

[283] Gascoine, D. and O'Connor and Company, 2006. Private voluntary standards within the wto multilateral framework. Report prepared for the Department for International Development, London, UK, p. 10, par 16.

And as Cascoine concludes: 'In any event, the perceived problem is not yet so significant as to warrant such a major step, especially since other initiatives, both within the framework of the WTO and amongst the private bodies concerned, may be able to ameliorate particular difficulties more conveniently'.[284]

One thing is clear: the applicability through Article 13 of the WTO/SPS-rules to private standards would turn the civil law systems of many countries upside down. It would, for example, make governments responsible for the 'deals' buyers and sellers make with one another.

6.6 Food for thought

It is apparent from the debate in the WTO (and other organisations) that there are real concerns regarding the relationship between public and private standards and market access, the development and the credibility of the SPS system. The proliferation of private standards creates a certain risk of undermining the progress made in regulating sanitary and phytosanitary measures through adoption and implementation of the SPS agreement. Nevertheless, the question remains whether a focus on the legal issue of Article 13 of the SPS Agreement will be effective enough to address these concerns.

A better approach might be to look for cooperation models between public and private standard setters, to develop more synergies between them, and also to look at the phenomenon of private standards in light of increasing public concerns about food safety. It is clear that increased public awareness is placing more pressure on government agencies to be more prescriptive and proactive in their regulation of the food industry. Given the scarcity of public sector resources and other concerns about the impact of regulation on competitiveness and the scale of the task at hand, there is also a growing interest in cooperation between public and private sectors to deliver safer food at lower (regulatory) cost.

To conclude, what is really needed is a regular dialogue between the public standard setting organisations (Codex, IPPC and OIE) and the private standard setters. Although some preliminary contacts have taken place, to date we still have two different worlds that speak different languages. There is therefore ample need for a better understanding of each others roles. This might really help the debate further.

[284] Gascoine, D. and O'Connor and Company, 2006. Private voluntary standards within the wto multilateral framework. Report prepared for the Department for International Development, London, UK, p. 7.

References

EU, 2002. Regulation (EC) No 178/2002 of the European Parliament and of the Council of 28 January 2002 laying down the general principles and requirements of food law, establishing the European Food Safety Authority and laying down procedures in matters of food safety. Official Journal of the European Union L 31, 1/2/2002: 1-24.

Gascoine, D. and O'Connor and Company, 2006. Private voluntary standards within the wto multilateral framework. Report prepared for the Department for International Development, London, UK. Available at: http://www.handelenduurzaamheid.nl/web/index.php?page=agenda.

Henson, S.J. and Humphrey, J., 2009. The impacts of private food safety standards on the food chain and on public standard-setting processes. Paper prepared for FAO/WHO, ALINORM 09/32/9D-Part II, Codex Alimentarius Commission, Rome, Italy.

Smith, G., 2009. Interaction of public and private standards in the food chain. OECD Food, Agriculture and Fisheries Working Papers, No. 15. OECD Publishing, Paris, France. DOI: 10.1787/221282527214.

Rau, M.L., 2009. Public and private standards and certification in agri-food trade. Small assignment report prepared for the Ministry of Agriculture, Nature and Food Quality, The Hague, the Netherlands.

Stanton, G.H. and Wolff, C., 2008. Private voluntary standards and the World Trade Organisation (WTO) Committee on Sanitary and Phytosanitary Measures. Fresh Perspectives 2008. Available at: http://www.agrifoodstandards.net/en/filemanager/active?fid=134.

Van der Meulen, B., 2010. The global arena of food law. Emerging contours of a meta-framework. Erasmus Law Review 3(4).

Van Exel, L.D., 2009. Public vs. Private Standards. Does GlobalGAP exceed the international standard requirements, and if so, to what extent? Report for the Permanent Mission of the Netherlands, Geneva, Switzerland.

WTO, 1995. Agreement on the Application of Sanitary and Phytosanitary Measures. WTO, Geneva, Switzerland.

WTO, 2007. Private Standards and the SPS Agreement, Note by the Secretariat, G/SPS/GEN/822. WTO, Committee on Sanitary and Phytosanitary Measures, Geneva, Switzerland.

WTO, 2010. Possible actions for the SPS Committee regarding SPS-related private standards. Committee on Sanitary and Phytosanitary Measures, G/SPS/W/247/Rev.2.

7. Private law making at the round table on sustainable palm oil

Otto Hospes

7.1 Introduction

Palm oil is used in nearly every kitchen or bathroom in the world. It is a key ingredient of margarines and fats and is also widely used for frying. Popular non-edible uses of palm oil include soaps and cosmetics. Palm oil is not only used at home but also in restaurants and during the mass production of fried potatoes, French fries, pies, pastries and doughnuts. Palm oil is the leading tropical vegetable oil of the world.

The production of palm oil is much more concentrated: Indonesia and Malaysia each produce more than 40% of the world production and, not surprisingly, these two countries are the two major palm oil exporters of the world. The main importing countries are the EU, China and India. Palm oil is a highly globalised commodity.[285]

The global and increasing demand of palm oil has led to a rapid growth in palm oil production and trade. This has generated income and employment to millions of people as well as foreign currency for palm oil producing countries. At the same time, the expansion of palm oil production has caused different kinds of problems. These include increasing food and income insecurity, land conflicts, and social-political unrest at community level. Next to these social problems, palm oil production at plantations and mills has induced negative environmental effects. These effects include deforestation, loss of biodiversity, carbon dioxide emissions through deforestation and the exploitation of peat soils, and pollution of watersheds due to the use of chemical fertilisers and pesticides and discharges from processors.[286] These effects and the many global-local links that connect production,

[285] Product board for margarines, fats and oils, 2010. Fact sheet palm oil.
United States Department of Agriculture, 2010. Indonesia: rising global demand fuels palm oil expansion. Commodity Intelligence Report, Foreign Agricultural Service, Washington, DC, USA. Available at: http://www.pecad.fas.usda.gov/highlights/2010/10/Indonesia/.

[286] McCarthy, J. and Cramb, R.A., 2009. Policy narratives, landholder engagement, and oil palm expansion on the Malaysian and Indonesian frontiers. The Geographical Journal 175(2): 112-123.
McCarthy, J., 2010. Processes of inclusion and adverse incorporation: oil palm and agrarian change in Sumatra, Indonesia. Journal of Peasant Studies 37(4): 821-850.
Danielsen, F., Beukema, H., Burgess, N., Parish, F., Bruhl, C.A., Donald, P.F., Murdiyarso, D., Phalan, B., Reijnders, L., Struebig, M. and Fitzherbert, E.B., 2009. Biofuel plantations on forested lands: double jeopardy for biodiversity and climate. Conservation Biology 23(2): 348-358.
McCarthy, J. and Zen, Z., 2010. Regulating the oil palm boom: assessing the effectiveness of environmental governance approaches to agro-industrial pollution in Indonesia. Law and Policy 32(1): 153-179.
Reijnders, L. and Huijbregts, M.A.J., 2008. Palm oil and the emission of carbon-based greenhouse gases. Journal of Cleaner Production 16(4): 477-482.

processing and consumption, have increasingly globalised the challenges of the palm oil sector.[287]

In the absence of intergovernmental initiatives and institutions to address and regulate these problems, the World Wildlife Fund (WWF) and Unilever started to explore the possibilities for organising a new form of private governance at the global level in 2002. One year later they organised the first RSPO conference. In 2004 they officially launched the Round Table on Sustainable Palm Oil (RSPO). The main purpose of this round table has been to develop a set of principles and criteria for sustainable production of palm oil on the basis of a series of worldwide multi-stakeholder consultations. In 2010 the eighth annual RSPO conference was held.

The RSPO is not the only initiative of non-governmental organisations and multinational companies to develop private sustainability law for the production of global commodities. In the 1990s and even more so in the 2000s similar initiatives have been taken to regulate sustainable production of forest, fisheries, soy, cotton, sugar, beef and aquaculture by private standards. WWF is a founding member of seven multi-stakeholder initiatives directed at standard setting for sustainable production of a global commodity.[288] Unilever is founding member of two initiatives. Each of these two initiatives concerns one of the two most traded oil crops: the RSPO and the Round Table on Responsible Soy (RTRS).

This chapter has two objectives. The first objective is to analyse the RSPO as a law making process. For this purpose, the normative contents, actors and instruments of the RSPO will be described in relation to market power, public law and state authorities. To assess whether the RSPO principles and criteria are (increasingly or decreasingly) *de jure* or *de facto* binding, I will consider the power of sustainability as a market driver and the role of state actors in the standard setting process. The second objective is to assess to what extent the RSPO standard setting process has contributed to the development of new public standards on sustainable palm oil. For this purpose, I want to describe the different ways in which the governments of Indonesia and the Netherlands have each responded to the establishment of the RSPO principles and criteria. I have selected Indonesia and the Netherlands because non-state actors from these countries are well represented in the RSPO membership. This makes one wonder how their governments who, like any other government, are not member of the RSPO, treat the RSPO and its global, private standard. The comparative review of the two countries will show that one cannot

[287] Teoh, C.H., 2010. Key sustainability issues in the palm oil sector: a discussion paper for multi-stakeholders consultations (commissioned by the World Bank Group). IFC, World Bank, Washington, DC, USA. Available at: http://www.ifc.org/ifcext/agriconsultation.nsf/AttachmentsByTitle/Discussion+Paper/$FILE/Discussion+Paper_FINAL.pdf.

[288] World Wide Fund for Nature (WWF), 2010. Certification and roundtables: do they work? WWF review of multi-stakeholder sustainability initiatives. WWF, Gland, Switzerland. Available at: http://assets.wwf.org.uk/downloads/wwf_certification_and_roundtables_briefing.

generalise about 'the government' when discussing the making and meaning of transboundary private food law.

7.2 The normative content of the RSPO

The opening paragraph of the Preamble of 'Principles and Criteria for Sustainable Production of Palm Oil' (2007) presents the key issue of the RSPO and the kind of activity that the RSPO wants to change and regulate:

> Sustainable palm oil production is comprised of legal, economically viable, environmentally appropriate and socially beneficial management and operations. This is delivered through the application of the following set of principles and criteria, and the accompanying indicators and guidance.

The key issue is about sustainability, which does not only consist of economic, environmental and social dimensions but also legal ones. The activity is about palm oil production.

The document (from now on named as: 'RSPO Principles and Criteria') lists eight principles (Table 7.1).[289] Three principles (2, 3 and 5) exclusively relate to one of the four dimensions of sustainability as referred to in the Preamble. The inclusion

Table 7.1. RSPO principles.

RSPO principles	Sustainability dimension
1. Commitment to transparency	-
2. Compliance with applicable laws and regulations	Legal
3. Commitment to long-term economic and financial viability	Economic
4. Use of appropriate best practices by growers and millers	Environmental, social
5. Environmental responsibility and conservation of natural resources and biodiversity	Environmental
6. Responsible consideration of employees and of individuals and communities affected by growers and mills	Social, legal
7. Responsible development of new plantings	Environmental, social, legal
8. Commitment to continuous improvement in key areas of activity	-

Source: RSPO 2007. RSPO Principles and Criteria for Sustainable Production of Palm Oil.

289 Round Table on Sustainable Palm Oil (RSPO), 2007. RSPO principles and criteria for sustainable production of palm oil, including indicators and guidance. Available at: http://www.rspo.org/files/resource_centre/RSPO%20Principles%20&%20Criteria%20Document.pdf.

of a legal dimension of sustainability is remarkable. Many sustainability debates and documents focus on economic, environmental and social dimensions, leaving out legal ones.

The framing of the legal criterion of sustainable palm oil is both comprehensive and vague. Palm oil is considered sustainable if the production complies with 'all applicable local, national and ratified international laws and regulations' and if 'use of the land for oil palm does not diminish the legal rights, or customary rights, of other users, without their free, prior and informed consent'. Annex 1 of the 'RSPO Principles and Criteria' lists applicable international laws and conventions, including 15 different ILO conventions, the UN Declaration on the Rights of Indigenous Peoples (2007) and the UN Convention on Biological Diversity (1992). The document provides no guidance on how to define and identify what are 'legal' or 'customary' rights, and on how to deal with possible tensions or contradictions between international, national and local laws.

Three other principles (4, 6 and 7) have been specified into a mix of environmental, social and/or legal indicators. The use of best practices by growers and millers (principle 4) includes health criteria. The commitment to transparency (principle 1) and the commitment to continuous improvement in key areas of activity (principle 8) are of a different nature: they specify criteria for information management and monitoring and can be seen as preconditions for auditing and enforcement.

The eight principles cannot be categorised into the categories of public law requirements for food businesses as discussed in the first chapter of this book: product, process, presentation and public power. On the one hand, not a single principle or indicator defines the bio-physical quality of palm oil as a product. On the other hand, all principles define what turns palm oil into sustainable palm oil.

To provide further guidance to the eight principles, they have been specified into a total of 35 criteria. For every criterion, different indicators are distinguished. The indicators of principles 4, 5 and 7 refer to (best) agricultural practices (like integrative pest management techniques, ground cover management, recycling and re-use of nutrients) whereas the indicators of principles 4 and 6 refer to Best Management Practices, safe working practices or non-discriminatory practices. For best practices related to storage of chemicals, the RSPO principles refer to the 'FAO or GIFAP Code of Practice'.

The issue of presentation is not covered in the 'RSPO Principles and Criteria' but in the document 'RSPO guidelines on communication and claims' (2009).[290] This document provides instructions and restrictions on the use of the RSPO logo in

[290] Round Table on Sustainable Palm Oil (RSPO), 2009. RSPO guidelines on communication and claims. Available at: http://www.americanpalmoil.com/publications/RSPO_cc_guidelines%20(oct%2009).pdf.

corporate, on-pack and about-product communication. Only RSPO members can receive written authorisation to apply the logo after they have agreed to comply with the rules set forth in the document.

In corporate communication, members may report their membership of the RSPO. They may use the RSPO logo and/or RSPO web address together with such a membership claim. In on-pack and about-product communication, members may use the RSPO logo and/or RSPO web address only if combined with an approved claim on the use or the advancement of sustainable palm oil. The RSPO has approved four supply chain certification systems: identity preserved, segregated, mass balance, and book and claim.[291] For each of them, communication rules and approved story telling have been specified. The proportion of plantations certified defines how producers should communicate about their RSPO membership. Three different claims have been approved (Table 7.2).

Table 7.2. How producers should communicate about their RSPO membership.

Producer status	Approved claim in corporate communication
RSPO member, no plantations certified	'[Company X] is a member of the RSPO'
RSPO member, some plantations certified	'[Company Y] is a member of the RSPO and x% of its production capacity has been RSPO-certified'
RSPO member, all plantations certified	'[Company Z] is a member of the RSPO and produces only RSPO certified sustainable palm oil'

Source: RSPO 2009, RSPO Guidelines on Communication and Claims.

Finally, the issue of public power is also not covered in the 'RSPO Principles and Criteria'. However, the document does include an implicit reference to public power. One of the criteria of principle 6 is:

> Any negotiations concerning compensation for loss of legal or customary rights are dealt with through a documented system that enables indigenous peoples, local communities and other stakeholders to express their views through their *own representative institutions* [italics, OH].

However, the indicators of this criterion do not explicitly refer to state authorities or public law to explain what is meant with 'own' representative institutions. Something similar can be concluded from another criterion of principle 6: 'There

[291] Round Table on Sustainable Palm Oil (RSPO), 2009. Supply chain systems: overview. Available at: http://www.rspo.org/sites/default/files/RSPO_Supply%20Chain%20Systems%20Overview.pdf.

is a mutually agreed and documented system for dealing with complaints and grievances, which is implemented and accepted by all parties'. The indicators of this criterion do not refer to state authorities or public law but to 'dispute resolution mechanisms that should be established through open and consensual agreements with relevant affected parties'.

For the RSPO categorisation of RSPO principles, criteria and indicators, different P's can be distinguished: plan, policy, procedure, practice. Of the total of 35 indicators, 17 refer to plans to be developed, 4 refer to policies to be established, and 10 to procedures to be put in place. There are 17 indicators that explicitly refer to agricultural, management or other kinds of (best) practices.

7.3 Principle(d) actors

The Preamble of the 'RSPO Principles and Criteria' does not explicitly mention who is to meet the principles and criteria. The production of palm oil can be defined in a narrow way to refer to palm oil growers but also to include millers, who produce palm oil out of fresh fruit bunches harvested from the oil palm trees. However, the naming of principles 4 and 6 clearly suggests that these principles concern both growers and millers. The specification of principles 1, 3, 5 and 8 into criteria and indicators reveals that these principles relate to both growers and millers. Principles 2 and 7 seem to concern growers only, also when reading the criteria and indicators.

The 'RSPO Principles and Criteria' are not specifically geared towards smallholders. The Preamble mentions that, 'The development of more detailed guidance for application of the principles and criteria by smallholders ... is still on-going'. The principles and criteria are basically meant for plantations, defined by ILO (Convention 110, article 1/1) as: an agricultural undertaking regularly employing hired workers... concerned with the cultivation or production of ... [inter alia] palm oil....', but commonly understood as large-scale production. Principles 5, 6 and 7 explicitly refer to 'plantations', 'plantation area' and/or 'plantation development'.

Whereas the producers of palm oil are the ones to adopt the principles and criteria and to get certified (see below), they have not played a leading role in the establishment of the RSPO, the membership, the standard setting process and the national interpretations of the global RSPO principles and guidelines.

The RSPO was an initiative of the WWF and Unilever. In 2002 they started to explore possibilities for establishing a business partnership model for sustainable palm oil. In 2004 the RSPO was registered as a foundation under Swiss law, starting with 10 members. In November 2008, membership consisted of 261 members. Membership had grown to 380 members in October 2010. Whereas the RSPO distinguishes seven categories of members, they can be regrouped into three main categories (Table 7.3):

value chain actors (palm oil processors, traders, consumer goods manufacturers, retailers, banks and investors), oil palm growers and civil society organisations (environmental and nature conservation organisations, social and development organisations). As per October 2010, membership consisted of 276 value chain actors, 84 oil palm growers and 20 civil society organisations (Table 7.3).

Table 7.3. Membership of the RSPO as per October 2010.

Type of member	Number	Percentage of membership
A. Value chain actors	276	72.6
Palm oil processors and traders	151	39.7
Consumer goods manufacturers	94	24.7
Retailers	23	6.1
Banks and investors	8	2.1
B. Oil palm growers	84	22.1
C. Civil society organizations	20	5.3
Environmental and nature conservation organizations	11	2.9
Social and development organizations (NGOs)	9	2.4
Total	380	100

Source:www.rspo.org as per October 19th of 2010.

Whereas palm oil is mainly produced in Indonesia and Malaysia, the membership of the RSPO is not limited to business or civil society organisations from these two countries. The RSPO is a truly international organisation, with membership coming from 42 countries all over the world. Malaysia (87) and Indonesia (75) do provide the largest number of members but companies and NGOs from the United Kingdom (66), the Netherlands (37), Germany (31) and United States (26) cannot be easily overlooked.[292]

The highest authority of the RSPO is the annual General Assembly of members. This assembly has the power to establish the principle guidelines for the general policy of the RSPO. The decisions are taken by a simple majority vote of the members present or represented. Oil palm growers represent a minority of the membership. Palm oil processors, traders and consumer goods manufacturers together form a majority of the membership. The General Assembly also has the power to elect the members of the Executive Board within their own sector and

[292] See: http://www.rspo.org (as per January 2011).

to make recommendations to the Executive Board in view of the establishment of any useful Committee or Working Group (as specified in the RSPO by-laws).[293]

In 2005 the General Assembly adopted a set of general principles and criteria for sustainable production of palm to be field tested for two years. The document was prepared by the RSPO Criteria Working Group (CWG), which has played a key role in the standard setting process. The working group consisted of nine representatives of producers, six supply chain and investors, three environmental organisations and three social organisations. Major Indonesian palm oil companies, however, were not represented. The category of producers included two semi-public bodies and two research institutes. Eight members of the RSPO, including four members of the Executive Board of the RSPO, were appointed as observers of the CWG. The consultative process of discussing draft principles and criteria was facilitated by ProForest, an environmental NGO.[294] After field-testing and further consultations, the 'Principles and Criteria for Sustainable Production of Sustainable Palm Oil' were adopted by the General Assembly in 2007.

After the adoption of the global and generic principles and criteria, 'national implementation and interpretation teams' were organised in seven palm oil producing countries. The national implementation and interpretation teams do not only consist of non-state actors but also of state actors: about one fifth to one third of the membership of these teams consists of representatives of ministries. The membership includes public or semi-public bodies that are supposed to articulate interests of the plantation sector, like the Indonesian Palm Oil Commission (IPOC) and the Malaysian Palm Oil Association (MPOA). Representatives from industry form between one and two thirds of the membership, whereas NGOs do not reach more than one fifth.

The main purpose of every national team has been to ensure that the implementation of the global principles and criteria is 'congruent or compatible with the norms, laws and values of countries, or sovereign states'. With this test in mind, each team distinguished 'major non-conformities' and 'minor non-conformities'. The first ones refer to indicators that are considered critical in a specific country; the latter ones refer to indicators that are not very appropriate or relevant in a specific country. At the same time, not all global indicators could be simply disregarded

[293] Round Table on Sustainable Palm Oil (RSPO), 2004. RSPO statues, by-laws and codes of conduct. Available at: http://www.rspo.org/?q=page/896.
[294] Round Table on Sustainable Palm Oil (RSPO), 2004. Minutes of the first meeting of the RSPO Criteria Working Group. Compiled by ProForest and Andrew Ng. Available at: http://www.rspo.org/files/resource_centre/CWG%201%20minutes.pdf.

or considered less important in the national context: 'At least 45% all indicators must be identified as compulsory'.[295]

7.4 Compliance and complaints

Without third party verification and certification, no public claims relating to compliance with the RSPO principles and criteria can be made. The third party has to be a RSPO approved independent certification body. The RSPO will use a mechanism for approving certification bodies that is based on accreditation against ISO/IEC Guide 65 ('General requirements for bodies operating product certification systems') and/or ISO/IEC Guide 66 ('General requirements for bodies operating assessment and certification/registration of environmental management systems'), where the generic accreditation is also supplemented by a set of specific RSPO certification process requirements. Certification bodies must be accredited by national or international accreditation bodies, such that their organisation, systems and procedures conform to ISO Guide 65 and/or ISO Guide 66. RSPO certification assessments need to be initiated by palm oil producers, by contacting one of the approved certification bodies. Growers will be assessed for certification once every five years, and if certified, will be annually assessed for continued compliance.[296]

Downstream processors or users of RSPO certified sustainable palm oil can claim the use (or support) of RSPO certified palm oil when they adhere to the intent and requirements of the RSPO Supply Chain Certification Systems. This is independently verified by an RSPO approved and accredited certification body. Palm oil can be traded through one of the four supply chain models that are approved by RSPO to be able to preserve the claim to sustainable production: identity preserved, segregation, mass balance, or book and claim.[297]

Complaints and grievances can be submitted by any interested party, where the interested party has a legitimate interest in, or is directly affected by, the operations of the organisation which has been assessed for compliance against the RSPO Criteria or by the certification decision. This includes complaints relating to the process and the outcome of a certification assessment or concerning other aspects relating to implementation of the RSPO certification systems. A complaint or grievance can be made either through the certification body's mechanism for complaints (which will include subsequent referral to the accreditation body, and

[295] Round Table on Sustainable Palm Oil (RSPO), 2007. RSPO principles and criteria for sustainable production of palm oil, including indicators and guidance. Annex 3. Available at: http://www.rspo.org/files/resource_centre/RSPO%20Principles%20&%20Criteria%20Document.pdf.

[296] Round Table on Sustainable Palm Oil (RSPO), 2011. 'How to be RSPO certified'. Available at: http://www.rspo.org/?q=page/510.

[297] Round Table on Sustainable Palm Oil (RSPO), 2009. Supply chain systems: overview. Available at: http://www.rspo.org/sites/default/files/RSPO_Supply%20Chain%20Systems%20Overview.pdf.

then to RSPO, if the complainant remains unsatisfied by the outcome), or directly to the RSPO Executive Board. In the latter situation, the RSPO Executive Board will then determine whether the complaint or grievance should firstly follow the certification body's mechanism, or whether it can be referred directly to the RSPO Certification Complaints Committee. The Executive Board will appoint the members of the complaints committee, which has to consist of at least four individuals from the following RSPO sectors: producers, supply chain and investors, social organisations and environmental organisations. The committee is to decide on the complaint by consensus. The RSPO Secretary General is responsible for the implementation of any follow up action as required, and for informing the parties to the complaint, in writing, of the decision, not later than ten days after date of decision.[298]

The RSPO has also organised a focal point for official complaints against RSPO members: a grievance panel (of five members) is to deliberate and to decide on the legitimacy of any given grievance or complaint made against RSPO members.[299] Potential complainants include non-members of the RSPO but also non-residents of the home country of the RSPO member to whom the complaint is addressed. Any RSPO member could be subjected to a complaint. However, complaints will most likely concern palm oil producers as they are the ones to get certified and to stick to the RSPO criteria. The head of the grievance panel used to be the president of the RSPO from Unilever. Recently, the Netherlands-based Oxfam Novib became the head of the grievance panel to handle complaints against members of the RSPO.

7.5 How voluntary are the RSPO principles and criteria?

Legally speaking, there is no obligation for palm oil producers to apply for RSPO certification. RSPO principles and criteria are not *de jure* binding. Generally speaking, palm oil producers have two options to sell their commodity to: they can sell to traders, processors and countries that want certified sustainable palm oil, or to traders, processors and countries that do not require CPO. Having noted this, the conclusion could be that RSPO principles and criteria are not *de facto* binding. However, this conclusion is too quickly drawn.

Whereas the market share of certified sustainable palm oil is still quite modest (estimated at only 7%of global palm oil production of 45 million tons in 2009), its growth from zero till 7% in a few years' time has been quite impressive. This growth figure and the more fundamental commitment of corporate and political

[298] Round Table on Sustainable Palm Oil (RSPO), 2007. RSPO certification systems. Final document approved by RSPO Executive Board. Available at: http://www.rspo.org/sites/default/files/RSPO%20P&C%20certification%20system.pdf.

[299] See: Round Table on Sustainable Palm Oil (RSPO), 2011. RSPO grievance procedure. Available at: http://www.rspo.org/files/resource_centre/RSPO%20Grievance%20Procedure.pdf.

leadership to the promotion of sustainable production, as the only way forward, have increasingly made palm oil producers wonder whether they have a choice at all. What if not only Europe but also China and India start to formulate guidelines, criteria or even directives for import of sustainable palm oil? 'Sustainability is the current hottest issue faced by the industry', said the chairman of the Indonesian Palm Oil Association (GAPKI) during his welcome address at the Indonesian palm oil conference in December 2010, describing sustainability as 'a new market driver'.

Market powers increasingly show their sustainability 'muscles'. Multinationals, retailers and banks announce that they will only purchase trade or invest in certified sustainable palm oil in a few years from now: as per 2015 Unilever only wants to buy sustainably produced palm oil. Retailer Ahold and agricultural investor Rabobank have made similar commitments. This way the application for certification feels more and more like an obligation.

Multinationals, traders, retailers and bankers cannot simply impose their will on producers but the majority of their membership certainly gives them extra weight. Decisions in the General Assembly are taken by simple majority vote. From the very beginning till now the producers have not been behind the steering wheel of the RSPO. The RSPO was an initiative of Unilever, together with WWF; the Executive Board has never been chaired by a representative of the producers but by Unilever. Major Indonesian palm oil companies were not represented in the RSPO Criteria Working Group. Unilever headed the RSPO Grievance Panel.

Next to the dominant role of market power, the importance of public law in the RSPO standard and the increasing role of state authorities in the standard process are salient. 'Compliance with applicable laws and regulations', including international, national and local ones, is one of the main principles of the RSPO. This way the RSPO reproduces and reinforces existing state laws. After the adoption of the global principles in 2007, representatives of ministries of palm oil producing countries were invited to participate in the 'national interpretation and implementation teams'. The main purpose of these teams is to make the global RSPO standard compatible and congruent with norms, laws or values of 'a country or sovereign state'. The participation of government officials in the process of national interpretation, which can be seen as a specific or new phase of the standard setting process, is quite remarkable: until 2007 state actors had not been invited or involved in the RSPO standard setting process; more generally speaking, state actors are simply assumed not to play an active role in the interpretation and adjustment of 'private' law to regulate sustainable production of global food commodities.

7.6 Governments as consultative cheerleaders or competitive law makers

Whereas *de facto* the RSPO principles and criteria may increasingly be binding, governments have insisted on the voluntary character of membership and certification by the RSPO. However, they have not done so for the same reasons. Their own reactions and initiatives differ much, as can be seen from the cases of the Netherlands and Indonesia.

The Netherlands' government is one of the greatest moral supporters of the RSPO. This has to do with two related things. First, the way in which the RSPO has been organised much reflects Dutch values on social order, emphasising the importance of multi-stakeholder consultations to address common threats or problems. The RSPO may even be seen as a Dutch cultural product and global expression of the '*poldermodel*', with representatives of business and civil society from different parts of the world collaborating not to prevent polders from flooding but forests from palm oil.

One of the great supporters of the RSPO is the Dutch Product Board for Margarines, Fats and Oils (MVO), a more local expression of the deeply rooted Dutch culture of multi-stakeholder consultation. Whereas the board presents itself as an industry association, it has been created by law and is endowed with statutory powers. As a statutory public body, the MVO is to serve the common interests of all business actors in the production chain of oils and fats which carry out activities in the Netherlands, and to facilitate deliberations with authorities and societal organisations. The MVO is a member of the RSPO, that actively contributes to discussions at annual meetings and is willing to take a leading role in working groups. Currently, the MVO chairs the working group on trade and traceability and for that reason is invited to meetings of the RSPO Executive Board.

Second, the Netherlands' government insists on the voluntary character of RSPO certification because it is hesitant to develop or support initiatives that could be seen as new trade barriers. Sustainability initiatives should be WTO-proof. From this perspective, the Netherlands wants the RSPO to remain a voluntary private sector initiative, characterised by non-binding rules. Together with the UK government, the Netherlands' government has convened meetings with palm oil leadership to commit them to 'continued rapid growth in the production and use of RSPO-certified palm oil'. At the 'Palm Oil Leadership Group Meeting' of July 2010 in London, a 'Sustainable Palm Oil Declaration' was formulated, acknowledging 'the RSPO as the fundamental and credible platform for defining acceptable sustainability standards and as the forum for multi-stakeholder engagement'.[300]

[300] Proforest, 2010. Palm oil leadership group meeting: global business of biodiversity conference. Excel conference centre, London, UK, 13 July 2010. Summary report.

The Indonesian government also emphasises the voluntary character of the RSPO standard. However, it finds the RSPO insufficient and has started to develop an own standard, which should be binding for all palm oil producers in Indonesia: the Indonesian Sustainable Palm Oil (ISPO). The standard was announced by the Indonesian Minister of Agriculture in his opening address of the eight round table RSPO conference of November 2010 in Jakarta. At the same conference, the chair of the Indonesian Palm Oil Commission (IPOC) further explained the plans of the minister: oil palm plantations will be classified into five categories according to Decree no. 17 of 2009 of the Ministry of Agriculture; the plantations that will be classified under categories I, II and III can seek certification under the Indonesian Certification System, using ISPO as a standard; independent certification bodies will assess the application and plantation company. The Ministry wants to offer companies that are already RSPO-certified, the possibility to get ISPO-certified as well. To be eligible for this, the company should also be classified under category I, II or III by the Ministry of Agriculture. In that case, the last audit report will be studied in a selective way, provided this audit 'complies with regulations'. The chair of the IPOC said that an ISPO approved certified body will only audit those criteria that 'do not follow Indonesian regulations'. However, it is still unclear how an ISPO certified body will determine what RSPO criteria do not follow Indonesian regulations, how such criteria will be audited and what 'weight' will be given to performance on these criteria.

Like the Dutch government, the Indonesian government wants food law or sustainability standards not to interfere with WTO law. Unlike the Dutch government, the Indonesian government is preparing notification of its own standard under the WTO. The chair of the IPOC referred to compatibility with 'Codex' and approval by the 'International Organization for Standardization' (ISO) during her presentation at the eighth RSPO conference. At the same time, the Indonesian Ministry of Agriculture explores the possibility of organising Multilateral Agreements with accreditation bodies in palm oil importing countries (EU member states, USA, etc.) to enable acceptance of the ISPO certificates. Finally, the Ministry seeks bilateral cooperation with buyers' countries.

7.7 Conclusion

The law making process of the RSPO can be divided into three phases. The first phase is the phase in which multinationals, retailers and international NGOs – mainly from buyers' countries – formulated principles and criteria for sustainable palm oil in producers' countries. This phase began with the first exploratory talks of WWF and Unilever on a business partnership model in 2002 and ended with the adoption of the sustainability principles and criteria at the General Assembly of the RSPO in 2007. One of the key principles is compliance with all applicable laws and regulations.

The second phase is the phase in which government officials of producers' countries were invited to develop a national interpretation of the global RSPO principles and criteria, together with representatives of business and civil society. Strictly speaking, one could argue that this was not a phase of standard setting. However, during this phase 'minor' and 'major' non-conformities were distinguished. Some RSPO indicators were defined as more relevant than others in a specific national context. Also, the members of the national teams made explicit references to state laws during the 'interpretation' of the RSPO principles and criteria. This all suggests that it is more appropriate to speak of a re-setting of the normative contents of the RSPO standard. The second phase began in 2007. For some producers' countries this phase has already ended with the approval of national interpretations by the Executive Board of the RSPO (in 2008 for Indonesia and Papua New Guinea and in 2010 for Malaysia and Columbia). Other producers' countries (Ghana, Thailand and Salomon Islands) are still working on the national interpretation.

The third phase is the phase in which governments of producers' countries start to develop their own standard for sustainable production of palm oil. In the case of Indonesia, it is not unlikely that the RSPO standard setting process and the involvement of government officials in the national interpretation have challenged the Indonesian government to develop an own standard. The government was excluded from the RSPO standard setting process and observed that Indonesian plantation owners were not represented in the RSPO Working Group on Principles and Criteria. They also noted that the approval of the Indonesian national interpretation of the RSPO standard did not prevent a series of incidents to cause bad press for Indonesian RSPO-certified plantation owners (like Wilmar and Sinar Mas, Indonesia's biggest palm oil producer) and food giants to stop buying from these growers. Greenpeace alleged that Sinar Mas has been responsible for widespread deforestation and peat lead clearance practices, which release vast amounts of carbon dioxide. The environmental NGO has submitted a complaint about PT Smart, which is part of Sinar Mas Group, to the RSPO Grievance Panel. Currently, this panel is headed by the Dutch development organisation Oxfam Novib.

Unlike the Dutch government, the Indonesian government has given up its role as cheerleader of the RSPO. It is now actively developing its own standard, and more importantly, trying to challenge the RSPO as the single, legitimate power to define sustainability law for the production of palm oil on its own territory. A similar initiative has been taken by the Malaysian government. Whether the new and active role of governments of producers' countries in the development of national standards, marks the beginning of the end of the RSPO as a global standard setting body and grievance panel, remains to be seen. Market power of global value chain actors and moral power of international NGOs should not be underestimated. They will not easily give up a global forum and standard where they can sell sustainability. National governments of producers' countries will not quickly do away with the RSPO but at the same time try to make global private law subordinate to national public law. The rivalling of private and public standards at different levels, may contribute to intensification, joining and widening of efforts to make production of palm oil more sustainable, but also to new 'shopping' behaviour, inefficiencies, confusion and frictions.

At the last RSPO conference, held in Jakarta in November 2010, the new initiatives of the Indonesian and Malaysian government were presented at a workshop entitled 'external developments'. The framing is telling: own, national initiatives are considered 'external' in the eyes of the international organisers of the RSPO conference. This looks like the worst invitation to the Indonesian or Malaysian government, who will not feel challenged to explore how legal diversity and mixing can be ingredients for sustainable food production.

References

Danielsen, F., Beukema, H., Burgess, N., Parish, F., Bruhl, C.A., Donald, P.F., Murdiyarso, D., Phalan, B., Reijnders, L., Struebig, M. and Fitzherbert, E.B., 2009. Biofuel plantations on forested lands: double jeopardy for biodiversity and climate. Conservation Biology 23(2): 348-358.

McCarthy, J., 2010. Processes of inclusion and adverse incorporation: oil palm and agrarian change in Sumatra, Indonesia. Journal of Peasant Studies 37(4): 821-850.

McCarthy, J. and Cramb, R.A., 2009. Policy narratives, landholder engagement, and oil palm expansion on the Malaysian and Indonesian frontiers. The Geographical Journal 175(2): 112-123.

McCarthy, J. and Zen, Z., 2010. Regulating the oil palm boom: assessing the effectiveness of environmental governance approaches to agro-industrial pollution in Indonesia. Law and Policy 32(1): 153-179.

Product board for margarines, fats and oils, 2010. Fact sheet palm oil.

Proforest, 2010. Palm oil leadership group meeting: global business of biodiversity conference. Excel conference centre, London, UK, 13 July 2010. Summary report.

Reijnders, L. and Huijbregts, M.A.J., 2008. Palm oil and the emission of carbon-based greenhouse gases. Journal of Cleaner Production 16(4): 477-482.

Round Table on Sustainable Palm Oil (RSPO), 2004a. Minutes of the first meeting of the RSPO Criteria Working Group. Compiled by ProForest and Andrew Ng. Available at: http://www.rspo.org/files/resource_centre/CWG%201%20minutes.pdf.

Round Table on Sustainable Palm Oil (RSPO), 2004b. RSPO statues, by-laws and codes of conduct. Available at: http://www.rspo.org/?q=page/896.

Round Table on Sustainable Palm Oil (RSPO), 2007a. RSPO certification systems. Final document approved by RSPO Executive Board. Available at: http://www.rspo.org/sites/default/files/RSPO%20P&C%20certification%20system.pdf.

Round Table on Sustainable Palm Oil (RSPO), 2007b. RSPO principles and criteria for sustainable production of palm oil, including indicators and guidance. Available at: http://www.rspo.org/files/resource_centre/RSPO%20Principles%20&%20Criteria%20Document.pdf.

Round Table on Sustainable Palm Oil (RSPO), 2009a. RSPO guidelines on communication and claims. Available at: http://www.americanpalmoil.com/publications/RSPO_cc_guidelines%20(oct%2009).pdf.

Round Table on Sustainable Palm Oil (RSPO), 2009b. Supply chain systems: overview. Available at: http://www.rspo.org/sites/default/files/RSPO_Supply%20Chain%20Systems%20Overview.pdf.

Round Table on Sustainable Palm Oil (RSPO), 2011a. 'How to be RSPO certified'. Available at: http://www.rspo.org/?q=page/510.

Round Table on Sustainable Palm Oil (RSPO), 2011b. RSPO grievance procedure. Available at: http://www.rspo.org/files/resource_centre/RSPO%20Grievance%20Procedure.pdf.

Teoh, C.H., 2010. Key sustainability issues in the palm oil sector: a discussion paper for multi-stakeholders consultations (commissioned by the World Bank Group). IFC, World Bank, Washington, DC, USA. Available at: http://www.ifc.org/ifcext/agriconsultation.nsf/AttachmentsByTitle/Discussion+Paper/$FILE/Discussion+Paper_FINAL.pdf.

United States Department of Agriculture, 2010. Indonesia: rising global demand fuels palm oil expansion. Commodity Intelligence Report, Foreign Agricultural Service, Washington, DC, USA. Available at: http://www.pecad.fas.usda.gov/highlights/2010/10/Indonesia/.

World Wide Fund for Nature (WWF), 2010. Certification and roundtables: do they work? WWF review of multi-stakeholder sustainability initiatives. WWF, Gland, Switzerland. Available at: http://assets.wwf.org.uk/downloads/wwf_certification_and_roundtables_briefing.

8. GlobalGAP smallholder group certification

Challenge and opportunity for smallholder inclusion into global value chains[301]

Margret Will

8.1 Challenge or opportunity? An introduction to GlobalGAP option 2 smallholder certification

8.1.1 Localising global market opportunities means localising global challenges

Overseas markets for fresh and processed high-value agricultural and especially horticultural produce are attractive outlets for many developing countries. Effective exploitation of such opportunities, however, entails the requirement of complying with public mandatory and private voluntary food safety and quality standards ruling market access. These requirements have been considerably tightened since the turn of the Millennium, mainly for two reasons: firstly, the harmonisation of national regulatory frameworks under the guidance of the World Trade Organization (WTO) aiming at facilitating the internationalisation of trade; and secondly, numerous food hazard incidences and food scandals that forced legislators in industrialised countries to revise their national or, in the case of the European Union, their supranational food laws.

Alongside tightening of legal provisions and restructuring of public risk management and food inspection systems, the principle of due diligence was established, placing the primary responsibility for food safety with the business operators along the food supply chain; that is, with importers in the case of foreign country supplies. As a consequence, trade and industry introduced self-regulatory schemes, ranging from legally mandated and publicly inspected hygiene management and quality assurance systems to private voluntary standards (PVS).

Aiming at managing supply side risks in response to due diligence obligations, a plethora of PVS emerged, often going beyond formal market access requirements. At the same time, PVS were introduced for marketing reasons: to homogenise product attributes with a view to making market transactions more transparent and cost-efficient; and to label products as a means to differentiating from the offer of competitors.

[301] This contribution is an abridged and slightly revised version of Will, M., 2010. Integrating smallholders into global supply chains: GlobalGAP option 2 smallholder group certification generic manual: lessons learnt in pilot projects in Kenya, Ghana, Thailand and Macedonia. GTZ Sector Project Agricultural Trade, Eschborn, Germany. This contribution is published with kind permission of the editor, the Deutsche Gesellschaft für Technische Zusammenarbeit (GTZ) GmbH, Eschborn, Germany.

In the light of rising concerns over food safety, food security, climate change, growing competition for arable land between food, feed and bioenergy crops, standards will in all probability gain further importance. Especially, if positive impacts can be expected on the sustainable use of water and soil resources, the conservation of biodiversity, the preservation of traditional knowledge and, last but not least, on productivity, farmers' welfare, workers' health and consumer protection. Against this background, public mandatory and private voluntary standards are of growing relevance to smallholders in developing countries. Already now, compliance with standards is not anymore an issue only for farmers exporting overseas but increasingly also for smallholders competing for shares in domestic and sub-regional markets.

GlobalGAP certification, for example, is required by a considerable number of retailers, especially supermarkets, worldwide. On the one hand, this represents a real threat for many farmers in developing countries, as non-compliance may lead to their exclusion. On the other hand, if appropriately managed, compliance does not only offer income opportunities but as well the chance to introducing sustainable agricultural practices with positive impacts on farm economics, the environment and social networks. Various stakeholders report that investments into good agricultural practices (GAP) and quality assurance systems along the food supply chain are justifiable and result in reasonable returns on investment, for example through productivity gains, reduction of production and transaction costs, improved market access, and ultimately, improved consumer protection. However, two questions still remain unanswered: to what extent are small-scale farmers in developing countries likely to be squeezed out of global value chains and which capacities need to be built to enable smallholders to seize opportunities from access to higher value markets ruled by PVS such as GlobalGAP.

In so far, the question of 'challenge or opportunity' may have to be changed into the question, on how the challenges posed by standards in general and private voluntary standards in particular can be turned into opportunities for small-scale farmers in developing countries. The following citations stand for the current discussion on 'challenge or opportunity' for smallholder inclusion into global value chains:

From challenge ...

> 'Farmers do perceive codes often as ... imposed externally ... they are required to follow to improve market access. This is understandable as it seems that

> codes often have been designed with insufficient involvement of farmers and in particular smallholder farmers reflecting on their typical environment'.[302]
> 'The ability of smallholders to access the high-value markets dominated by supermarkets has declined dramatically since the implementation of GlobalGAP'.[303]
> 'One obstacle standing in the way of more extensive dissemination of the certification ... are the associated costs for small-scale farmers'.[304]
> 'Private standards will not go away. Therefore the solution is evolution and adaptation of standards rather than demanding their abolishment'.[305]
> 'The actual challenge consists of tapping the advantages of GlobalGAP (good agricultural practices) in such a way that long-term economic sustainability is ensured for goods produced by small-scale farmers...'[306]

... to opportunity!

> 'In this globalised world, GlobalGAP certification provides an opportunity to smallholder groups to play on an equal ground with other bigger suppliers in the world, and be connected to international buyers'[307]
> 'The most successful GlobalGAP-compliant smallholder schemes are highly committed to a commercial farming approach, well organised in strongly-managed producer groups, and linked to a large, well-resourced export company'[308]

[302] Opitz, M., Potts, J. and Wunderlich, C., 2007. Closing the gaps in GAPS: a preliminary appraisal of the measures and costs associated with adopting commonly recognised 'good agricultural practices' in three coffee growing regions, appendices. International Institute for Sustainable Development (iisd) and Sustainable Coffee Partnership, Winnipeg, Canada, p. 67.
[303] Legge, A., Orchard, J., Graffham, A., Greenhalgh, P., Kleih, U. and MacGregor, J., 2009. Mapping different supply chains of fresh produce exports from Africa to the UK. In: Borot de Batisti, A., MacGregor, J. and Graffham, A. (eds.) Standard bearers: horticultural exports and private standards in Africa. International Institute for Environment and Development (IIED), Winnipeg, Canada, pp. 40-44.
[304] BMZ [Federal Ministry for Economic Cooperation and Development], 2008. Introduction of voluntary social and ecological standards in developing countries, summary version of the evaluation. Evaluation Report 043, BMZ, Bonn, Germany, p. 15.
[305] Wainwright, H. and Labuschagne, L., 2009. Private standards: a personal perspective from a training service provider. In: Borot de Batisti, A., MacGregor, J. and Graffham, A. (eds.) Standard bearers: horticultural exports and private standards in Africa. International Institute for Environment and Development (IIED), Winnipeg, Canada, pp. 154-157.
[306] BMZ [Federal Ministry for Economic Cooperation and Development], 2008. Introduction of voluntary social and ecological standards in developing countries, summary version of the evaluation. Evaluation Report 043, BMZ, Bonn, Germany, p. 15.
[307] Enomoto, R., 2009. GlobalGAP Smallholder Support Kit. GlobalGAP Tour 2009 Good Agricultural Practice, Nairobi, Kenya, 16 September 2009.
[308] Graffham, A. and Cooper, J., 2009. Making GlobalGAP smallholder-friendly: can GlobalGAP be made simpler and less costly without compromising integrity? In: Borot de Batisti, A., MacGregor, J. and Graffham, A. (eds.) Standard bearers: horticultural exports and private standards in Africa. International Institute for Environment and Development (IIED), Winnipeg, Canada, pp. 83-88.

> 'While the existing literature documents how both the non-recurring and the recurring costs of compliance with GlobalGAP can be considerable, our results demonstrate that the returns on the associated investments in terms of export sales growth are considerable. ... we might reasonably expect appreciable 'knock-on' benefits to small producers'[309]

8.1.2 The GlobalGAP standard

GlobalGAP (formerly EurepGAP) has emerged as the leading private voluntary standard for the access of agricultural products to major import markets. Initiated in 1997 by retailers united in the Euro-Retailer Produce Working Group (EUREP), the pre-farm-gate standard Good Agricultural Practices covers all on-farm processes from inputs through farming until the product leaves the farm.

While 'retailer and supplier representatives are equally responsible for decision making in the different GlobalGAP committees' (GlobalGAP, 2009), GlobalGAP with support of the UK Department for International Development (DfID) and the Deutsche Gesellschaft für Technische Zusammenarbeit (GTZ) GmbH, initiated an 'Africa Observer' project in 2007 with a view to also representing the interests of African smallholders in standard setting processes. The Africa Observer furthermore collects global best practices to facilitate standard implementation by small-scale farmers worldwide.[310]

In addition to its relevance for accessing major export markets, GlobalGAP gains importance in developing countries' domestic markets. Kenya may serve as an example, where the private Fresh Produce Exporters Association of Kenya (FPEAK) promotes KenyaGAP, which is fully benchmarked with GlobalGAP. The Kenyan supermarkets recently started establishing preferred supplier schemes that oblige farmers to comply with KenyaGAP. Even if the shares of supermarkets in the commercialisation of horticultural produce in most parts of Africa are still small and will not grow fast, without question, there is a trickle-down effect of GlobalGAP into the local market.

8.1.3 The GlobalGAP option 2 group certification

With a view to easing compliance for small-scale farmers, GlobalGAP offers options for group certification (for an overview of different certification options see Table 8.1). The advantage compared to individual certification (option 1) is that under group certification (option 2), qualified staff within the farmer group (optionally

---

[309] Henson, S., Masakure, O. and Cranfield, J., 2009. Do fresh produce exporters in sub-Saharan Africa benefit from GlobalGAP certification? International Food Economy Research Group, InFERG Working Paper No. 2_FT, University of Guelph, Guelph, Canada, p. 23.

[310] See http://www.africa-observer.info.

Table 8.1. GlobalGAP certification options.

	Individual certification	**Group certification**
GlobalGAP	option 1	option 2
Benchmarked scheme	option 3	option 4

	An individual farmer owns the certificate. Verification of compliance through one external inspection per year.	A farmer group owns the certificate. Verification through Quality Management System, internal inspections and audits plus one external inspection and audit per year.

a subcontracted service provider) acts as internal auditor of the individual group members. The task of external certification is hence reduced to examine the proper working of the group's Internal Quality Management and Control System (ICS) accompanied by an external inspection of only a random sample of farmers. This 'system check', which nevertheless remains the major recurrent cost factor, allows the external Certification Body to certify the entire group rather than each individual farmer.

Compared with individual certification, a group certificate under option 2 implies some advantages: auditing costs and centralised investments (e.g. pesticide store) can be shared among group members; exchange of information and capacity building can be delivered more straightforward through the groups; and, the motivation to comply is boosted by the groups' peer pressure on members since failure of one member would affect the entire group. It can well be assumed that group certification is more feasible for small-scale producers and can hence contribute to reducing the risk of smallholders' exclusion from (global) value chains.

For achieving GlobalGAP option 2 group certification, four key requirements have to be met by the farmer groups[311]:

- The group has to operate an *Internal Quality Management and Control System (ICS)* consisting of (a) a Quality Management System (QMS) with a documented management structure and a written control and procedures manual as well as (b) a Central Administration and Management of the group responsible for the implementation of the control and sanction system across all member farms.

[311] Günther, D. and Neuendorff, J., 2006. EurepGAP smallholder manual: understanding EurepGAP. Powerpoint presentation held at the occasion of the Kick-off workshops.

- Since the ICS does not replace the internal self-inspection, each farmer registered as group member has to implement a *Farmer Internal Self Assessment* based on the GlobalGAP checklist and to be undertaken by the farmer himself. The findings must be available for review by either the internal or the external inspector.
- Based on the GlobalGAP checklist, the Farmer Group has to realise a *Farmer Group Internal Control* aimed at inspecting each registered farm and all declared produce handling sites at least once per year. This audit can be realised by qualified staff within the Farmer Group (Internal Farmer Group Inspector) or can be subcontracted to an external verification body as long as it is different from the external Certification Body.
- The Farmer Group registers with a GlobalGAP approved Certification Body (CB)[312] and signs a sub-licence agreement with the CB. The *External Verification* undertaken by the CB consists of two parts: the audit of the ICS ('System Check') and the inspection of a random sample of farms.

8.1.4 The challenges

Even though small-scale farmers contribute major shares to fresh produce destined for export and for the local processing industry in many developing countries and even if they derive significant levels of income in return, smallholders are especially challenged with achieving GlobalGAP certification. The main concern is that the costs of compliance render smallholder production unfeasible. As a consequence, customers who previously bought from small-scale farmers, may switch to either sourcing from larger farms or from fully-integrated own production. The special challenge of option 2 certification is to build the necessary technical and managerial capacities within farmer groups and to generate the financial means for realising necessary investments and perhaps subcontracting service providers capable of supporting the operation of the ICS.

8.2 Challenge and opportunity! The GlobalGAP smallholder pilot project

8.2.1 The objective of the pilot project

The overall objective of the smallholder pilot project was to identify ways, in which the GlobalGAP standard can become more inclusive for smallholder farmers in developing countries and to assist GlobalGAP to develop new and adjust existing technical standards and tools appropriate for smallholder certification. The pilot project was designed to develop a generic 'GlobalGAP smallholder QMS manual' serving smallholder producer groups to lower the costs for creating their own

[312] List of approved CB available online at: *http://www.globalgap.org*.

Internal Quality Management and Control System (ICS) as the main prerequisite for getting certified under GlobalGAP Option 2.

Furthermore, the 'GlobalGAP Smallholder QMS Manual' was pilot-tested in four countries, thereof two in Africa and one country each in Asia and South East Europe, with the purpose of:

- identifying critical aspects of success and failure in GlobalGAP option 2 group certification and developing practical solutions;
- adapting the generic manual to local conditions to develop a simple public-domain local shareware QMS manual readily usable for adaptation by smallholder groups; and
- developing standard training packages and qualifying local public and private service providers for training and auditing of farmer groups.

Finally, the generic 'GlobalGAP smallholder QMS manual' was made available as public shareware.[313]

8.2.1 The partners of the pilot project

The GlobalGAP smallholder pilot project was jointly implemented by the sector project Agricultural Trade of the Deutsche Gesellschaft für Technische Zusammenarbeit (GTZ) GmbH and the Gesellschaft für Ressourcenschutz (GfRS) in cooperation with GlobalGAP (by then still EurepGAP) and various development programmes implemented in the four pilot countries by the GTZ (Germany), the UK Department for International Development (DfID), the United States Agency for International Development (USAID) and the World Bank financed Horticulture Export Industry Initiative (HEII in Ghana).

8.2.2 The generic GlobalGAP QMS smallholder manual

The 'smallholder manual' serves as a practical guidance on how to make the complex system requirements for certification as laid down in the GlobalGAP general regulations for fruit and vegetables manageable for smallholder groups. The generic guidance document provides a systemised compilation of all information relevant for farmer groups to prepare and achieve GlobalGAP certification under Option 2.

More precisely, the 'smallholder manual' introduces the GlobalGAP scheme, explains the certification process, discusses critical success factors for group certification and imparts practical guidance to group representatives on how to

[313] See: GTZ, 2010. GlobalGAP smallholder QMS set-up guide – How to establish QMS in your group. Deutsche Gesellschaft für Technische Zusammenarbeit, Eschborn, Germany. Based on: GlobalGAP IFA Ver. 3.02 QMS checklist_Mar 08, English Version, 2010.

establish and document an Internal Control System in full compliance with the general regulations of GlobalGAP. The manual provides a template of a handbook for a Quality Management System (QMS) including standard operating procedures and recording forms modelled for a fictional farmer group.

Not to be understood as a one-size-fits-all solution, the applicant smallholder groups can use this practical guide for creating their own QMS by adapting procedures and forms to the specific situation of their groups and members. The manual does not give guidance on how to implement the GlobalGAP standard at farm level, but leaves this task intentionally to the group managers, to expert farmers, agricultural extensionists or other specialists who are familiar with the specific crops as well as the local production systems and farming practices.

To support smallholder groups in designing their own group-specific QMS, the smallholder QMS guidelines had to be adapted to the prevailing conditions of the pilot groups. The development of these local generic QMS manuals was guided by the following tasks:

- definition of methods for risk assessments;
- definition of the organisational and administrative set-up;
- development of a quality policy;
- development of standard operating procedures;
- development of recording forms (templates, e.g. model contracts and control sheets).

After validation by international experts and the certification body, the draft local generic QMS manuals were discussed with the end users (growers, exporters) in the four pilot countries whose feedback was taken into consideration for the development of the respective final versions.

8.2.3 The four pilot projects in Africa, Asia and Europe

The draft version of the 'smallholder manual' was tested with several smallholder farmer groups in Kenya, Ghana, Thailand and Macedonia. In all four countries, pilot testing of the manual complemented activities of development programmes implemented by GTZ, DfID, USAID and the World Bank and some countries' Ministries respectively, aiming at integrating smallholder farmers into food supply chains (usually referred to as 'value chains' in the development context).

Pilot testing was implemented in a stepwise action-oriented approach to provide for evaluation and reflection loops allowing the project partners to adapt the implementation to the progress of farmer groups and group instructors involved in advice and training. In general, the implementation of the pilot projects was guided by the following sequence of actions:

- group profiling and selection of farmer groups according to pre-established eligibility criteria (see below);
- assignment of a local coordinator for managing the pilot project and instructors for assisting the farmer groups in adapting the groups' QMS manuals and ICS to their situation;
- kick-off workshop implemented by GTZ with development partners, the local coordinator, the farmer groups' managers, the instructors aimed at introducing the stakeholders to the smallholder manual;
- group work implemented by the management of the farmer groups with assistance by the group instructors and accompanied by a training needs assessment and tailor-made training courses;
- mid-term review jointly implemented by GTZ, GlobalGAP, development partners, the local coordinator, the farmer groups' managers and the instructors to review the draft group QMS manuals;
- implementation of the ICS by the groups (mainly training of farmers, internal inspections, pesticide testing, produce handling) to test, review and finalise the documented group QMS manuals;
- final workshop jointly implemented by GTZ, development partners, the local coordinator, the farmer groups' managers and the instructors to feed the practical experiences into the final version of the documented QMS of the farmer groups to be submitted to the certification body for approval.

The project used the following eligibility criteria for the selection of the pilot groups:
- the group shall be a smallholder group as defined below;
- the group wishes to apply for certification under option 2 of GlobalGAP;
- the group shall be a legal entity;
- the group must be large enough to sustain an ICS (minimum of 30 to 50 smallholders);
- the group shall have sufficient human and financial resources to maintain the ICS;
- the group shall share all experiences in an open and transparent way internally and with the technical advisors who will assist them in adapting their ICS documentation.

The ICS guidance document of the EU Commission (2000) defines smallholder farmer groups as follows:
- The cost of individual certification is disproportionally high in relation to the sales value of the product sold (higher than 2% of sales).
- Farm units are mainly managed by family labour.
- There is homogeneity of members in terms of geographical location, production system, size of holdings and common marketing system.
- Only smallholder farmers shall be members of the group. Under certain conditions, larger farmers, processors and exporters can be part of the structure of the group.

The characteristics of the four pilot projects are briefly described in Table 8.2. A detailed description of the four pilot projects has been published.[314]

Table 8.2. Characteristics of the four GlobalGAP smallholder pilot projects.

Kenya	
Crops and target markets	french beans (fresh and canned) destined for exports
Number of farmer groups	9 farmer groups, thereof 6 certified by October 2006
Partners	
• Private sector	455 farmers, 3 exporters, 1 PMO[1], service providers
• Public sector	Ministry of Agriculture extension services, research centre
• Dev. partners	GTZ, DfID
A glance on the way forward	'The process of benchmarking KenyaGAP to EurepGAP [GlobalGAP] is the unique opportunity for our small-holders to demonstrate that they are world class producers'.[2]
Ghana	
Crops and target markets	pineapples (fresh and fresh-cut) destined for exports
Number of farmer groups	6 farmer groups; all certified by mid 2007
Partners	
• Private sector	142 farmers; no linkage to exporters or other buyers
• Public sector	Ministry of Food and Agriculture with different departments
• Dev. partners	GTZ, USAID, HEII (World Bank financed)
A glance on the way forward	'Ghana must face the challenges of quality ... This development must be seen as part of a full-service marketing proposition, which would stem from its current position of low-cost spot supplier to the European ... market'.[3]
Thailand	
Crops and target markets	asparagus exported to Japan/Taiwan/EU; durian exported to China/ASEAN
Number of farmer groups	6 farmer groups, thereof 3 certified by 2007
Partners	
• Private sector	240 farmers, 1 exporter/2 collectors, 1 association
• Public sector	1 university as service provider
• Dev. partners	GTZ
A glance on the way forward	'We need to think about how we can focus on globalisation, yet remain locally accountable'.[4]

[314] Will, M., 2010. Integrating smallholders into global supply chains: GlobalGAP option 2 smallholder group certification generic manual: lessons learnt in pilot projects in Kenya, Ghana, Thailand and Macedonia, GTZ Sector Project Agricultural Trade, Eschborn, Germany.

Table 8.2. Continued.

Macedonia	
Crops and target markets	fruit and vegetables (pepper); partly destined for exports
Number of farmer groups	2 farmer groups, both certified in 2006
Partners	
• Private sector	125 farmers, 1 consultant
• Public sector	Ministry of Agriculture (MoA)
• Dev. partners	GTZ, USAID
A glance on the way forward	In Macedonia, the 'greatest food system weaknesses [are rooted in the] inadequate supply of uniform quality primary products for the fresh and processed product markets'.[5]

[1] PMO = Produce Marketing Organisation or Primary Marketing Organisation.
[2] Shah, H., 2005. The global partnership for safe and sustainable agriculture. In: EurepGAP global report, p. 8.
[3] Danielou, M. and Ravry, C., 2005. The rise of Ghana's pineapple industry: from successful takeoff to sustainable expansion. ESSD Africa Working Paper Series No. 93, p. 42.
[4] Thai Deputy Prime Minister Kosit Pupienmrat, 2007, http://www.globalgap.org/cms/front_content.php?idart=372&idcat=46&lang=1&client=1.
[5] Scott, W., 2006. Linking producers to increased horticultural exports from Macedonia. Presentation at the Regional Consultation on Linking Producers to Markets, Cairo, Egypt, 29 January to 2 February 2006.

8.3 Turning challenges into opportunities: conclusions from the GlobalGAP smallholder pilot project

8.3.1 Results intended ... results achieved

By building technical and managerial capacities of farmers, group managers and service providers as well as institutional capacities of farmer groups to jointly manage GlobalGAP option 2 certification, the pilot project attained significant impacts with regard to economic viability, environmental sustainability and social advancement (Table 8.3). Seventeen pilot farmer groups achieved certification, representing 75% of all groups having participated in the pilot project.

While some weaker smallholders dropped out of the programme (Table 8.4) there is evidence from a recent post-pilot evaluation in Ghana that the majority of pilot farmers increased their incomes through increased productivity, reduced production costs and rejects and that the membership in several pilot groups increased (trickle-down effect).

As a further result, the generic 'smallholder manual' has been revised in the light of the experiences gained from the pilot projects. The manual is now available as

Table 8.3. Positive results according to sustainability criteria.

	Achievements
Economic viability	• Productivity increased; input costs and rejects reduced; income stabilised/ increased[1] • Crop risk (yield fluctuations) reduced through application of GAP/better quality inputs • Market/price risks (volatility) reduced for groups that are integrated into supply chains • Access also improved to local, regional upmarket segments not requiring certification • Financial assets generated enabling farmers to re-invest and to access (trade) credits • Infrastructure improved on farms and at group level • Capacities of service providers (extension/training) improved • Access to a more reliable supply of required qualities improved for PMOs/ exporters
Environmental sustainability	• Protection of natural resources improved through reduced use of chemicals • Efficiency of the use of natural resources increased through reduced waste (rejects)
Social advancement	• Management/technological capacities for compliance/QMS management built among farmers and group managers; with positive effects on wider farm/group management • Farmer groups strengthened (growth in membership e.g. in Ghana; even if this is also owed to longer-term (post-pilot) assistance in farmer group development) • Supply chain governance[2] has become more transparent and trustful with more equitable relationships assuring fairer compensation for smallholders • Farm worker and family health improved through improved handling of pesticides and better knowledge on food safety and hygiene issues

[1] Evidence mainly anecdotic, apart from the on-going post-pilot evaluation in Ghana, which confirms assumptions made in numerous recent publications.

[2] Supply chain governance: power relations, information exchange/transparency and distribution of gains between operators along the supply chain.

a public shareware[315]. Additionally, the lessons learnt in elaborating the generic QMS manual and implementing the pilot projects have fed into the GlobalGAP

[315] See: GTZ, 2010. GlobalGAP smallholder QMS set-up guide – How to establish QMS in your group. Deutsche Gesellschaft für Technische Zusammenarbeit, Eschborn, Germany. Based on: GlobalGAP IFA Ver. 3.02 QMS checklist_Mar 08, English Version, 2010.

Table 8.4. Negative unintended results.

Economic viability	Unsatisfactory first-time certification/retention rates[1] in some pilot projects, mainly due to: • weak groups (partly in Kenya) • wrong groups not (yet) requiring certification (durian group in Thailand) • wrong expectations about price premiums paid for certified produce • inadequate approaches for building required farmer and group capacities • insufficient time for training of smallholder farmers not meeting their requirements • insufficient/lacking group-exporter linkages, partly too donor-driven • questionable ownership and hence sustainability (especially in Thailand)
Social advancement	• Dropout of weaker group members prior to the first post-pilot certification in most cases (e.g. -14% in Ghana); however, this decrease was at least in Ghana more than compensated by an increase in group membership of +52% one year later (retention rate and membership are expected to grow further).

[1] Retention rate: rate of satisfaction with and renewal of the certification.

Africa Observer project, which represents the interests of the smallholder farmer community in and beyond Africa, and the GlobalGAP smallholder task force. As a major result, initial steps have been taken to make the standard as well as the conformity assessment more smallholder-friendly.

8.3.2 Critical success factors

The following citation of Graffham and Cooper[316] perfectly depicts the lessons, which have as well been learnt in the pilot projects: 'The most successful GlobalGAP-compliant smallholder schemes are highly committed to a commercial farming approach, well organised in strongly-managed producer groups, and linked to a large, well-resourced export company'.

The results achieved illustrate the ambiguous effects inherent to smallholder group certification: compliance with GlobalGAP can advance the integration of responsive farmer groups into global supply chains but may drive weaker small-scale farmers

[316] Graffham, A. and Cooper, J., 2009. Making GlobalGAP smallholder-friendly: can GlobalGAP be made simpler and less costly without compromising integrity? In: Borot de Batisti, A., MacGregor, J. and Graffham, A. (eds.) Standard bearers: horticultural exports and private standards in Africa. International Institute for Environment and Development (IIED), Winnipeg, Canada, pp. 83-88.

out of export markets. 'In principle, smallholders ... recognise the need to adhere to international market standards because this enhances trade confidence with their buyers. In particular, GlobalGAP has shown to enhance market access, especially for producers in [countries] where official control (government) systems are under-developed and hence less assuring. ... [However,] the main concerns on GlobalGAP ... are the high cost of certification and the complexity of implementation'.[317]

There is no simple solution to these two critical issues. Rather, the following success factors illustrate the need for a situation-specific (country, supply chain, groups, products) system's approach to promoting smallholder group certification (see also Table 8.5):

1. *Smallholder capacities (group performance and farmers' skills)*: 'The social cohesion of the ... groups involved in group certification is one key success factor for sustainable certification and concurrently one of the core challenges for group certification ... Hence, capacity development is essential. A common need must be identified within the group ... This can be common marketing, common production, common training – but never certification as the only purpose ... The most important challenges identified are capacity development at both individual farmer level and farmer group level ... and strong engagement in group cohesion ... Therefore, for sustainable group certification the focus ... [needs to be] placed on group selection and group development'.[318]
2. *Incentives, financial viability and market prospects*: For up-scaling group certification, it is necessary to identify (a) realistic incentives based on a cost-benefit analysis along the entire supply chain to convince supply chain operators to committing required resources; (b) viable solutions for sustainable financing of initial investments and recurrent costs of compliance; and (c) buyers (traders, exporters, processors or PMOs) capable to assuring access to markets, committed to contracting smallholder farmer groups and willing to supporting certification through embedded services.
3. *Supply chain governance (fair partnership and industry commitment)*: Given that 'GlobalGAP certification can hardly be achieved by smallholder groups alone ..., intense support from exporters [or other customers] is one core element of success. [Furthermore,] ... the advantages of strong exporter support are that standard implementation is performed quickly and effectively'.[319] In the same line of thinking, the International Institute for Environment and Development (IIED) and the Natural Resources Institute (NRI) argue that there is a '... need to rethink the concept of 'costs'. On average, farmers pay 14% of recurrent costs associated with [GlobalGAP] and exporters (and/or) donors pay the rest.

[317] Mbithi, S., 2009. GlobalGAP – report on smallholder taskforce consultations – smallholder ambassador. Nairobi, 21st June 2009. GlobalGAP Africa Observer, p. 2.
[318] GTZ, GlobalGAP and Marktkontor Obst und Gemüse Baden e.V., 2008. GlobalGAP group certification: a challenge for smallholders in Europe and developing countries. International Workshop, 28- 29 April 2008, Frankfurt am Main, Germany, p. 7.
[319] *Ibid.*, p. 6.

Rather than labelling exporter investments as unsustainable, it can be argued that this illustrates a healthy and functioning system with the two private sector investors sharing the costs and benefits as part of a sustainable business model.'[320]

4. *Adaptation to smallholder capacities (standard complexity and certification costs)*: Experience shows that the main challenges are more related to certification than to compliance as such. Hence, it will be essential to find ways of adapting the costs of certification and the implementation of the Control Points and Compliance Criteria (CPCC) to the capacities of smallholder groups without compromising essential GlobalGAP requirements. 'On the cost, it is true that the main part (at least 85%) ... is audit fees paid to CBs [Certification Bodies], with the rest being standard fees. In order to enhance more compliance amongst smallholders, it will be necessary to initiate discussions with CBs on how credible but cost effective audit structures can be implemented. On complexity, apart from reducing CPCCs, GlobalGAP will need to work with farmers ... and other organisations to translate/simplify implementation of the standard'.[321] In this context, the establishment of AfriCert, a Kenya-based/Ghana-represented CB is worth mentioning. As a local CB, AfriCert cannot only bring down costs but its inspectors also have a better understanding of how CPCCs can be reached under local conditions.
5. *Support services and framework conditions*: Last but not least, the public and private sectors play a crucial role in, firstly, establishing (public task) and lobbying for (private task) enabling framework conditions (policies, legislation, enforcement, appropriate infrastructure) and, secondly, offering appropriate, competent, affordable and accountable financial and non-financial services.

8.3.3 Improving the smallholder-orientation of option 2 certification

Until recently, discussions on the effects of standards on small-scale farmers centred on the risks of exclusion, nurtured by figures from Kenya showing dramatic declines in smallholder participation in vegetable exports. But evidence is growing that compliance with GlobalGAP can as well enhance competitiveness of smallholders and support their participation in global supply chains. Even if data are not yet available, reports on increased productivity as well as reduced input costs and rejects are not any more only anecdotic. These findings of the pilot project can be confirmed by recent studies reporting for example that '... the returns on the associated investments in terms of export sales growth are considerable. ... Given that firms in the survey tended to procure a significant proportion ... from small

[320] IIED [International Institute for Environment and Development] and NRI [Natural Resources Institute], undated. Costs and benefits of EurepGAP compliance for African smallholders: a synthesis of surveys in three countries. Fresh Insights 13, p. 20.

[321] Mbithi, S., 2009. GlobalGAP – report on smallholder taskforce consultations – smallholder ambassador. Nairobi, 21st June 2009. GlobalGAP Africa Observer, p. 2.

Table 8.5. Critical success factors in GlobalGAP option 2 group certification.

Smallholder capacities sufficient to achieve group certification:	Market prospects promising for certified smallholder supplies:	Incentives for the commitment of supply chain partners:
Group performance: • Group cohesion • Capacities for ICS/QMS management • Exporter/PMO linkages Farmers' skills: • Technological skills (e.g. GAP) • Financial assets	Market potential: • Market volume and growth • Lucrative price segments • Significance of smallholder shares in overall supplies Distribution systems: • Supply chain organisation • Transport/logistics capacities	Income generation: • Profitability for all supply chain partners • Fair distribution of gains along the supply chain (win-win) Broad understanding of benefits: • Cost cuts, productivity gains • Worker and consumer health
Issue: How to up-scale capacity building to achieve broad impact	Issue: How to up-scale marketing skills of smallholders	Issue: Which cost-benefit ratio irrespective of pilot conditions
Supply chain governance conducive for smallholder integration:		Financial viability/financing for sustaining certification:
Fair partnership: • Mutual trust/contract-abiding • Fair distribution of profits and costs along the supply chain • Industry commitment: • Contracting farmer groups • Supporting certification (e.g. through embedded services)	Promising product features for GlobalGAP certification[1]: • High-value • Low-cost • Labour-intensive • Existing demand	Access to finance via: • Industry embedded service • Private financial sector • Government/donor funds Low-cost solutions for: • Initial investments (infrastructure, training, certification) • Recurrent compliance costs
Issue: How to promote CSR for fair partnerships in local firms		Issue: Which solutions for smallholder low-risk attitudes

outgrowers, ... we might reasonably expect appreciable 'knock-on' benefits to small producers'.[322]

[322] Henson, S., Masakure, O. and Cranfield, J., 2009. Do fresh produce exporters in sub-Saharan Africa benefit from GlobalGAP certification? International Food Economy Research Group, InFERG Working Paper No. 2_FT, University of Guelph, Guelph, Canada, p. 23.

Table 8.5 continued.

Adaptation to smallholder capacities (precondition for up-scaling):	Support services competent and affordable for farmer groups:	Framework conditions enabling smallholder compliance:
Adaptation of the standard: • Simplified traceability • Simplified record-keeping • Risk-based CCPCs Certification costs: • Solutions for CB costs • Revision of standard fees	Embedded services for: • Training/advice/supervision • QMS management • Financing 3rd party public/private services: • Extension/training • Laboratories/CBs • Financing	Enabling policies and legislation: • Conducive sector policies • Adequate food/company laws Reliable/unbiased enforcement: • inspections, contracts, etc. Appropriate infrastructure, e.g.: • roads, logistic centres • public utilities (water/ power)
Issue: How to assure food safety while adapting to smallholders	Issue: How to build viable and accountable service systems	Issue: Which strategic partners to advocate for necessary changes

[1] This list of promising product features for GlobalGAP certification does not imply that other products cannot or should not be certified. However, the first three product features represent typical competitive advantages of smallholder over large-scale production systems. The fourth product feature may not be relevant in cases where supply chain operators intend to diversify products or markets respectively and prospects are promising for creating demand.

> 'While standards compliance (or non-compliance) can bring about changes that harm the livelihoods of the poor, advantages may accrue to those able to participate in evolving supply chains. This can certainly apply to smallholders, especially those operating in suitable locations with adequate infrastructure and in the context of effective producer organisations and long-term relationships with buyers. Smallholder farmers can frequently adopt the necessary technical measures and investments to comply with emerging standards. A key challenge is thus to reduce, through collective action, the transaction costs associated with monitoring and certifying compliance. Public policy and investment can make a difference in the pattern of 'winners' and 'losers''.[323]

[323] Jaffee, S., 2005. Food safety and agricultural health standards: challenges and opportunities for developing country exports. The Worldbank, Washington, DC, USA. January 10, p.112.

Concluding, group certification is not an easy process and the pilot project has shown that smallholders cannot achieve certification on their own. The main issue is not whether small-scale farmers are able to achieve certification and integrate into up-market supply chains; the question is rather how the weak service sector and business environment can be compensated for so as to enable smallholders to achieve certification without compromising supply chain competitiveness.

Experience suggests that (a) standard regulations have to be adapted to the actual risks of smallholder production; (b) technical and managerial capacities need to be built; and (c) initial investments as well as recurrent costs of compliance need to be made affordable for smallholders or, at least partly, be borne by business partners reaping fruit from smallholders' low-cost, labour-intensive production or by other stakeholders committed to integrating smallholders into food supply chains.

Selected initiatives supporting better smallholder-orientation of GlobalGAP Option 2 ...

- GlobalGAP Africa Observer/Smallholder Ambassador project. A GlobalGAP initiative supported by DfID and GTZ for strengthening the representation of smallholder interests in standard setting processes.[324]
- International Workshop 'GlobalGAP Group Certification. A challenge for smallholders in Europe and developing countries'.[325]
- GlobalGAP Tour 2009. Initiatives for assuring stakeholder participation in standard revision.[326]
- GlobalGAP Smallholder Support Kit. A GlobalGAP initiative for making the best practices developed in capacity building for QMS development and management as well as good agricultural practices available to the public.
- Strengthening the national quality infrastructure through the establishment of local Certification Bodies. A GTZ (Ghana) and AfriCert (Kenya) initiative for opening a branch office of AfriCert in Ghana with a view to reducing certification costs.
- Up-scaling KenyaGAP in the local market. An initiative of the Fresh Produce Exporters Association of Kenya (FPEAK) in cooperation with supermarkets for upgrading local market supplies to the ('domesticated') GlobalGAP-benchmarked KenyaGAP standard.[327]

... and the key role of standard owners

A regular update of the generic manual by the standard owner to follow both, the revision of the GlobalGAP standard and developments in production systems

[324] See *http://www.africa-observer.info/index.html.*
[325] See *http://africa-observer.info/docs_sub1.html.*
[326] See *http://www.tour2009.org.*
[327] See *http://www.fpeak.org/code.html.*

and local framework conditions will be of essential help for any smallholder farmer group interested to establish a Quality Management System for option 2 certification.

8.4 GlobalGAP: challenge and opportunity! Conclusions and recommendations

While the pilot cases clearly suggest that small-scale farmers are capable of achieving GlobalGAP certification under option 2, it became obvious that the viability and sustainability of group certification depends on several key conditions: together with technical capacities, due attention needs to be paid to developing group structures, management capacities and leadership skills in order to achieve and sustain group cohesion, an indispensable precondition for maintaining group certification; simplified handouts have to be developed to adapt the generic QMS to the capacities of smallholder farmers; depending on the previous technical knowledge, management skills and group performance, sufficient time needs to be provided to enable farmer groups to adopt relevant technologies as well as attitudes necessary for achieving compliance. A major issue in the promotion of GlobalGAP certification is that the benefits are difficult to predict while costs incur immediately. However, aspiring to promote certification, it is necessary to inform farmers on probable benefits and costs of group certification to assist them take informed business and investment decisions. This especially holds true for resource-poor and risk-averse smallholders.

8.4.1 Conclusions

The following conclusions can be drawn from the lessons learnt in the four pilot projects:

In well-managed schemes, opportunities outweigh challenges of GlobalGAP certification. The pilot projects in Kenya, Ghana, Thailand and Macedonia confirm that costs of compliance with GlobalGAP are usually considerable. But at the same time, certification offers a wide range of economic and social benefits to smallholder farmer groups and their trade partners seeking to maintain and expand market shares in an ever-more competitive global market. In addition to the broad range of economic impacts attained, which directly or indirectly translate into financial benefits, farmer groups also realised considerable non-financial benefits. As reported in other studies, benefits of GlobalGAP 'include the production of quality produce, improved field hygiene, better knowledge of pesticide use, and wider farm management benefits. In truth, many of these so-called non-financial benefits are quantifiable; access to trade credit or higher quality inputs will improve farm efficiency and yields'.[328]

Furthermore, 'many farmers said that they were using [GlobalGAP] records to understand their financial viability and run their farms more commercially. Proper handling of pesticides and improved food safety and hygiene had health benefits on the farms, and in addition most farmers said that they had transferred hygiene messages to the homestead with positive implications for family health. ... Further benefits have been gained through supply chain relationships that might accompany [GlobalGAP] certification. For example, contracts enable some [smallholder groups] to access trade credit through designated input sellers for seeds, fertiliser or chemicals'.[329]

There is an obvious need to change from a smallholder-centred to a supply chain-oriented perspective of costs and benefits. To assess the overall benefits of GlobalGAP certification, costs and benefits need to be measured along the entire supply (value) chain since costs and returns on investments into compliance accrue at all nodes of the supply chain. Just like their suppliers, PMOs, exporters and retailers have to invest into managing quality with major costs incurred when procuring from fragmented supplier networks, namely the costs for training, advice and supervision of smallholder groups.

It is obvious that, where firms procure from smallholders, a fair share of returns on smallholder compliance should trickle down to the farmers as incentive to renew the certification and maintain long-term and reliable contract relations. In many cases, this fair share of returns on investments is already realised at the farm level through the increase in yields and reduction of input costs thanks to the application of GAP; in other cases, where the returns on smallholders' investments into certification are mainly or only realised at the downstream end of the supply chain, PMOs, exporters and in some instances retailers may consider paying price premiums for certified produce (e.g. in the case of asparagus in Thailand). In cases where demand outstrips supplies of certified produce, buyers may also pay premiums (e.g. pepper in Macedonia). However, reality shows that the potential for premium prices paid to smallholder farmers usually pales in comparison to the benefits achieved through the income increased thanks to increased yields, reduced input costs and reliable market access.

The complexity of the mission that aims at integrating smallholders into (global) supply chains requires a system's approach coordinating diverse stakeholders. In summary,

[328] IIED [International Institute for Environment and Development] and NRI [Natural Resources Institute], undated. Costs and benefits of EurepGAP compliance for African smallholders: A synthesis of surveys in three countries. Fresh Insights 13, p. 11.
[329] *Ibid.*

the GlobalGAP Smallholder Pilot Project achieved its intended results, even if it took longer than initially planned. This was mainly due to the ignorance of the comprehensive needs for capacity building and change of attitudes among farmers (technical and entrepreneurial), groups (general group and specific QMS management, leadership, cohesion and joint commercial activities) and their buyers (contractual relations and embedded services). Furthermore, the need to orient certification efforts to market opportunities; the importance of reducing transaction costs through the promotion of consistent and reliable supplier-buyer linkages; and the complexity of the supply chain systems, into which the smallholder groups have to integrate were initially largely underestimated. Last but not least, the efficiency and effectiveness of support structures (private and public service systems) and conducive framework conditions (policies, laws and regulations as well as economic and social infrastructure) are likewise determinant for success or failure.

As the lessons learnt suggest, smallholder group certification is subject to multifaceted challenges. Accordingly, quite complex strategies have to be developed depending on the prevailing situation with regard to the access requirements in target markets, the governance structures in the supply chain, the features of the smallholder groups, the performance of the service sector as well as the in-country and export market framework conditions.

There is a need to define and coordinate the different roles of the private sector, the public sector and development partners. Obviously, the private sector, the public sector and development partners play critical roles in promoting a better market-integration of smallholders without compromising supply chain competitiveness. Roles and responsibilities have to be defined to foster synergies and avoid inefficient isolated approaches: 'The private sector drives the organisation of value chains that bring the market to smallholders and commercial farms. The state – through enhanced capacity and new forms of governance – corrects market failures, regulates competition, and engages strategically in public-private partnerships to promote competitiveness in the agribusiness sector and support the greater inclusion of smallholders and rural workers'.[330]

Private sector leadership will be essential: with farmers committing resources to the certification process and owning the QMS and with PMOs, exporters and processors respectively playing a key role in supporting their supplier networks. Despite the important responsibilities the public sector and development partners assume in fostering smallholder integration, both have to abstain from distorting existing private initiatives and commitment since they cannot replace the actual business partners of smallholders in the supply chain. Public-private partnerships

[330] The World Bank, 2007. World development report 2008 – Agriculture for development. The World Bank, Washington, DC, USA, p. 8.

(PPP), though, can serve as an appropriate tool for promoting smallholder inclusion into global supply chains by adding public (governments, donors) value to private initiatives.

8.4.2 Highlights on some key recommendations

In Chapter 2 'Quasi-states? The Unexpected Rise of Private Food Law', Lawrence Busch argues that 'the neoliberal project of limiting the role of the state has led to the rise [of] a wide range of 'quasi-states' consisting of individual firms, industry groups, and private voluntary organisations, each pursuing their own aims and interests through the production of private codes, laws, rules, and regulations. Whether some or all of these quasi-states are able to achieve legitimacy and develop democratic modes of governance remains to be seen.'

Busch further argues that 'The state-like character of TSRs [Tripartite Standards Regimes], as well as their democratic deficit, poses a number of ... problems ... These include questions about accountability, effectiveness, transparency, innovation, fairness, and legitimacy.' The following recommendations for up-scaling option 2 certification are guided by the line of thinking of Bush and the criteria proposed in his chapter.[331] Given the importance for success or failure of investments into any efforts for achieving compliance or certification respectively, viability has been added to the list of criteria proposed by Bush.

Viability can be achieved through e.g.:
- the selection of products that promise to be viable for smallholder farming and of producer groups capable of upgrading (farm/group management, production technologies);
- the facilitation of bridging the finance gap between short-term costs of compliance/certification and medium to long-term return on investments;
- the support by governments, donors or non-governmental organisations (e.g. subsidies for compliance and/or certification, bank guarantees), however only as a solution of last resort.

Effectiveness can be achieved through e.g.:
- the assurance of smallholder compliance (not necessarily certification) through the realisation of cost-benefit-'plus', working as incentives;
- the recognition of increased productivity, reduced production and transaction costs, market access, etc. as benefits/incentives (don't count (only) on price premiums!);

[331] For more detailed recommendations see Will, M., 2010. Integrating smallholders into global supply chains: GlobalGAP option 2 smallholder group certification generic manual: lessons learnt in pilot projects in Kenya, Ghana, Thailand and Macedonia. GTZ Sector Project Agricultural Trade, Eschborn, Germany, pp. 32-33.

- the integration of embedded services into the supplier-customer business relation with traders/processors providing advisory, training, input, credit services to their suppliers.

Innovations (increased adoption rates of innovations) can be achieved through e.g.:
- the ex-ante assessment of realistic benefits (essential for achieving commitment);
- the development of capacities (individual farmers' and group management capacities);
- the upgrading of local infrastructure (e.g. access roads, telecommunication, water, power).

Fairness can be achieved through e.g.:
- the development of capacities for fair distribution of costs and benefits along the chain;
- the development of negotiation skills of smallholders (e.g. prices, outgrower contracts);
- the promotion of corporate social responsibility (CSR) among downstream customers;
- the granting of commensurate producer prices.

Transparency can be achieved through e.g.:
- the consideration of smallholder capacities and requirements in standard development and standard revision (GlobalGAP assures smallholder participation through e.g. the Africa Observer/Smallholder Ambassador project and worldwide stakeholder consultations).

Legitimacy can be achieved through e.g.:
- the recognition of country-specific standards to facilitate spill-over to local/regional markets in parallel to full benchmarking (e.g. KenyaGAP domestic scope).

Accountability can be achieved through e.g.:
- the development of co-regulation systems combining industry self-regulation (responsibility of food operators) with government risk-based control systems (e.g. as initiated in Kenya through the joint private-public effort to up-scale KenyaGAP).

8.4.3 Final conclusions

With a view on the topic of this publication 'Private food law – non-regulatory dimensions of food law', the following citation may give guidance on the role of private voluntary standards (PVS) such as GlobalGAP: 'It is easy to get tangled in the standards and eco-labels as a fixed end in themselves. It would be wiser to realise that they can better be understood as a starting point for improved efficiencies, better quality, and an increased awareness of social and environmental issues.'[332]

It should be added that PVS are a matter of fact and are needed, since public law is not complete as many fields are not regulated given the complicated process of harmonising the interests and regulations of 27 Member States in the European Union.[333] Furthermore, public inspection cannot control the entire industry and every final product, neither in industrialised nor in developing countries. The EU General Food Law therefore provides for the following principles that oblige the private sector to take responsibility for food safety: (a) the farm to table approach; (b) the primary responsibility of food operators; (c) the traceability concept; and (d) coherent concepts for food safety including imported food (due diligence).

In this setting, PVS such as GlobalGAP already start serving as a vehicle for upgrading currently largely inefficient supply chains in developing countries. While the main incentives for the private sector are to achieve competitiveness in local, regional and international markets, the public sector strives for smallholder integration, consumer protection and assurance of public health.

References

BMZ [Federal Ministry for Economic Cooperation and Development], 2008. Introduction of voluntary social and ecological standards in developing countries, summary version of the evaluation. Evaluation Report 043, BMZ, Bonn, Germany. Available at: http://www.bmz.de/en/publications/type_of_publication/evaluation/evaluation_reports_since_2006/BMZ_Eval-043e_web.pdf.

Danielou, M. and Ravry, C., 2005. The rise of Ghana's pineapple industry: from successful takeoff to sustainable expansion. ESSD Africa Working Paper Series No. 93. Available at: http://www.worldbank.org/afr/wps/wp93.pdf.

Enomoto, R., 2009. GlobalGAP Smallholder Support Kit. GlobalGAP Tour 2009 Good Agricultural Practice, Nairobi, Kenya, 16 September 2009. Available at: http://www.globalgap.org/cms/upload/Documents/temp_upload/TOUR2009/Presentations/Enomoto_GlobalGAP_Presentation_Nairobi.pdf.

Graffham, A. and Cooper, J., 2009. Making GlobalGAP smallholder-friendly: can GlobalGAP be made simpler and less costly without compromising integrity? In: Borot de Batisti, A., MacGregor, J. and Graffham, A. (eds.) Standard bearers: horticultural exports and private standards in Africa. International Institute for Environment and Development (IIED), Winnipeg, Canada, pp. 83-88. Available at: http://www.iied.org/pubs/pdfs/16021IIED.pdf.

[332] Van Dingenen, K., Koyen, M-L., Koekoek, F-J., Pierrot, J. and Giovannucci, D., 2010. European and Belgian market for certified coffee. Idea Consult on behalf of the Belgian Development Agency/BTC Trade For Development, 62 pp.

[333] Freely quoted after: D. Gorny, Personal communication (18th International EFLA [European Food Law Association] Congress: Private food law – Non-regulatory dimensions of food law, Amsterdam/The Netherlands, 16-17 September 2010).

GTZ, 2010. GlobalGAP smallholder QMS set-up guide – How to establish QMS in your group. Deutsche Gesellschaft für Technische Zusammenarbeit, Eschborn, Germany. Based on: GlobalGAP IFA Ver. 3.02 QMS checklist_Mar 08, English Version, 2010. Available at: http://www.gtz.de/en/dokumente/gtz2010-en-globalgap-smallholder-qms-set-up-guide.pdf.

GTZ, GlobalGAP and Marktkontor Obst und Gemüse Baden e.V., 2008. GlobalGAP group certification: a challenge for smallholders in Europe and developing countries. International Workshop, 28- 29 April 2008, Frankfurt am Main, Germany. Available at: http://www.africa-observer.info/documents/Report_GlobalGAP_workshop.pdf.

Günther, D. and Neuendorff, J., 2006. EurepGAP smallholder manual: understanding EurepGAP. Powerpoint presentation held at the occasion of the Kick-off workshops.

Henson, S., Masakure, O. and Cranfield, J., 2009. Do fresh produce exporters in sub-Saharan Africa benefit from GlobalGAP certification? International Food Economy Research Group, InFERG Working Paper No. 2_FT, University of Guelph, Guelph, Canada. Available at: http://www.inferg.ca/workingpapers/WP2_FT.pdf.

IIED [International Institute for Environment and Development] and NRI [Natural Resources Institute], undated. Costs and benefits of EurepGAP compliance for African smallholders: a synthesis of surveys in three countries. Fresh Insights 13. Available at: http://www.agrifoodstandards.net/en/filemanager/active?fid=128.

Jaffee, S., 2005. Food safety and agricultural health standards: challenges and opportunities for developing country exports. The Worldbank, Washington, DC, USA. January 10. Available at: http://siteresources.worldbank.org/INTRANETTRADE/Resources/Topics/Standards/standards_challenges_synthesisreport.pdf.

Legge, A., Orchard, J., Graffham, A., Greenhalgh, P., Kleih, U. and MacGregor, J., 2009. Mapping different supply chains of fresh produce exports from Africa to the UK. In: Borot de Batisti, A., MacGregor, J. and Graffham, A. (eds.) Standard bearers: horticultural exports and private standards in Africa. International Institute for Environment and Development (IIED), Winnipeg, Canada, pp. 40-44. Available at: http://www.iied.org/pubs/pdfs/16021IIED.pdf.

Mbithi, S., 2009. GlobalGAP – report on smallholder taskforce consultations – smallholder ambassador. Nairobi, Kenya, 21st June 2009. GlobalGAP Africa Observer, available at: http://www.africa-observer.info/documents/Summary_of-Smallholder-recommendations-june_09.pdf.

Opitz, M., Potts, J. and Wunderlich, C., 2007. Closing the gaps in GAPS: a preliminary appraisal of the measures and costs associated with adopting commonly recognised 'good agricultural practices' in three coffee growing regions, appendices. International Institute for Sustainable Development (iisd) and Sustainable Coffee Partnership, Winnipeg, Canada. Available at: http://www.iisd.org/pdf/2008/closing_gap_appendices.pdf.

Scott, W., 2006. Linking producers to increased horticultural exports from Macedonia. Presentation at the Regional Consultation on Linking Producers to Markets, Cairo, Egypt, 29 January to 2 February 2006. Available at: http://www.globalfoodchainpartnerships.org/cairo/presentations/WilliamScott.pdf.

Shah, H. (2005): The global partnership for safe and sustainable agriculture. In: EurepGAP Global Report 2005, p. 8.

The World Bank, 2007. World development report 2008 – Agriculture for development. The World Bank, Washington, DC, USA. Available at: http://siteresources.worldbank.org/INTWDR2008/Resources/2795087-1192111580172/WDROver2008-ENG.pdf.

Van Dingenen, K., Koyen, M-L., Koekoek, F-J., Pierrot, J. and Giovannucci, D., 2010. European and Belgian market for certified coffee. Idea Consult on behalf of the Belgian Development Agency/BTC Trade For Development, 62 pp. Available at: http://www.befair.be/site/download.cfm?SAVE=21535&LG=1.

Wainwright, H. and Labuschagne, L., 2009. Private standards: a personal perspective from a training service provider. In: Borot de Batisti, A., MacGregor, J. and Graffham, A. (eds.) Standard bearers: horticultural exports and private standards in Africa. International Institute for Environment and Development (IIED), Winnipeg, Canada, pp. 154-157. Available at: http://www.iied.org/pubs/pdfs/16021IIED.pdf.

Will, M., 2010. Integrating smallholders into global supply chains: GlobalGAP option 2 smallholder group certification generic manual: lessons learnt in pilot projects in Kenya, Ghana, Thailand and Macedonia. GTZ Sector Project Agricultural Trade, Eschborn, Germany. Available at: http://www.gtz.de/en/dokumente/gtz2010-en-globalgap-group-certification.pdf.

9. Towards the self-regulation code on beer advertising in Italy

Steps on the long lasting path of competition/ co-operation of public and private food law

Ferdinando Albisinni

9.1 The peculiar relation between innovation and food law

Innovation and food law share a peculiar relation, characterised by reciprocal strong influence and interference.[334] Technological innovation, mainly from 19th century and starting with the revolutionary ideas of the famous French chef Appert – as it is well known – has radically changed the techniques applied to food, in all phases from production, to processing, to distribution. As a consequence, Food law is often under strong pressure to find adequate regulatory answers to the challenges of technological innovations. In Europe, the Novel Foods legislation,[335] the introduction of rules on the traceability of beef after the explosion of the 'mad cow' crisis,[336] and the general adoption of the precautionary principle,[337] may be cited among the many significant examples of an approach of the food legislator, aimed to answer to needs and demands coming from innovation and not finding sufficient regulatory tools in the already existing legislation. But the impact of technological innovation is not limited to the substantive aspects of food, as it deals largely – in an always more pervasive way – to immaterial aspects, first of all those related to the communication on the market, thus forcing the food legislator to deal both with substantive and immaterial issues in regulating food production and marketing.

On the other side, juridical innovation is not only an answer to techniques; it is also by itself an expression of the elaboration of new and original models and institutions, through an experimental approach, which may develop in what

[334] See Jongen, W.M.F. and Meulenberg, M.T.G. (eds.), 2005. Innovation in agri-food systems. Wageningen Academic Publishers, Wageningen, the Netherlands.
On the peculiar relation between innovation and food law see Albisinni, F., 2009. Strumentario di diritto alimentare Europeo. Utet, Torino, Italy.

[335] Regulation (EC) No 258/97 of the European Parliament and of the Council of 27 January 1997 concerning novel foods and novel food ingredients. Official Journal of the European Union L 43:. 1-6.

[336] Council Regulation (EC) No 820/97 of 21 April 1997 establishing a system for the identification and registration of bovine animals and regarding the labelling of beef and beef products. Official Journal of the European Union L 117: 1-8.

[337] See Article 7 of Regulation (EC) No 178/2002 of the European Parliament and of the Council of 28 January 2002 laying down the general principles and requirements of food law, establishing the European Food Safety Authority and laying down procedures in matters of food safety. Official Journal of the European Union L 31: 1-24.

some German scholars effectively defined as *Rechtsreform in Permanenz.*[338] In this complex and articulated process, which enriches the armoury of legal tools with new additions, and in the same time enhances consolidation and simplification of existing rules,[339] the traditional borders between public and private law are assuming new contents. Protected interests typically classified as 'public goods' - like consumer's protection and fair competition - increasingly rely on new tools which utilise private models, like contracts, voluntary acceptance of rules and standards, and civil compensation as enforcement tool instead of criminal or administrative sanctions. In the field of control and surveillance, certification bodies of private nature are chosen to perform duties which are traditionally considered 'public'.[340]

9.2 Private regulatory law

Along this path, Private Regulatory Law is anticipating and/or integrating Public and State Law. In the specific field of communication on the market, the traditional answers of European legal systems relied largely on public law. In Italy, criminal sanctions, already established in the *Penal Code* of 1930,[341] punish fraud in trademarks and commercial signs,[342] fraud in trade,[343] sale of non-genuine food,[344] and more recent legislation introduced specific provisions of criminal sanctions in case of illicit use of international and European PDO and PGI.[345] Other sanctions, provided by the general law on safety rules for food products and materials,[346] punish with a fine, originally of criminal nature and lately weakened to administrative nature, the use of misleading labelling and advertising.[347] In

[338] See, with reference to company law, Noack, U. and Zetzsche, D.A., 2005. Corporate governance reform in Germany: the second decade. Center for Business & Corporate Law Research Paper Series No. 0010, Düsseldorf, Germany.

[339] See, e.g. Council Regulation (EC) No 1234/2007 of 22 October 2007 establishing a common organisation of agricultural markets and on specific provisions for certain agricultural products (Single CMO Regulation). Official Journal of the European Union L 299: 1-149.

[340] See, e.g. Article 5 of Regulation (EC) No 882/2004 of the European Parliament and of the Council of 29 April 2004 on official controls performed to ensure the verification of compliance with feed and food law, animal health and animal welfare rules. Official Journal of the European Union L 165: 1-141.
Article 11 of Council Regulation (EC) No 510/2006 of 20 March 2006 on the protection of geographical indications and designations of origin for agricultural products and foodstuffs. Official Journal of the European Union L 93: 12-25.
Article 118(7) of Council Regulation (EC) No 1234/2007 of 22 October 2007 establishing a common organisation of agricultural markets and on specific provisions for certain agricultural products (Single CMO Regulation). Official Journal of the European Union L 299: 1-149.

[341] Royal.Decree No. 1389, 19 October 1930. Italian legislation, in original and updated text, is published at http://www.normattiva.it.

[342] *Ibid.* at Article 514.

[343] *Ibid.* at Article 515.

[344] *Ibid.* at Article 516.

[345] Article 517-*quater*, added to the *Penal Code* by L. No. 99, 23 July 2009.

[346] L. No. 283, 30 April 1962.

[347] *Ibid.* at Article 8.

both cases, nevertheless, there was lack of specific attention for protecting 'goods' different from *bona fides* in marketing, labelling and business communication.

Even when dealing with immaterial aspects of food products, law (public law) was looking to the substantive aspects of the products *per se* considered, and to the fairness of communication, advertising and labelling by comparison to what is promised to the consumer, neglecting other relevant aspects pertaining to general expectations of consumer and to social behaviour. This gap has been filled by private regulatory rules, looking to new needs and to new disciplinary areas, and implementing new tools of regulation and enforcement.

9.3 The IAP – Institute of self-regulation in Marketing Communication (1963)

The first significant experience in Italy of a self-regulatory system on marketing communication was set up in 1963, on the basis of the following principles:

> '(...) having established that the function of advertising is equally in the interest of enterprises and consumers, having acknowledged the requirement that the principles of morality and loyalty underlying the profession are enhanced, (...) as a further concrete impediment to marginal degenerative forms of advertising, *commits advertisers, artists, producers and the media, through their respective associations to work out and collect under a single moral Code of Advertising the rules that are to govern all advertising activities in defence of entrepreneurial activities and of the fundamental interests of consumers;* moreover it commits such associations to appropriate themselves of such Code and ensure its application by their members.'[348]

The IAP – Istituto dell'Autodisciplina Pubblicitaria, as regulated in its own Statute, is a non-profit organisation. It aims to collect in one single voluntary organisation, together with all subjects operating along the food chain, also artists, producers and the media, adopting an inclusive approach, which does not limit competences for a fair market communication to producers and sellers of the products, but assigns responsibilities in the process of establishing, maintaining, and respecting adequate standards, to all subjects in any way engaged in communication. The shift is from the liability approach of public law, aimed to prohibit and sanction illicit acts, and therefore addressed mainly to producers and resellers of substantive products, to a responsibility approach characteristic of private regulatory law, aimed to assure voluntary (and, whenever possible, preventive) compliance and therefore addressed to all subjects who in any way play a role (i.e. assume responsibility) in the communication process.

[348] Agreement approved at the Conference of Ischia on October 3 to 6, 1963.

Amongst the main tasks of IAP are the formulation and updating of the rules of the Code of Self-Regulation, and the establishing of a Jury and a Review Board, having the task to decide 'according to the provisions of the Code'[349] on any form of advertising, both prior to and after the publication. The Code of Self-Regulation was first published on 12 May 1966 and has been constantly updated. At present time it is in force the 50th edition, effective from 18 January 2010.[350] Decisions are published, posted on the website and filed in the database of the IAP.[351] The media through which marketing communication is disseminated, *'which directly or through their trade associations accept the self-regulatory code, even if not involved in the proceedings before the Jury'*, are obliged to observe its decisions;[352] therefore going well beyond the borders and limits usually assigned to public law decisions. Through this way, the Code marked a significant innovation regarding subjects involved, roles assigned, effects of decisions.

Even with reference to the content of regulation, the Code marked a step in the path, broadening the discipline of fair market communication, from rules limited to specific illicit behaviours (as happened under the criminal provisions of the *Penal Code*) to a more general consideration of misleading marketing communication, including overall fairness in trade and taking into account behaviour, values and habits of the consumers.

Inter alia, provisions of the Code:

- Hold a broad and open definition of what is covered by the Code.

  'Preliminary and General Rules – (e) Definitions.
  The term 'message' refers to any form of public presentation of the product and therefore includes the outer packaging, wrapping, labelling, etc.'

  'Article 2 – Misleading marketing communication
  Marketing communication must avoid statements or representations that could mislead consumers, including omissions, ambiguity or exaggeration that are not obviously hyperbolical, particularly regarding the characteristics and effects of the product, prices, free offers, conditions of sale, distribution, the identity of persons depicted, prizes or awards.'
- Expressly mention ideal values, which obtain further protection, added to the protection of economic interests already recognised by public law.

  'Article 10 – Moral, Civil, and Religious Beliefs and Human Dignity
  Marketing communication should not offend moral, civil and religious beliefs. Marketing communication should respect human dignity in every form and expression and should avoid any form of discrimination.'

[349] Article 32 of the Code of Self-Regulation.
[350] Published at: http://www.iap.it.
[351] *Ibid.* at Article 40.
[352] *Ibid.* at Article 41.

- Operate distinct evaluation of messages by reason of their destination to adult consumers or to children and young people.

  'Article 11 – Children and young people

  Special care should be taken in messages directed to children and young people or to which they may be exposed. Such messages should avoid material that could cause psychological, moral or physical harm, and should not exploit the credulity, inexperience or sense of loyalty of children or young people.

  In particular, such marketing communication must not suggest:
  - violating generally accepted rules of social behaviour;
  - acting dangerously or seeking exposure to dangerous situations;
  - that failure to posses the promoted product means either their own inferiority or their parents' failure to fulfil their duties;
  - that the role of parents and educators is inadequate in supplying healthy nutritional advice;
  - adopting poor eating habits or neglecting the need for a healthy lifestyle;
  - soliciting other people to purchase the promoted product.

  The portrayal of children and young people in marketing communication must avoid playing on the natural sentiments of adults towards the young.'
- Include specific rules for specific categories of products, among which spirits, cosmetics, food supplements and alia.

  'Article 22 – Alcoholic Beverages

  Marketing communication concerning alcoholic beverages should not be in contrast with the obligation to depict styles of drinking behaviour that project moderation, wholesomeness and responsibility. This principle aims to safeguard the primary interest of the population in general, and of children and young people in particular, in a family, social and working environment safeguarded from the negative consequences of alcohol abuse.'.

We may therefore reasonably assume that, in the Italian legal experience, the voluntary creation of IAP, and its operation using tools proper of private regulatory law, largely anticipated lines and themes of regulations, which have found place in public law only much later.

9.4 Legislative reforms in the 1990's: cooperative competition between public and private law

The last decade of 20th century has seen significant legislative reforms in Italy in the areas of market regulation and of commercial communication, mainly under the influence (directly or indirectly) of the European Community. The institution of the Anti-trust Authority in 1990[353] gave express and formal recognition to the

[353] L. No. 287, of 10 October 1990, establishing the AGCM – Autorità Garante della Concorrenza e del Mercato.

European principles against agreements in any form having the effect of restriction or distortion of competition, against abuses of dominant position, and in favour of effective free competition between undertakings. Two years later, in 1992, the implementation of European directives led to the introduction of new rules on labelling, presentation and advertising of food products,[354] and on deceptive advertising.[355]

Those reforms crossed and interacted each other, and their application offered an original framework to rules on commercial communication, especially with reference to food products. The D. Leg.vo No. 74/1992 on deceptive advertising had to identify competences and procedure for the effective implementation of the new regulation. The Anti-trust Authority created in 1990, with its recognised skill and qualification, appeared to the Italian legislator of 1992 to be the natural choice for an independent body called to investigate and decide on the compliance with the newly introduced rules.[356] The underlining (and largely shared) idea was that advertising is a decisive part of market competition and that assuring non deceptive advertising is essential both to protect consumers, and to guarantee full and fair competition among undertakings.

On the basis of the new competence received, the Anti-trust Authority largely expanded the field of its review, adopting a very broad definition of advertising and assuming that within this expression must be included any form of commercial communication, with any means, therefore including labels.[357] So arguing, the Italian Anti-trust Authority expressed by way of interpretation as general rule, even in the field of public law, the principle, originally introduced by IAP in the field of private regulatory law, that the discipline of commercial communication in the market may not be divided in separate chapters, subject to different rules, on the basis of means and media utilised, but must be considered in an unified and uniform perspective.[358] Moreover, the Anti-trust Authority affirmed the principle (lately accepted also by the courts of justice) that the existence of trade marks duly registered under the laws applicable at the time of registration does not allow the trade mark owner to reproduce this mark on the labels, when this use may induce the consumer to a deceptive understanding of the characteristics of the product. This principle has been recognised primarily in the area of Food Law. With specific reference to the true origin of a food product, the Anti-trust Authority in a long and coherent series of decisions declared that the true origin

[354] D. Leg.vo No. 109, of 27 January 1992, implementing Directives 79/112/EEC, 89/395/CEE, 89/396/EEC.

[355] D. Leg.vo No. 74, of 25 January 1992, implementing Directives 84/450/EEC and 97/55/EC.

[356] Article 7 of D. Leg.vo No. 74/1992.

[357] See dec. No. 1078, 21 April 1993, *Oleificio Viola*, at http://www.agcm.it.

[358] On competence and powers of the AGCM in cases of consumers' protection, with specific reference to food law, see Astazi, A., 2008. Pratiche commerciali scorrette nell'ambito dei contratti del consumo alimentare e tutela dei consumatori. Revista di Diritto Alimentare 2(2): 1-9.

of olive oil is that of the place where olives have been picked and that a trade mark containing a reference to a geographic place cannot be used on the label of a bottle of olive oil when the olives come from a geographic place different from that mentioned in the trade-mark.[359]

With these conclusions, the Anti-trust Authority, operating as a powerful independent Authority with effective decision-making powers, anticipated criteria which will find official legislative recognition in Europe only later, through Regulation (EC) No. 1019/2002 of 13 June 2002.[360] The reciprocal influence and interference within rules, criteria and arguments among private regulatory bodies, public bodies, and formal rules of law, expanded in those cases from a domestic dimension to a larger European sphere.

The above mentioned D. Leg.vo No. 74/1992, in addition to the public law competences attributed to the Anti-trust Authority to investigate and sanction cases of deceptive advertising, introduced a relevant and original example of co-operative competition mechanism between public and private law (and between public and private institutions and bodies). Article 8 stated that, in case of deceptive advertising, interested parties may turn to self-regulation voluntary bodies and in this case may agree to postpone any application to the Anti-trust Authority until the private body adopts its final decision. Under the same Article 8, even when the application is previously deposited before the Anti-trust Authority, any party may ask the Authority to suspend its procedure waiting for the decision of the private body. In this hypothesis the Authority may suspend the procedure for a term up to thirty days.

This rule has been afterwards confirmed with a larger application area by the *Consumers' Code.*[361] Article 27-*ter*[362] of the Code states that consumers and competitors, even through their associations or organisations, may agree in general terms with trade operators and producers that, before making any petition to the Anti-trust Authority or to the courts of justice, they will turn to self-regulation voluntary bodies to stop unfair commercial practices (including but not limited to deceptive advertising). Along the steps and with the peculiarities mentioned

[359] See among others, dec. No. 4970, 30 April 1997, *Bertolli-Lucca;* dec. No. 5562, 18 December 1997, *Olio Carapelli Firenze;* dec. No. 5563, 18 December 1997, *Olio Carli Oneglia;* dec. No. 5564, 18 December 1997, *Olio Monini Spoleto;* dec. No. 5713, 19 February 1998, *Olearia del Garda;* dec. No. 7619, 13 October 1999, *Cooperativa agricola Trevi*, published at http://www.agcm.it. For a comment of the decisions of the AGCM see: Albisinni, F., 2010. Etichettatura dei prodotti alimentari. In: Albisinni, F. (ed.) Diritto alimentare. Mercato, Sicurezza. Wolters Kluwer, Milano, Italy.

[360] Commission Regulation (EC) No 1019/2002 of 13 June 2002 on marketing standards for olive oil. Official Journal of the European Union L 155: 27-31. Amended by Commission Regulation (EC) No 182/2009 of 6 March 2009 amending Regulation (EC) No 1019/2002 on marketing standards for olive oil. Official Journal of the European Union L 63: 6-8.

[361] D. Leg.vo No. 206, 6 September 2005.

[362] Introduced by D. Leg.vo No. 146, 2 August 2007, article 1.

above, largely moving from experiences and cases in Food Law, public and private regulatory bodies have therefore been expressly recognised in the Italian *Consumers' Code* as an arena of co-operative competition, offering to undertakings and consumers a large armoury of possible tools and remedies, concurrent but harmonised in their content and in their operational procedures.

9.5 The Beer Advertising Code: private regulation as tool to expand and anticipate consumer protection

The new *Beer Advertising Code*, adopted in 2010 by the Italian Brewers Association (see Alessandro Artom's detailed discussion of the *Code* in Chapter 10 of this volume), locates itself within this long lasting path of competition/co-operation of public and private food law, getting inspiration from previous private and public experiences and in the meantime introducing some provisions having innovative character. In the *Code*[363] special attention is given to young people and to minors and more generally to the declared fundamental goal to avoid any sort of commercial communication, which could in any way induce consumers to drink in a 'non-responsible way'.

The inspiration of the *Code* is expressly derived from Directive No. 1989/552/EEC[364], of 3 October 1989, on television broadcasting activities, whose Article 15 states, with reference to alcoholic beverages:

> 'Article 15.
> Television advertising for alcoholic beverages shall comply with the following criteria:
> (a) it may not be aimed specifically at minors or, in particular, depict minors consuming these beverages;
> (b) it shall not link the consumption of alcohol to enhanced physical performance or to driving;
> (c) it shall not create the impression that the consumption of alcohol contributes towards social or sexual success;
> (d) it shall not claim that alcohol has therapeutic qualities or that it is a stimulant, a sedative or a means of resolving personal conflicts;
> (e) it shall not encourage immoderate consumption of alcohol or present abstinence or moderation in a negative light;
> (f) it shall not place emphasis on high alcoholic content as being a positive quality of the beverages.'

[363] Published at http://www.assobirra.it.
[364] Council Directive 89/552/EEC of 3 October 1989 on the coordination of certain provisions laid down by Law, Regulation or Administrative Action in Member States concerning the pursuit of television broadcasting activities. Official Journal of the European Union L 298: 23-30.

Those provisions has been later amended[365]:

- by Directive No. 97/36/EC, of 30 June 1997, which extended to '*teleshopping*' the application of the mentioned rules;
- and by Directive No. 2007/65/EC, of 11 December 2007, which introduced Article 3(*e*), stating at par. 1(*e*):

> *'audiovisual commercial communication for alcoholic beverages shall not be aimed specifically at minors and shall not encourage immoderate consumption of such beverages'.*

The entire Directive of 1989 has been finally repealed by Directive No. 2010/13/EU, of 10 March 2010, concerning the provision of audiovisual media services, which confirmed the above mentioned provisions.[366]

Comparison of the provisions of these Directives with the content of the *Beer Advertising Code* clearly shows the innovative approach of the *Code*. In the directive of 1989 the scope of regulation was initially limited to television advertising, only later extended to teleshopping and recently to general audiovisual commercial communication, while we have seen that already back in 1963 the IAP (whose model inspired the *Beer Advertising Code*) adopted an inclusive and comprehensive approach considering any form of commercial communication as subject to the Code of Self-regulation. And even with the last amendments, the public European legislation above mentioned, on advertising of alcoholic beverages, is applicable only to 'audiovisual commercial communication', with a scope much narrower than that of the private *Beer Advertising Code*. Moreover the *Code*, on the basis of the experience of IAP, inter alia:

- pays special attention to the mode and form of the commercial communication, e.g. adopting specific rules regarding the use of comics or animated figures, considered as addressed to young people by their nature and therefore evaluated on the basis of strict criteria;
- adopts a very broad definition of 'dangerous activities' which in any case cannot be associated with the use of beer, including in the definition even simple domestic activities;

[365] Directive 97/36/EC of the European Parliament and of the Council of 30 June 1997 amending Council Directive 89/552/EEC on the coordination of certain provisions laid down by law, regulation or administrative action in Member States concerning the pursuit of television broadcasting activities. Official Journal of the European Union L 202: 60-70.
Directive 2007/65/EC of the European Parliament and of the Council of 11 December 2007 amending Council Directive 89/552/EEC on the coordination of certain provisions laid down by law, regulation or administrative action in Member States concerning the pursuit of television broadcasting activities. Official Journal of the European Union L 332: 27-45.

[366] See Article 9(e) and Article 22 of Directive 2010/13/EU of the European Parliament and of the Council of 10 March 2010 on the coordination of certain provisions laid down by law, regulation or administrative action in Member States concerning the provision of audiovisual media services (Audiovisual Media Services Directive). Official Journal of the European Union L 95: 1-24.

- excludes any association of beer with the use of any vehicle, on the assumption that all vehicles require in any case full attention from the driver.

Summing up, even on the basis of a brief examination of this new private *Code* and comparison with public rules applicable to the same products, the consideration which may be shared is that the private *Code* expands and anticipates consumer protection through recourse to self-regulation, i.e. to a regulation which is introduced and managed by the same subjects who are called to apply it. The duty to respect is the result of voluntary acceptance, as part of the collective participation to the Brewers Association. It is a case of private food law, which is distinguished by its nature of being collective food law, whose strength and effectiveness is mainly based in its being a shared regulation, not imposed by a stronger contracting party (as happens in some instances of private standards fixed by powerful market players), but created by the same parties that assume the obligation to respect it.

9.6 Some open questions

Even after placing the self-regulation *Code* within the comprehensive concept of private collective food law above mentioned, some relevant questions remain open. Which kind of formal (or informal) relation may be construed between private and public regulation in the final evaluation of commercial communication on food? Which effects on the liability/responsibility of food producers at public law, may have the compliance with private established *Code* rules? In particular, may a food producer evoke as protection against public law sanctions the circumstance that the discussed commercial communication has been verified by the jury of the *Code* prior to its use? Which is the status of the jury? Is it the same of private arbitrators, which cannot be held responsible for errors at law? Or must it be compared to certification bodies, responsible for not having checked the violation of applicable rules? In an area of regulation, like Food Law, where rules of responsibility may in some cases operate even in absence of any fault criteria,[367] any possible answer to those questions involves crucial issues, both for the producers involved and for the viability of the system proposed.

In the past century legal scholars analysed, along opposite perspectives, the competition between modern State monopoly of law sources[368], and social experiences vindicating a territory outside the State regulation[369]. In this century we have to deal with a multilevel and pluralistic situation, with many competing and

[367] See, e.g. Article 19 of Regulation (EC) No. 178/2002 on the duty to withdraw, and see Court of Justice, 23 November 2006, C-315/05, *Lidl Italia*, on the liability of distributors.
[368] See the fundamental contribution of Kelsen, H., 1945. General theory of law and state. Harvard University Press, Cambridge, MA, USA.
[369] It is sufficient here to remember: Romano, S., 1945 [1918]. L'ordinamento giuridico. 2nd ed. Firenze, Italy.

overlapping sources of regulation, with still no clear rules of effective supremacy and with a sort of pendulum among different sources.

References

AGCM, 1993. Decisions of 21 April 1993, No. 1078, *Oleificio Viola.*

AGCM, 1997a. Decisions of 30 April 1997, No. 4970, *Bertolli-Lucca.*

AGCM, 1997b. Decisions of 18 December 1997, No. 5562, *Olio Carapelli Firenze.*

AGCM, 1997c. Decisions of 18 December 1997, No. 5563, *Olio Carli Oneglia.*

AGCM, 1997d. Decisions of 18 December 1997, No. 5564, *Olio Monini Spoleto.*

AGCM, 1998. Decisions of 19 February 1998, No. 5713, *Olearia del Garda.*

AGCM, 1999. Decisions of 13 October 1999, No. 7619, *Cooperativa agricola Trevi.*

Albisinni, F., 2009. Strumentario di diritto alimentare Europeo. Utet, Torino, Italy.

Albisinni, F., 2010. Etichettatura dei prodotti alimentari. In: Albisinni, F. (ed.) Diritto alimentare. Mercato, Sicurezza. Wolters Kluwer, Milano, Italy.

Astazi, A., 2008. Pratiche commerciali scorrette nell'ambito dei contratti del consumo alimentare e tutela dei consumatori. Revista di Diritto Alimentare 2(2): 1-9.

Court of Justice, 2006. C-315/05, *Lidl Italia*, 23 November 2006.

EU, 1989. Council Directive 89/552/EEC of 3 October 1989 on the coordination of certain provisions laid down by Law, Regulation or Administrative Action in Member States concerning the pursuit of television broadcasting activities. Official Journal of the European Union L 298: 23-30.

EU, 1997. Council Regulation (EC) No 820/97 of 21 April 1997 establishing a system for the identification and registration of bovine animals and regarding the labelling of beef and beef products. Official Journal of the European Union L 117: 1-8.

EU, 1997. Directive 97/36/EC of the European Parliament and of the Council of 30 June 1997 amending Council Directive 89/552/EEC on the coordination of certain provisions laid down by law, regulation or administrative action in Member States concerning the pursuit of television broadcasting activities. Official Journal of the European Union L 202: 60-70.

EU, 1997. Regulation (EC) No 258/97 of the European Parliament and of the Council of 27 January 1997 concerning novel foods and novel food ingredients. Official Journal of the European Union L 43: 1-6.

EU, 2002. Commission Regulation (EC) No 1019/2002 of 13 June 2002 on marketing standards for olive oil. Official Journal of the European Union L 155: 27-31.

EU, 2007. Directive 2007/65/EC of the European Parliament and of the Council of 11 December 2007 amending Council Directive 89/552/EEC on the coordination of certain provisions laid down by law, regulation or administrative action in Member States concerning the pursuit of television broadcasting activities. Official Journal of the European Union L 332: 27-45.

EU, 2002. Regulation (EC) No 178/2002 of the European Parliament and of the Council of 28 January 2002 laying down the general principles and requirements of food law, establishing the European Food Safety Authority and laying down procedures in matters of food safety. Official Journal of the European Union L 31: 1-24.

EU, 2004. Regulation (EC) No 882/2004 of the European Parliament and of the Council of 29 April 2004 on official controls performed to ensure the verification of compliance with feed and food law, animal health and animal welfare rules. Official Journal of the European Union L 165: 1-141.

EU, 2006. Council Regulation (EC) No 510/2006 of 20 March 2006 on the protection of geographical indications and designations of origin for agricultural products and foodstuffs. Official Journal of the European Union L 93: 12-25.

EU, 2007. Council Regulation (EC) No 1234/2007 of 22 October 2007 establishing a common organisation of agricultural markets and on specific provisions for certain agricultural products (Single CMO Regulation). Official Journal of the European Union L 299: 1-149.

EU, 2009. Commission Regulation (EC) No 182/2009 of 6 March 2009 amending Regulation (EC) No 1019/2002 on marketing standards for olive oil. Official Journal of the European Union L 63: 6-8.

EU, 2010. Directive 2010/13/EU of the European Parliament and of the Council of 10 March 2010 on the coordination of certain provisions laid down by law, regulation or administrative action in Member States concerning the provision of audiovisual media services (Audiovisual Media Services Directive). Official Journal of the European Union L 95: 1-24.

Italian government, 1930. R.D. No. 1389, 19 October 1930, Approvazione del testo definitivo del Codice penale.

Italian government, 1962. L. No. 283, 30 April 1962, Disciplina igienica della porducione e della vendita delle sostanze alimentari e delle bevande.

Italian government, 1990. L. No. 287, 10 October 1990, Norme per la tutela della concorrenza e del mercato.

Italian government, 1992a. D. Leg.vo No. 109, 27 January 1992, Attuazione della direttiva 89/395/CEE e della direttiva 89/396/CEE concerni l'etichettatura, la presentazione e la pubblicità dei prodotti alimentari.

Italian government, 1992b. D. Leg.vo No. 74, 25 January 1992, Recante attuazione della direttiva 84/450/CEE in materia di pubblicità ingannevole.

Italian government, 2005. D. Leg.vo No. 206, 6 September 2005, Codice del consumo.

Italian government, 2007. D. Leg.vo No. 146, 2 August 2007, Attuazione della direttiva 2005/29/CE relativa alle pratiche commerciali sleali tra imprese e consumatori nel mercato interno e che modifica le direttive 84/450/CEE, 97/7/CE, 98/27/CE, 2002/65/CE, e il Regolamento (CE) n. 2006/2004.

Italian government, 2009. L. No. 99, 23 July 2009, Disposizioni per lo sviluppo e l'internazionalizzazione delle imprese, nonché in materia di energia.

Jongen, W.M.F. and Meulenberg, M.T.G. (eds.), 2005. Innovation in agri-food systems. Wageningen Academic Publishers, Wageningen, the Netherlands.

Kelsen, H., 1945. General theory of law and state. Harvard University Press, Cambridge, MA, USA.

Noack, U. and Zetzsche, D.A., 2005. Corporate governance reform in Germany: the second decade. Center for Business & Corporate Law Research Paper Series No. 0010, Düsseldorf, Germany.

Romano, S., 1945 [1918]. L'ordinamento giuridico. 2nd ed. Firenze, Italy.

10. Self-regulation code on beer advertising

Alessandro Artom

10.1 Introduction

The Italian Code on Beer Advertising provides an interesting example of private food law addressing communication practices of food business operators. The idea of adopting a self-regulation code on beer advertising comes from the need to improve on public law, including both national and EU law, through the private instrument of self-regulation and the voluntary respect by all parts concerned.

The aim of such private activity is to integrate public law in this field with the adoption of simple and fast rules, voluntarily adopted by the companies, to properly market alcoholic products, such as beer, and to promote those products to the consumer in ways encouraging safer consumption.

In this regard, the rules of the Italian self-regulation Code represent a considerable progress in the field of commercial advertising's self-regulation, because both regulatory and procedural provisions have been conceived not only to regulate brewers and distributors' marketing activities, but also to make consumers more aware of important social matters, related to the consumption of alcoholic beverages. In addition, the self-regulation makes the consumer able to use free, fast and efficient instruments of private justice, in order to prevent immediately the spread of unfair advertising.

The Code is therefore the instrument by which the companies associated to Assobirra[370] show, on one hand, their will to be bound by the principles commonly shared by the brewers of Europe, and on the other hand, the need to educate the public on a responsible consumption of alcohol. This principle constitutes the common ground adopted by all European beer producers.

This chapter is structured as follows. Section 10.2 introduces the principles that have been elaborated in the Code, directly but also via the more general Italian Code of Self-Regulation of Marketing Communication. Section 10.2.1 addresses principles derived from EU legislation. Section 10.2.2 presents seven operational standards that the association of Brewers of Europe has agreed upon which are operationalised in the Code. Section 10.3 analyses the content of the Code. This section starts in 10.3.1 with a discussion of purpose and obligations, followed by

[370] Assobirra – Associazione degli industriali della Birra e del Malto. Assobirra is the Italian association of beer and malt producers/distributors, based in Rome (viale di Val fiorita n. 90). Assobirra is associated with Confindustria (Confederation of Italian industry) and with the European Association Brewers of Europe (BOE), embracing in full the ethical code of the latter. It carries out institutional, promotional and technological development tasks.

Section 10.3.2 on conduct rules. Procedural rules are set out in Section 10.3.3 and an instrument of compliance assistance called 'copy advice' in Section 10.3.4. Section 10.3.5 presents cases illustrating how the Code may work in practice. The final section, Section 10.4, concludes this chapter with a discussion of the relevance of the Code in the wider context of private food law.

10.2 Underlying principles

10.2.1 The European principles

On the European level, the principles concerning the alcohol advertising were set for the first time in 1989 by the Directive no. 552[371] and then confirmed in the Directive 2007/65/CE concerning television broadcasting activities, which has been recently transposed into Italian law by the Decree 15/3/2010, no. 44.[372] Now, those principles are laid down in the Directive no. 13 of 10 March 2010, concerning television advertising and teleshopping for alcoholic beverages[373] and the Italian Beer Advertising Code fully complies with the Directive's purposes.

First of all, the new Directive stresses the importance of the self-regulation instruments in the field of alcoholic products' advertising. In recital no. 44, they state that self-regulation instruments can play an important role in delivering a high level of consumer protection and that Member States should recognise this role as a complement to the legislative and judicial and/or administrative

[371] Council Directive 89/552/CEE of 3 October 1989 on the coordination of certain provisions laid down by Law, Regulation and Administrative Action in Member States concerning the pursuit of television broadcasting activities. Official Journal of the European Union L 298, 17.10.1989, 23-30.

[372] Decreto Legislativo 15/03/2010, no. 44 implementing Directive no. 2007/65/CE on the coordination of certain provisions laid down by Law, Regulation and Administrative Action in Member States concerning the pursuit of television broadcasting activities. Gazzetta Ufficiale n. 73 del 29.03.2010.

[373] Directive 2010/13/EU of the European Parliament and of the Council of 10 March 2010 on the coordination of certain provisions laid down by law, regulation and administrative action in Member States concerning the provisions of audiovisual media services (Audiovisual Media Services Directive). Official Journal of the European Union L 95, 15.4.2010, 1-24.

mechanisms in place.[374] In addition, the recital no. 89 of the Directive foresees that it is necessary to lay down strict criteria relating to the television advertising of alcoholic beverages, in order to safeguard the consumer against unfair commercial communication.[375]

These criteria are laid down in Article 22 of the Directive[376] and they have been implemented in our self-regulation Code on Beer Advertising.

In particular, the advertising of beer may not be aimed at minors or depict minors consuming these beverages; it shall not link the consumption of alcohol to enhanced physical performance or to driving; it shall not create the impression that the consumption of alcohol contributes towards social or sexual success; it shall not claim that alcohol has therapeutic qualities or that it is a stimulant, a sedative or a means of resolving personal conflicts; it shall not encourage immoderate

[374] Directive 2010/13/EU, recital no. 44: 'In its Communication to the European Parliament and to the Council on Better Regulation for Growth and Jobs in the European Union, the Commission stressed that a careful analysis of the appropriate regulatory approach is necessary, in particular, in order to establish whether legislation is preferable for the relevant sector and problem, or whether alternatives such as co-regulation or self-regulation should be considered. Furthermore, experience has shown that both co-regulation and self-regulation instruments, implemented in accordance with the different legal traditions of the Member States, can play an important role in delivering a high level of consumer protection. Measures aimed at achieving public interest objectives in the emerging audiovisual media services sector are more effective if they are taken with the active support of the service providers themselves. Thus self-regulation constitutes a type of voluntary initiative which enables economic operators, social partners, non-governmental organisations or associations to adopt common guidelines amongst themselves and for themselves.
Member States should, in accordance with their different legal traditions, recognise the role which effective self-regulation can play as a complement to the legislative and judicial and/or administrative mechanisms in place and its useful contribution to the achievement of the objectives of this Directive. However, while self-regulation might be a complementary method of implementing certain provisions of this Directive, it should not constitute a substitute for the obligations of the national legislator. Co-regulation gives, in its minimal form, a legal link between self-regulation and the national legislator in accordance with the legal traditions of the Member States. Co-regulation should allow for the possibility of State intervention in the event of its objectives not being met. Without prejudice to formal obligations of the Member States regarding transposition, this Directive encourages the use of co-regulation and self-regulation. This should neither oblige Member States to set up co-regulation and/or self-regulatory regimes nor disrupt or jeopardise current co-regulation or self-regulatory initiatives which are already in place within Member States and which are working effectively'.

[375] Directive 2010/13/EU, recital no. 89: 'It is also necessary to prohibit all audiovisual commercial communication for medicinal products and medical treatment available only on prescription in the Member State within whose jurisdiction the media service provider falls and to lay down strict criteria relating to the television advertising of alcoholic products'.

[376] Directive 2010/13/EU, Article 22: 'Television advertising and teleshopping for alcoholic beverages shall comply with the following criteria: (a) it may not be aimed specifically at minors or, in particular, depict minors consuming these beverages; (b) it shall not link the consumption of alcohol to enhanced physical performance or to driving; (c) it shall not create the impression that the consumption of alcohol contributes towards social or sexual success; (d) it shall not claim that alcohol has therapeutic qualities or that it is a stimulant, a sedative or a means of resolving personal conflicts; (e) it shall not encourage immoderate consumption of alcohol or present abstinence or moderation in a negative light; (f) it shall not place emphasis on high alcoholic content as being a positive quality of the beverages'.

consumption of alcohol or present abstinence or moderation in a negative light and, finally, it shall not place emphasis on high alcohol content as being a positive quality of the beverages.

Our Code reflects also the provisions of the Italian Code of self-regulation of Marketing Communication, that is a private standard voluntarily adopted in Italy since 1966 by all producers, advertising agencies and media, concerning the advertising in all fields. Article 22 of this Code refers to alcoholic beverages and affirms that the marketing communication concerning such beverages should not be in contrast with the principle to invite a responsible consumption of alcohol. This principle aims to safeguard the primary interest of the consumers in general, especially children and young people, in a family, social and working environment protected from negative consequences of alcohol abuse.[377]

Regarding beer, in application of Article 22, the jury has decided some cases. One of the most interesting cases recently decided by the jury on alcohol advertising was about a TV advertising of 'Birra Nastro Azzurro' of April 2008.[378]

This TV advertising shows a cargo ship full of beer boxes that leaves the harbour. The captain does not realise that a group of young people are going after the boat, on board dinghy, because he's listening to the music with earphones. The guys manage to reach the boat, start a big party where all the people on board dance and drink beer all night long. At sunrise, the cargo ship arrives to the harbour of destination and from the pier one guy ask the captain where the Italian party is. The captain is surprised and realises what happened during the night: the guys drank all beers and wrote 'Thanks Italy!' on the cargo ship. The captain and the boy laugh and drink the last two beers. The TV advertising finishes with a claim 'Nastro Azzurro. C'è più gusto a essere italiani' and the image of some empty bottles of beers.

---

[377] Italian Code of Self-Regulation of Marketing Communication, Article 22 'Alcoholic beverages': 'Marketing communication concerning alcoholic beverages should not be in contrast with the obligation to depict styles of drinking behaviour that project moderation, wholesomeness and responsibility. This principle aims to safeguard the primary interest of the population in general, social and working environment safeguarded from negative consequences of alcohol abuse. In particular, such marketing communication should not: – encourage an excessive, uncontrolled, and hence harmful consumption of alcoholic beverages; – depict an unhealthy attachment or addiction to alcohol, or the belief that resorting to alcohol can solve personal problems; – target or refer to minors, even only indirectly, or depict minors consuming alcohol; – associate the consumption of alcoholic beverages with the driving of motorised vehicles; – encourage the belief that that the consumption of alcoholic beverages promotes clear thinking and enhanced physical and sexual performance, or that the failure to consume alcohol implies physical, mental or social inferiority; – depict sobriety and abstemiousness as negative values; – induce the public to disregard different drinking styles associated with specific features of individual beverages and the personal conditions of consumers; – stress high alcoholic strength as being the principal feature of a beverage'(official English translation of Istituto dell'Autodisciplina Pubblicitaria – Advertising Self-Regulation Institute. Available at: http://www.iap.it/en/code.htm).

[378] Decision no. 54/2008 of 26.5.2008 – Comitato di Controllo vs *Birra Peroni spa*.

On this advertising, the Jury, appointed by the Review Board, stresses the strong psychological impact that marketing communication on alcoholic beverages could have on consumers, if alcohol is depicted as a means to satisfy unreasonable needs and behaviours. On the other hand, Article 22 of the Code (CAP) on alcoholic beverages doesn't have the function to give a teaching direction to advertising agency's creativity. In the TV advertising examined, the jury considers that the story shown put the event on unreal basis because it's unlikely that a group of young people attacks a cargo ship, makes a noisy party without any reaction from the captain. In addition the TV advertising shows only few bottles of beer inside a box full of ice. Consequently there are no messages that suggest an irresponsible consumption of alcohol. For these reasons, the jury has considered that the message complies with Article 22 of CAP.

The specific improvement made by our Code is that to emphasise a moderate and responsible consumption of beer, as an alcoholic beverage. Despite the importance of such rules, there was no specific and independent regulation on beer advertising in Europe. Therefore Brewers of Europe, with the adoption of Seven Operational Standards, has satisfied this need.

10.2.2 Seven Operational Standards of the 'Brewers of Europe'

The Association of the Brewers of Europe[379] currently represents 27 national brewing associations and producers of around 95% of the beer brewed in the EU and is a founding member of the EU Alcohol and Health Forum and is committed to being part of the solution when it comes to tackling alcohol misuse.

The Brewers of Europe has adopted uniform criteria for the institution in the Member States of self-regulation systems for the commercial communication of beer.[380] The Seven Operational Standards, agreed upon by the Brewers of Europe in 2007, have been adopted in order to optimise, in the field of the beer advertising, the effectiveness of the national mechanisms of self-regulation and to guarantee responsible commercial communications in the EU.[381] These particular guidelines would have to be implemented by April 2010 and Italy has fulfilled the task trough the adoption of the Code. At the moment, we have implemented all the Seven Operational Standard, fixed by the Brewers of Europe.

---

[379] Website: www.brewersofeurope.org.

[380] On May 26th 2010 the association of the Brewers of Europe has published a report on the Seven Operational Standards: 'Responsible beer advertising through self-regulation'. The report offers, both from an EU-wide and a national perspective, an overview of the background, baseline, progress and next steps in relation to the full implementation of the Seven Operational Standards by the membership of the Brewers of Europe.

[381] Operational Standards for National Self-Regulatory Action Plan: (1) Full code coverage; (2) Increased code compliance; (3) Impartial judgements; (4) Fast procedure; (5) Effective sanctions; (6) Consumer awareness; (7) Own-initiative compliance monitoring.

The first Standard deals with 'code coverage'. The objective is that all the commercial communications on beer, regardless of their form and source, have to be covered by the Code, as well as all brewers and all distributors and all practitioners (for example advertising agencies, promotional commercial agencies, points of sale involved on promotional operations).

The second Operational Standard deals with 'code compliance'. It's essential that there is the maximum compliance with both the letter (that's to say, formal rules) and the spirit (which are the principles not formally written down) of the code, in order to prevent that irresponsible commercial communications on beer reach the public sphere. In addition, it's important that the 'copy advice' is provided on a confidential and free basis, making this mechanism easily available to every beer producer.

The third Operational Standard requests an impartial judgement in the commercial communications. The private judicial body is composed by three independent and qualified members including the chairman. 'Independent' means that the individual should be independent not just from the company whose advertisement is being investigated, but also independent from the brewing and advertising industries as a whole.

The fourth Operational Standard requests a fast procedure, in order to have a speedy decision if the advertising is in breach of the Code and consequently the Jury may order to interrupt immediately the advertising or promotional activities in breach.

The fifth Operational Standard deals with 'Sanctions'. It is necessary that sanctions act as a deterrent to prevent brewers from launching unfair communications or promotional activities. The strength of such sanctions must show both the regulator and the consumer that self-regulation is an efficient and effective system.

The sixth Operational Standard stresses the importance to guarantee and increase the consumer's awareness of the self-regulatory system and its functioning. Such knowledge is reachable by the publication of the jury decisions.

The last Operational Standard deals with the monitoring, time to time, of the Code by the judicial body: the jury. Such system has to be based on planned and systematic check of the jury's activity, as well as on a continue up-dating of the code through a regular review.

10.3 The Code

10.3.1 Purposes and obligations

The Italian self-regulation Code on Beer Advertising is composed of two parts: the first one deals with general purposes, the obligations undertaken by all Italian beer producers associated to Assobirra, the field of application and the definitions commonly accepted; the second one is about conduct rules, procedural and judicial rules.

Concerning the purposes, the Code aims to ensure that any kind of marketing communication on beer (that's to say all media, TV, radio, cinema, web, press, public bills, promotional activities in the points of sale ... and including labels and packaging) must comply with the principle of 'responsible drinking'.

The subjects bound by the Code are all the Assobirra's associated companies (producers and distributors) and their marketing agencies, including advertising agencies, advertising dealers and media advertising managers and the point of sale of HoReCa (Hotel restaurant café sale channel) and large distribution dealers, where promotional activities are made.

In particular, the brewers and the practitioners voluntarily undertake to observe the code and its rules, to ensure compliance by their members, to efficiently disseminate the rulings of the jury and to adopt appropriate measures towards members who fail to comply with/or repeatedly disregard the jury's decisions.

In the part dedicated to definitions, the key terms of the Code are explicated. The principal terms contain definitions for: acceptance clause, agency, Assobirra, beer, code, copy advice, hazardous activities, jury, marketing communication, minors, point of sale, sanctions and secretariat.[382]

One of the most important definitions refers to the acceptance clause. In this regard, Italian brewers undertake to ensure that their members and associates shall include a special clause in the contracts subscribed with their advertising agencies, specifying acceptance of the provisions of the Code, its Regulations and

[382] Some of the most relevant definitions: Agency – Advertising agencies, advertising and marketing advisors, exclusive dealer advertising, advertising managing agents and promotional commercial agencies, points of sale involved on promotional operation in Italy; Point of sale – Location of supply and/or sale of beer including sales free of charge, to the consumer that belongs to HoReCa channel or to the large organised distribution or to exhibitions opened to public, to promote beer, in the latter case that such promotion will be made directly by the brewer itself, with or without distribution of free beer samples; Sanctions – The sanctions imposed by the jury to the beer company are as follows: (a) at the first stage the interruption of marketing communication on beer i.e. desist order should marketing communication submitted for examination appear to clearly violate one or more articles of the self-regulation code, the jury may order to the beer company and agencies to desist from publishing it, and the condemnation of jury's costs of procedure; (b) at the second stage, in case of non-compliance of the desist order, a fine will imposed through a compulsory contribution from a minimum of € 1000 to a maximum of € 10,000 devolved to scientific research, in the field of the prevention of alcohol abuse accompanied by the publication of the jury's decisions on Assobirra's website; Secretariat – The jury's secretariat is the independent office competent to receive the consumers petition on marketing communication about beer. The jury's secretariat, will examine the consumers petition and will perform the inquiring tasks for the jury. The jury's secretariat is the competent office to realise the copy advice.

the rulings of the jury including publication of the latter, as well as acceptance of final desist orders issued by the jury. The desist order consists in the immediate interruption of marketing communication on beer.

Another important definition refers to the Copy Advice. In fact, it consists of an advisory and confidential service given from the jury's secretariat to the brewers companies or relative agencies. The legal advice delivered by the jury's secretariat concerns the compliance or the non-compliance with the Code of a scheduled advertising campaign. The advertising is submitted in advance and in a confidential way by the brewers producers or agencies before its broadcast or publication.

Finally, the jury is defined. The jury is judging body and it is composed of three independent and qualified members chosen among experts in law, consumer affairs and communication. The expert in law (lawyer and/or university professor in law) holds the office of chief of justice, the expert of communication holds the office of reporter, the expert of consumerism is the third private judge.

10.3.2 Conduct rules

The crucial point of the Code is represented by the conduct rules, in accordance to the provisions of the mentioned Article 22 of Directive no. 13 of 2010. These rules are included in articles from 1 to 8 of the Code and their purpose is to address any kind of marketing communication and/or promotional activities towards principles of moderation and responsibility in the alcohol consumption, in relation to the standards established by the Brewers of Europe.

According to Article 1, the marketing communication about beer must not encourage excessive or irresponsible consumption of beer, nor present abstinence or moderation in a negative way; it must not be connected with violent, aggressive or anti-social behaviour and it must not show people who appear to be intoxicated or in any way present intoxication in a positive way.

In this regard, Assobirra has focused its attention on the social matters connected with an irresponsible consumption of alcohol and it has promoted a specific information and awareness campaign concerning the abuse of alcoholic beverages, in order to make the public more aware of the importance of a moderate and responsible consumption in our society.

The program is called 'Guida tu la vita. Bevi responsabilmente' (Drive your life. Drink responsible)[383] and it focuses on specific situations, like alcohol and driving, alcohol and pregnancy, alcohol and young people.

[383] There is also a website dedicated to the program, http://www.beviresponsabilmente.it.

On the matter dealing with alcohol and pregnancy, Assobirra wants the gynaecologists and women to be more aware of the risks connected with alcohol consumption during the pregnancy, especially for the future child. The Italian beer producers initiative on this problem has been recognised by the EU Alcohol & Health Forum as the first initiative promoted on this field in Europe by an alcohol company.

Regarding young people, Assobirra promotes some initiatives in cooperation with the Italian private radio, called 'Radio 105', one of the most popular radio station among young Italians. Young people are asked to talk together about alcohol and promote a correct behaviour in relation to alcohol consumption.

In the Code special attention is dedicated to minors. According to Article 2, the marketing communication about beer must not be aimed at minors, it must not show minors consuming beer, it must not be promoted in media, programs or at events where the majority of the audience is composed by minors. In particular, marketing communication must not address to minors through graphic pictures and/or cartoons on beer and it should not use testimonials whose normal working activity is addressed to minors.[384]

Article 3 of the conduct rules deals with driving. The advertising about beer must not associate, directly or indirectly, the consumption of beer with the act of driving vehicles of any kind. In this regard, the mentioned Assobirra's advertising campaign on alcohol, 'Drive your life. Drink responsible', is focused on the problems connected with alcohol and driving.

This clause of the Code is in accordance with the new provisions recently adopted in the Italian legislation. From August 13th it is in force the new Driving code (codice della strada), L. 29 July 2010 no. 120.[385] According to the Article 53, it is forbidden to sell alcoholic beverages in the highway stops between 2:00 and 6:00 a.m., in order to fight road accidents caused by alcohol consumption. For bars, discos, pubs and clubs the prohibition is valid between 3:00 and 6:00 a.m., according to Article 54 of the new legislation. A fixed level of 0.5 grams per litre is considered 'driving under the influence of alcohol' and forbidden by the new Driving code.

The other conduct rules of the Code refer to hazardous activities, therapeutic properties, alcohol content, performances and promotional activities. In particular, Article 4 establishes that marketing communication about beer must not associate

[384] Recently the new Decree 15.3.2010 no. 44 on audiovisual media services has increased minors' protection. In particular, Article 9 bans TV transmissions that could damage the physical, psychological and moral development of minors or show scenes of violence or pornography. Article 9 also introduces a protection system against any kind of transmission that could damage minors.

[385] L. 29.7.2010 n. 120 published in Gazzetta Ufficiale n. 175 del 29.7.2010.

consumption of beer with the performance of hazardous activities (all the human activities that involve a particular attention and/or physical effort), nor portray the act of consumption prior to or during such activities.

On therapeutic properties, Article 5 says that marketing communication must not lead to believe that beer has properties of preventing, treating or curing human diseases. Article 6 stresses the importance that beer advertising must not create any confusion as to the nature and strength of beer and it must not present the high alcoholic content of beer as a positive quality of the beverage or as a good reason for choosing it.

In accordance to Article 7, the beer advertising must not connect the consumption of beer to better physical abilities and it must not lead people to believe that consumption of beer enhances social or sexual success.

The last conduct rule, Article 8, underlines that promotional activities on beer in the point of sale must not drive to irresponsible or anti-social behaviour or encourage an excessive consumption of beer. In addition, the distribution of free beer samples must not be made out of the points of sale and no sampling of beer must be offered to minors.

10.3.3 Procedural rules

The competent bodies for beer advertising and promotional activities are the jury and the secretariat. The first one is the judging body and the second one is the independent office competent to receive the consumer's petitions and the requests of copy advice. The articles of the Code dedicated to the bodies and the procedures are from 9 to 18.

The procedure of monitoring starts with a petition, through which the consumer or the consumer's association make the jury's secretariat aware of an advertising which is probably infringing on the Code.

The secretariat evaluates the petition and in the term of 5 workdays it can act in three alternative ways:

- It can accept the petition and send it to the jury with a brief report and, simultaneously inform both the consumer and the beer company that the procedure is started. The beer company is invited to present pleadings to the jury in the successive 5 workdays.
- The secretariat can refuse the petition with a brief explanatory note to the consumer.
- It can consider the petition incomplete and so ask the consumer further information and, in case of refusal, the secretariat files the petition in the successive 10 workdays.

In case of acceptance of the consumer's petition, the jury within 10 workdays judges the submitted marketing communication and after the evaluation of pleadings eventually submitted by the beer company, it can decide in two alternative ways:

- It can order the company to stop immediately the advertising, which is called 'desist order', warning that in case of non-compliance a sanction will be imposed.
- It can consider the company's pleadings worthy to be discussed before the jury in a public hearing in the successive 10 workdays.

During the discussion before the jury, the secretariat will report on the case. The petitioner consumer and the beer company may explain their opinions on the matter, also through a nominee named for the purpose. At the end of the discussion, the jury takes a decision. The jury's decisions have a binding effect not only for the company involved in the procedure introduced by the consumer, but also for all the agencies and the points of sale which have accepted the Code's provisions with the acceptance clause. The decisions are also published on Assobirra's website.

10.3.4 Copy advice

The request of copy advice is examined confidentially by the secretariat with the story board, the graphic production completed by texts of marketing communication on beer. In the term of 7 workdays the secretariat gives its legal advice to the asking beer company or advertising agency. In particular, the secretariat can provide in three alternative ways:

- It can approve the marketing communication and if a consumer makes a petition on that advertising, the Secretariat will not accept the petition, in accordance to its previous copy advice, unless the consumer's petition concerns the ways of spread used for the advertising (for example: spread in programs for minors);
- It can find some infringements to the Code and order the beer company to remove them and to prepare a written commitment in the successive 5 workdays. If the commitment is received correctly and in time by the Secretariat, the marketing communication will be approved in the modified version;
- It can refuse the submitted marketing communication, considering it in breach of the Code. In case of successive broadcast or publication of this advertising, the Secretariat will alert immediately the Jury. In this case, the beer company involved can argue the copy advice before the Jury, which decides in the successive 10 workdays, in a private session.

10.3.5 Cases

To better understand the functioning of the procedures followed by the jury and the secretariat, it seems advisable to illustrate two practical cases:

- Case A is about a TV advertising.
- Case B is about promotional activities at points of sale.

Case A

The first case refers to a TV beer advertising broadcasted during the afternoon, a particular and protected time slot which is typically aimed at minors. According to Article 13 Italian L. 125/2001,[386] the advertising of alcoholic beverages is forbidden in programs aimed at minors, especially in the 15 minutes that come before and after the transmission of such programs.

First of all, the consumer communicates the secretariat (by mail or e-mail) the advertising he wants to submit to the jury, explaining his reasons and attaching the required documents. Then the jury's secretariat examines the petition sent by the consumer and in the term of 5 workdays ...

- It can accept the petition and send it to the jury with a report and, at the same time, inform both the consumer and the beer company that the procedure is started. The beer company is invited to present pleadings to the jury in the successive 5 workdays.

OR

- It can, on the contrary, refuse the petition with a brief explanatory note to the consumer.

OR

- Finally, it can consider the petition incomplete and so ask the consumer further information and, in case of refusal, the secretariat files the petition in the successive 10 workdays.

When the consumer's petition is sent to the Jury, the Jury within 10 workdays judges the marketing communication and after the evaluation of pleadings eventually submitted by the beer company

- It can order the company to stop immediately the TV beer advertising in question ('desist order'), warning that in case of non-compliance a sanction will be imposed as a compulsory contribution to be devolved to the scientific research, accompanied by the publication of the jury's decision on Assobirra website and the condemnation of Jury's costs of procedure.
- On the contrary, the jury can consider the company's pleadings worthy to be discussed before the jury in the successive 10 workdays.

During the discussion before the jury, the secretariat will report on the case. The petitioner consumer and the beer company may explain their opinions on the matter, also through a nominee named for the purpose. At the end of the discussion, the jury takes a decision.

[386] L. 30.03.2001, no. 125 on alcohol problems (Gazzetta Ufficiale no. 90 del 18.04.2001).

Case B

The second case deals with promotion and sampling within points of sale with distribution of leaflets to the consumer which lead to an irresponsible consumption of beer because of their content and pictures. In this case the consumer warns the Secretariat of the promotional activity at the point of sale and he explains his reasons, supporting them with the required documents. If the procedure starts before the jury, the secretariat asks the point of sale to send a brief report concerning the ways used for the promotion impeached, in the term of 5 workdays.

After having examined the pleadings submitted by the beer company and by the point of sale, the Jury in the term of 10 workdays...

- Can order the company and the point of sale to stop immediately the promotional activities in breach with the Code, warning that in case of non-compliance of such order, a sanction will be imposed as a compulsory contribution to be devolved to the scientific research, accompanied by the publication of the jury's decision on Assobirra website and the condemnation of Jury's costs of procedure.
- On the contrary, the jury can consider the company's pleadings and the brief sent by the point of sale worthy to be discussed before the jury in the successive 10 workdays.

During the discussion before the jury, the secretariat will report on the case. The petitioner consumer and the beer company may explain their opinions on the matter, also through a nominee named for the purpose. At the end of the discussion, the Jury takes a decision.

Finally, it is important to remember that whenever the Jury is involved in such cases, it has to inform the Italian Antitrust Authority (AGCM)[387] for the purpose of foreclosure and of stay of proceedings foreseen by the Italian Consumer Code, Article 27 ter par. 3,[388] and by the Regulation on the inquiry procedures concerning the unfair business-to-consumer commercial practices, Article 20.

10.4 The Code as Private Food Law

The Italian self-regulation Code on Beer Advertising shows a form of private food law not designed to regulate conduct within the food chain like most of the other examples presented in this book, but to regulate the communication

[387] Website: http://www.agcm.it.

[388] Decreto Legislativo 06.09.2005, no. 206 (Gazzetta Ufficiale n. 235 del 08.10.2005). Article 27-ter (Self-Regulation): paragraph 3 – when a procedure starts before a self-regulation body, the parties can decide not to refer to the Authority until the final decision of the body, or they can ask the Authority to suspend the proceeding, whenever the it has been introduced by another person entitled to do so, waiting for the self-regulation body to decide.

towards consumers. By consequence, it is less embedded in contractual relations regarding the sale and purchase of food products and more in agreement on the topic as such. The Code operationalises principles derived from public law sources – EU directives – as well as from civil law sources – the Brewers of Europe. For its enforcement and dispute settlement, it does not fully rely on the general instruments of civil procedure, but it is endowed with its own jury empowered to take decisions accepted by stakeholders as binding, but also to provide compliance assistance in the form of 'copy advice'. The obvious advantage of combining litigation and compliance assistance in one hand, is that it brings expertise and legal certainty to the 'copy advice'. The obvious disadvantage is that in providing 'copy advice' the jury limits its room for manoeuvre to decide upon consumer complaints before it has heard the consumer's arguments.

References

Asobirra, 2008. Decision no. 54/2008 of 26.5.2008 – Comitato di Controllo vs *Birra Peroni spa*.

Brewers of Europe, 2010. Responsible beer advertising through self-regulation – seven operational standards. Available at: http://www.brewersofeurope.org/docs/flipping_books/responsible_beer_ad_2010/index.html.

EU, 1989. Council Directive 89/552/CEE of 3 October 1989 on the coordination of certain provisions laid down by Law, Regulation and Administrative Action in Member States concerning the pursuit of television broadcasting activities. Official Journal of the European Union L 298, 17.10.1989, 23-30.

EU, 2010. Directive 2010/13/EU of the European Parliament and of the Council of 10 March 2010 on the coordination of certain provisions laid down by law, regulation and administrative action in Member States concerning the provisions of audiovisual media services (Audiovisual Media Services Directive). Official Journal of the European Union L 95, 15.4.2010, 1-24.

Italian government, 2001. Decreto Legislativo 30.03.2001, n. 125. Legge quadro in materia di alcol e di problemi alcolcorrelati. Gazzetta Ufficiale n. 90 del 18.04.2001.

Italian government, 2005. Decreto Legislativo 06.09.2005, n. 206. Codice del consumo. Gazzetta Ufficiale n. 235 del 08.10.2005.

Italian government, 2010a. Decreto Legislativo 15/03/2010, no. 44 implementing Directive no. 2007/65/CE on the coordination of certain provisions laid down by Law, Regulation and Administrative Action in Member States concerning the pursuit of television broadcasting activities. Gazzetta Ufficiale n. 73 del 29.03.2010.

Italian government, 2010b. Decreto Legislativo 29.7.2010 n. 120. Disposizioni in materia di sicurezza stradale. Gazzetta Ufficiale n. 175 del 29.7.2010.

11. Franchising strengthens the use of private food standards

Esther Brons-Stikkelbroeck

11.1 Introduction

The distinctive character of a business becomes more important every day, as well as concepts of sustainability and authenticity. Businesses are distinguishing from each other by offering sustainable products or ecologically sound business methods and fair trade labelled products.[389] Some of them are setting the standard not only for themselves, but also for their competitors in trade, that way creating a level playing field. They formulate codes of conduct or standards to be applied not only within their own business organisation, but also within the organisations of their suppliers and partners, within the entire supply chain. What is the influence of these supply chains, especially of the business method 'franchising', for the supposedly growing impact of private standards in the food sector?

The emphasis of this chapter is on the use of private food standards within franchise organisations and the influence of franchising according to Dutch Law on the effect of private standards. The first question to be answered is the meaning of the words private food standards. What are private standards and to what extent and with what purpose are they used? Having determined that, several cooperation agreements within supply chains are addressed and narrowed down to the ones that are supposed to have the most control over the execution of private food standards. How do franchise organisations operate in respect to private food standards and how do they handle with compliance-issues? Do franchise organisations strengthen the use of private food standards?

11.2 Private food standards

Global brand producers and retailers increasingly require their suppliers and also their cooperation partners to comply with certain social, environmental and safety norms.[390] These norms are referred to as 'private standards'. Private standards are not the same as technical regulations and national, regional or international voluntary standards such as might be encountered in trading with any partner. Private standards focus on social, safety and environmental issues and

[389] Franchise organisation Subway for instance increased its success by using health claims like: 'eat fresh live green' obliging its franchisees to comply with this claim.
[390] Before coming to terms with the matter of private standards, a supplier will need to fulfill some basic qualifications in order to qualify as a potential supplier to these brands and retailers. The enterprise must respect local and national legislation, needs certain scale and capacity, quality, competitive prices, volume and timely delivery and should comply with technical regulations and management qualifications. The same qualifications can be applied to cooperation partners such as franchisees.

are required by brand producers and retailers when they source their products. These standards come in various shapes and sizes. Standards may be applicable to the production site or the product itself.

Private standards can be divided into several categories. There are consortia standards, which are often developed by a sector-specific consortium (i.e. BRC or GlobalGAP[391]). There are civil society standards, established as an initiative by a non-profit organisation usually as a response to concerns over social and environmental conditions (e.g. MSC,[392] FSC,[393] GreenPalm or RSPO[394]) and there are company-specific standards, which are developed internally and apply to the whole supply chain of a company (i.e. codes of conduct). All of them are becoming increasingly popular tools to address sustainability issues across all types of organisations.

Franchisors like IKEA for instance require compliance with their code of conduct.[395] Suppliers of the franchisor IKEA and all franchisees are requested to adhere to the requirements outlined in such a code of conduct. As a result, compliance is often a prerequisite for having a business relationship with such franchise organisation. Others, like the Dutch franchise formula WAAR, require their franchisees to sell only labelled product (Fair Trade) and use only certificated shop interior (FSC-wood).[396] Some make it their goal, like General Mills who states: 'Our goal is to be among the most socially responsible food companies in the world'.

While private sector firms and consortia have been the driving force behind the formulation of management and product standards, the multitude of private standards and retailer requirements have a growing impact. There are probably a variety of reasons for this growing impact.

Firstly, there is more awareness about standards and regulations in general, also because of the availability of and easy access to information. Secondly, the building-up of health and safety concerns in industrialised countries (such as food safety, chemicals, allergens, working conditions, etc.) resulted in an environment where not only the government regulations have become stricter, but the retailers/supermarket chains have started to drive the trend for stringent standards due to consumer awareness. Consumers in developed economies are showing growing concerns about the social and environmental conditions prevailing

[391] Global Good Agricultural Practice.

[392] Marine Stewardship Council.

[393] Forest Stewardship Council.

[394] The Roundtable on Sustainable Palm Oil.

[395] IKEA has formulated formal social and environmental requirements for its suppliers in the IWAY standard, which is applicable to all external suppliers and service providers. IKEA includes these standards in its basic contracts. If non-compliance is encountered IKEA will stop delivery immediately and without adjustment the contract may be terminated.

[396] WAAR is a hard franchise formula that has accepted the fair trade principles. Besides the sales of fair trade products, WAAR has formulated specific environmental requirements for the suppliers of the shop interior from its franchisees.

in countries participating in the supply chains of products that are sold into their markets. As cases of severe breaches of workers' rights, human rights violations and environmental degradation caused by corporate activity reach the public, consumer confidence in the responsible conduct of the major brands and retailer's decreases. Furthermore the desirability of differentiation in the market place in relation to social and environmental production criteria also increases.

These developments have contributed to the response from large brands and retailers who are setting stricter standards within their supply chain with the aim of improving the social and ecological performance of developing producers along the whole length of these chains. In essence, many multinationals use private standards as an instrument of supply chain management, franchisors use private standards to influence their franchisees and as a mechanism to gain marketing advantage over rivals. Of course, reputation and brand protection, global sourcing, differentiation in the marketplace, and control and rationalisation of supply have been important drivers for private standards.

Private standards are known within vertical agreements as well, either expressed directly or indirectly. Directly, by requiring standards or conditions for the production or handling of food in order to meet the requested quality standards. This being part of the specific characteristics of the supply chain or supplied product. Indirectly, by stimulating the use of regulatory requirements, in order to guarantee food safety or a healthy food production. For instance by requiring the use of skilled personnel to accomplish hygienic working methods. The question is whether this cooperation form and the various methods of it are of any influence to the effect of private food standards.

11.3 Vertical agreements and franchising

Vertical agreements are agreements or concerted practices entered into between two or more undertakings each of which operates, for the purposes of the agreement or the concerted practice, at a different level of the production or distribution chain, and relating to the conditions under which the parties may purchase, sell or resell certain goods or services.[397] These conditions often contain vertical restraints, some of which are prohibited under EU competition rules.[398]

[397] EU, 2002. Commission Regulation (EC) No 1400/2002 of 31 July 2002 on the application of Article 81(3) of the Treaty to categories of vertical agreements and concerted practices in the motor vehicle sector. Official Journal of the European Union L 203, 01/08/2002: 30-41 and EU, 2010. Guidelines on Vertical Restraints. Official Journal of the European Union C 130, 19/05/2010: 1-46.

[398] Article 101 of the Treaty on the Functioning of the European Union applies to vertical agreements that may affect trade between Member States and that prevent, restrict or distort competition. Article 101 provides a legal framework for the assessment of vertical restraints, which takes into consideration the distinction between anti-competitive and pro-competitive effects. Article 101(1) prohibits those agreements which appreciably restrict or distort competition, while Article 101(3) exempts those agreements which confer sufficient benefits to outweigh the anti-competitive effects.

It is generally recognised that vertical restraints may have positive effects such as, in particular, the improvement of the quality of goods and services. After all, when a company has no market power, it can only try to increase its profits by optimising its distribution processes for instance by optimising its quality. In a number of situations vertical restraints may be helpful in this respect since the usual arm's length dealings between supplier and buyer, determining only price and quantity of a certain transaction, can lead to a sub-optimal level of investments and sales.

A recognised and justified use of vertical restraints is the positive effect of quality standardisation: 'a vertical restraint may help to create a brand image by imposing a certain measure of uniformity and quality standardisation on the distributors, thereby increasing the attractiveness of the product to the final consumer and increasing its sales. This can for instance be found in selective distribution and franchising'.[399]

There are several types of vertical agreements. The growing impact of private standards is best to be seen within the so-called qualitative selective distribution agreement and franchising agreement.

Selective distribution agreements restrict the number of authorised distributors on the one hand and the possibilities of resale on the other. It is a different kind of distribution method than exclusive distribution. The difference is that the restriction of the number of distributors does not depend on the number of territories but on selection criteria linked in the first place to the nature of the product. These selection criteria could be found in private standards. Selective distribution is almost always used to distribute branded final products. Private standards used in this kind of setting is therefore mostly limited to indirect requirements, such as the use of certain transport methods or hygienic working methods.

Franchise agreements are a species of distribution agreements;[400] a commercial cooperation agreement between independent entrepreneurs. According to the definition stated within EU competition rules, franchise agreements contain licenses of intellectual property rights relating in particular to trade marks or signs and know-how for the use and distribution of goods or services. In addition to the license, the franchisor usually provides the franchisee during the life span of the agreement with commercial or technical assistance. The license and the assistance are integral components of the business method being franchised. The

[399] Guidelines on Vertical Restraints (the guidelines describe the vertical agreements that generally do not fall within Article 101 of the Treaty on the Functioning of the European Union: agreements of minor importance, agreements between small and medium-sized firms and agency agreements).
[400] The existence of a distribution component is not necessary. The franchise agreement can also be limited to the use of a business concept, without transferring services from franchisor through franchisee to purchasers.

franchisor is in general paid a franchise fee by the franchisee for the use of the particular business method. Franchising may enable the franchisor to establish, with limited investments, a uniform network for the distribution of its products. In addition to the provision of the business method, franchise agreements usually contain a combination of different vertical restraints concerning the products being distributed, in particular selective distribution and/or non-compete and/or exclusive distribution or weaker forms thereof.

The purpose of a franchise agreement is the expansion of a well tested, proven and commercially successful business concept through franchisees, without large investments by the franchisor as the owner of the business concept. The investments are instead paid for by the franchisee who is looking for a way to make his business more profitable and is willing to do so by using this proven business concept.

Franchise is not an unambiguous word. People use the word 'franchise' or the phrase 'franchise business concept' in different manners to denote the concept of franchise. This is to be explained because of a non-legal use of the word. In that perspective franchise can be divided into three different business methods: distribution franchise, services franchise and industrial/production franchise.[401]

Distribution franchise is the classical method of franchising. These agreements aim to distribute goods that are being sold by the franchisor to franchisees and through them to consumers or other users of the products. Franchisor can be the one selecting and buying the product before selling them to the franchisees, he can be the manufacturer of the products himself or he could be the one selecting the suppliers. The first franchise organisation often mentioned is the organisation behind the distribution of Singer sewing machines. The sale of the product by franchisor (or at his advice) is combined with certain services, such as the knowhow of the best possible selection of products.

Services franchise is a method for a quick expansion of a business concept within the services industry without the necessity of large investments. The services provided for within the method are leading. They can be accompanied by the sale of products; however selling those products is meant to be supporting. Probably the products can even be provided by others than franchisor as long as those suppliers comply with the standards set forth by franchisor. In this kind of franchise the knowhow within the franchise organisation is of the utmost importance.

Industrial or production franchise is the method in which franchisor gives a licence to franchisee to be able to produce and sell products under franchisor's label or

[401] Franchise can be divided into more business methods than the three mentioned, however those three form a fair picture in the perspective of this chapter. In order to outline only the basics if have left the aspect of master franchising unspoken.

trade mark. The purpose of this agreement is to bring production and commercial sales closer to each other.

Franchise has been able to develop mainly through economic values more than through legislation. In the Netherlands there are still no specific franchise laws. Franchise agreements are to meet the requirements of general contract laws and, if applicable, the agency and distribution regulations. Greece, Germany, Austria, Finland and Sweden offer more or less the same arrangement. The USA, Canada, Belgium, France, Spain, Italy and Switzerland meanwhile all have specific legislation on the subject of franchising, most of them however limited to information requirements to be given prior to the agreement.

Generally there are two legally relevant documents, the franchise agreement and the information disclosed prior to the signing of this agreement. The franchise agreement is designed to assure that all of the franchisees within an organisation are treated equitably. The expectations must be uniform throughout the system. The franchise agreement is a document that is signed by both parties upon completion of the deal to do business together. It provides outlines on how the franchisor expects the franchisee to run the operation.

As mentioned before there are legal systems that require a franchisor to provide all potential franchisees with relevant business information merged in a so called 'Pre-contractual Information Document (PID)' or a 'Franchise Disclosure Document (FDD)'. These are documents presented prior to the final agreement. The potential franchisee has the opportunity to review this information before making a final determination as to whether or not to move forward in becoming a franchisee within the organisation. The type of information provided by the document entails comprehensive details about the company's background and history. Included in this document is the disclosure of any lawsuits or bankruptcies that has occurred within the franchise organisation. Also incorporated in this document is a host of financial data as well as distribution channel information. Any confidentiality restrictions are also disclosed, as far as what a franchisee may and may not discuss with others. The laws of the Netherlands do not require franchisors to disclose such information in general. However if a franchisor provides information, the information should be reliable.

The requirements of general contracts law in the Netherlands also contain competition rules. Those mostly economic values are of great interest to both franchisor and franchisee. Both parties act in close and ongoing collaboration and the success of the franchise organisation depends on mutual trust and transparency. Nevertheless the franchise agreement creates an unequal relationship between the franchisor and its franchisees. The franchisees economically depend on the franchisor and on the continued existence of the exclusive distribution relationship with the franchisor and within the franchise chain.

In order to streamline the consequences of the contractual freedom and in order to avoid government regulation to be incorporated, the franchise sector has been self-regulating. This self-regulation is best to be seen in the standards laid down in the European Code of Ethics for Franchising from the European Franchise Association as they are incorporated by all affiliated national franchise federations.[402] The force in law of these national codes of conduct, especially in relation to the customers of a franchisee, is questionable (as is for most other codes of conduct).[403] Their economic value on the other hand is of great importance as most of the franchise organisations incorporate the code of conduct in their franchise agreement. The Dutch Franchise association (NFV) commits its members to comply with the code of conduct and offers an enforcement mechanism through mediation.

One of the essential obligations of the franchisee mentioned in this code of conduct is to '*devote its best endeavours to the growth of the franchise business and to the maintenance of the common identity and reputation of the franchise network*'.[404] The common identity and reputation of the franchise network and the way in which this identity and reputation should be maintained, is often outlined in a franchise (operations) manual. The manual contains everything the franchisees need to know to successfully operate their business. It helps the franchisor ensure consistency from one franchisee to another and protects the business' profitability and good name.

At a minimum, the manual sets quality standards and provides a coherent framework to ensure uniformity across the entire franchise network. The franchise operations manual will list the agreed upon standards and procedures for the franchise. Each franchisee will be expected to carry out business according to these standards. If a franchisee fails to abide by these rules, he or she may be subject to serious consequences including loss of the right to operate the franchise or even litigation. This in itself stipulates the value of franchise agreements for the increase of standards.

An example of how franchising works (mentioned in the guidelines)[405] furthermore clarifies this value of franchise for the growing impact of standards:

> A manufacturer has developed a new format for selling sweets in so-called fun shops where the sweets can be coloured specially on demand from the consumer. The manufacturer of the sweets has also developed the machines to colour the sweets. The manufacturer also produces the colouring liquids.

[402] Europese Erecode inzake franchising, Nederlandse Franchise Vereniging (NFV).

[403] Van der Heijden, M.J., 2011. 'Supplier codes of conduct' en mensenrechten in een keten van contracten; Over enige vermogensrechtelijke implicaties van gedragscodes met betrekking tot mensenrechten en milieu in contractuele relaties. Contracteren 2011 (1): 3-22.

[404] European Code of Ethics for Franchising, European Franchise Federation (EFF).

[405] Article Nr. 191 of EU, 2010. Guidelines on Vertical Restraints. Official Journal of the European Union C 130, 19/05/2010: 1-46.

> The quality and freshness of the liquid is of vital importance to producing good sweets. The manufacturer made a success of its sweets through a number of own retail outlets all operating under the same trade name and with the uniform fun image (style of lay-out of the shops, common advertising, etc.). In order to expand sales the manufacturer started a franchising system. The franchisees are obliged to buy the sweets, liquid and colouring machine from the manufacturer, to have the same image and operate under the trade name, pay a franchise fee, contribute to common advertising and ensure the confidentiality of the operating manual prepared by the franchisor. In addition, the franchisees are only allowed to sell from the agreed premises, to sell to end users or other franchisees and are not allowed to sell other sweets. The franchisor is obliged not to appoint another franchisee nor operate a retail outlet himself in a given contract territory. The franchisor is also under the obligation to update and further develop its products, the business outlook and the operating manual and make these improvements available to all retail franchisees. The franchise agreements are concluded for a duration of 10 years.
> (...)
> Most of the obligations contained in the franchise agreements can be deemed necessary to protect the intellectual property rights or maintain the common identity and reputation of the franchised network and fall outside Article 101(1).[406] The restrictions on selling (contract territory and selective distribution) provide an incentive to the franchisees to invest in the colouring machine and the franchise concept and, if not necessary to, at least help maintain the common identity, thereby offsetting the loss of intra-brand competition. The non-compete clause excluding other brands of sweets from the shops for the full duration of the agreements does allow the franchisor to keep the outlets uniform and prevent competitors from benefiting from its trade name. It does not lead to any serious foreclosure in view of the great number of potential outlets available to other sweet producers. The franchise agreements of this franchisor are likely to fulfil the conditions for exemption under Article 101(3) in as far as the obligations contained therein fall under Article 101(1).

The example shows us that there are several moments in which a franchisor can impose private standards upon the franchisees within the franchise network or were standards used have effect on consumers or purchasers throughout the franchise chain.

In the example the franchisor is also the manufacturer of the liquids and machines. He can make sure that the coloured sweets that are produced within the franchising network comply with certain standards. The operating manual used within the

[406] Of the Treaty on the Functioning of the European Union.

network is, besides off course the franchise agreement itself and the general terms and conditions used for the sale of the products, the most common place to constrain conditions upon the use of the trade marks i.e. the know-how, such as compliance with private standards in order to guarantee uniformity or quality of goods and services.

What is mentioned in the example on the 'restrictions on selling' equally applies to the use of private standards within a network. It provides an incentive to the franchisees to invest in the network and the franchise concept and, if not necessary, to at least help maintain the common identity, thereby offsetting the loss of intra-brand competition.

The example mentions the obligation of the franchisee to *'contribute to common advertising'.* This shows us the effect of indirect regulations, for instance the franchisor that is using a communication campaign in order to teach its consumers or purchasers all about the essence of hygiene and food.

Another incentive for applying to the standards set within the franchise network could be found in the topic of legal responsibilities on products, although the opinions differ.[407] This incentive could work from the perspective of the franchisor as well as from the perspective of the franchisee. From the perspective of the franchisor because liability and/or claims could harm his know-how and the reputation developed by him or could lead directly to liability by fault in cases where the franchisor is the manufacturer of the products involved (distribution franchise). The franchisee is an independent entrepreneur who is, notwithstanding the commercial and technical support he receives by franchisor, liable by fault if he does not comply with food safety, even if the franchisor is the manufacturer of the products. From both sides because insurance companies demand the entire franchise network to comply with the desired behaviour.

Most likely, however, the real incentive for parties to comply with standards is the (fear of) reputation (damages) of the franchise network. This incentive applies to all parties, franchisor and franchisees. The connection between franchisor and the franchisees is important. Franchisor will do its utmost to maintain the interests of the franchise network and its know-how and reputation. That's what franchisor has invested in. Franchisee has an interest in the protection of his franchise rights and therefore will comply with the agreement and the franchise obligations in order to safeguard those franchise rights. Within those franchise networks where the adherence to the requirements is a binding provision, non-compliance will immediately lead to the loss of the business relationship with the franchise organisation.

[407] Havinga, T., 2010. Draagt aansprakelijkheidsrecht bij aan de voedselveiligheid. Recht der Werkelijkheid 31 (1): 6-27.

The difficulty of maintaining standards in franchise chains is the result of an inability of the individual franchisee to expropriate the full benefit that accrues to his investment in improving for instance quality. While his improvement of quality contributes to the overall reputation of the chain and therefore to the revenues generated in each of the chains markets, the franchisee can extract at the most only his contribution to the revenues in the franchised single outlet that he owns. Because of this gap between the private benefit to an individual franchisee and the total benefit to the owner of the chain many franchisors select to own or closely monitor a significant selection of their companies' retail outlets in an attempt to guarantee that adequate services are indeed provided. Others add some kind of bonus or malus system to their agreements in order to make franchisees 'voluntarily' comply with quality standards.

11.4 Conclusion

The continuous development of distribution chains, such as a franchise, occurs parallel with the harmonisation and benchmarking of private standards as a response to the overwhelming growth in their number and variety. Harmonisation and benchmarking of private standards is especially welcome as this greatly simplifies compliance and leads to costs savings for both buyer and the suppliers. Franchising will help realise this, at least for those franchisors that have a strong formula. They will use standards in order to gain share in the competitive market and they will oblige the franchisees to act accordingly. They formulate standards to be applied within their own business organisation and the organisations of their suppliers and partners, therefore within the entire supply chain and they supervise compliance. Through the use of private standards parallel with the use of standard franchise agreements within food organisations customers will eventually benefit from the strengthened use of private food standards.

References

EU, 2002. Commission Regulation (EC) No 1400/2002 of 31 July 2002 on the application of Article 81(3) of the Treaty to categories of vertical agreements and concerted practices in the motor vehicle sector. Official Journal of the European Union L 203, 01/08/2002: 30–41.

EU, 2008. Treaty on the Functioning of the European Union (consolidated version). Official Journal of the European Union C 115, 09/05/2008: 47-199.

EU, 2010. Guidelines on Vertical Restraints. Official Journal of the European Union C 130, 19/05/2010: 1-46.

Van der Heijden, M.J., 2011. 'Supplier codes of conduct' en mensenrechten in een keten van contracten; Over enige vermogensrechtelijke implicaties van gedragscodes met betrekking tot mensenrechten en milieu in contractuele relaties. Contracteren 2011 (1): 3-22.

Havinga, T., 2010. Draagt aansprakelijkheidsrecht bij aan de voedselveiligheid. Recht der Werkelijkheid 31 (1): 6-27.

12. On the borderline between state law and religious law

Regulatory arrangements connected to kosher and halal foods in the Netherlands and the United States[408]

Tetty Havinga

12.1 The developing supply of halal foods

In 2006, the Dutch supermarket chain Albert Heijn introduced halal meat products in some of its shops to better serve Muslim customers. Immediately, animal rights organisations protested strongly against the selling of meat from animals that had been slaughtered without being stunned first. They also pointed out that non-Muslim customers might unknowingly buy this meat and launched a campaign to remove it from the supermarket.[409] In response, Albert Heijn switched to another halal certification scheme that allows reversible electrical stunning prior to the killing of the animal.[410] This resulted in protests and warnings from Muslims not to eat this halal meat because it was not really halal.[411] On the other hand, a complaint was filed with the Dutch Advertising Standards Authority against Albert Heijn for making an unjust claim that its halal meat was kind to animals.[412]

The above case shows that halal food, and religious slaughter in particular, is a contentious subject in the Netherlands that involves complicated issues such as the substance of the religious requirements for halal food, the reliability of halal certificates, animal welfare in religious slaughter and the role of public authorities in relation to halal food.

Within the wider subject of food regulation, the regulation of halal food - food that is permitted for faithful Muslims because it is in accordance with Islamic dietary laws - is particularly interesting. Because of the obvious parallels, this

[408] A previous (almost similar) version of this chapter was published as: Havinga, 2010. Regulating halal and kosher foods: different arrangements between state, industry and religious actors. Erasmus Law Review 3(4): 241-255.

[409] Duizenden tegen halal vlees AH. De Volkskrant, 25 October 2006; Radar en Albert Heijn ruziën over halal-vlees. Available at: http://www.nu.nl/economie/860223; Haal halalvlees uit de schappen. De Volkskrant, 26 October 2006.

[410] See: http://www.ah.nl/halal; http://www.evmi.nl/nieuws, reports of 16, 24 and 31 October and 3 November 2006; AH: halalvlees toch van verdoofde dieren. de Volkskrant, 31 October 2006, AH stopt verkoop halal-vlees. AD, 30 October 2006.

[411] Moslims: Halal vlees Albert Heijn is niet halal', Elsevier, 3 November 2006.

[412] Albert Heijn claimt ten onrechte diervriendelijk halal. Available at: http://www.wakkerdier.nl/persbericht/447; Albert Heijn trekt halal-claim in. Distrifood, 10 December 2008; AH trekt claim 'diervriendelijk halal' in. Available at: http://www.maghreb.nl/2008/12/12.

chapter compares the regulation of halal food to the regulation of kosher food. The objective of this chapter is to analyse the division of roles between state actors, the food industry, certification agencies and religious authorities in regulatory arrangements connected with halal and kosher food in the Netherlands and the United States.

The chapter starts with a brief introduction to the issue of halal and kosher food and its regulation. Both the Netherlands and the United States have special arrangements for religious slaughter and the labelling of halal and kosher food. The next sections deal with the regulation of halal and kosher food in the Netherlands and the United States, relying on a review of literature and internet sources. The subsequent sections describe the system of kosher certification in the Netherlands, halal certification in the Netherlands, the regulation of religious slaughter in the Netherlands, the regulation of kosher food in the United States and the regulation of religious slaughter in the United States. The next section compares the regulation in the Netherlands with the regulation in the United States. The final section searches for an explanation for the different position of the Dutch and US public authorities on the regulation of halal and kosher food.

12.2 Regulating halal and kosher food

In addition to Islam, many other religions also forbid certain foods or have specific requirements related to food. There are Jewish, Islamic, Hindu and Buddha dietary laws. Food has always been the subject of taboos and obligations. Which food we prefer and what we consider fit for (human) consumption differs depending on the place and time we live and the faith we adhere to. Religious dietary laws are important to observant Jewish and Muslim populations, although not all the faithful comply with the religious dietary laws.

Islamic dietary laws determine which foods are permitted for Muslims. *Halal* means permitted, whereas *haram* means prohibited. Several foods are considered harmful for humans to consume and are forbidden. This is expressed by the prohibition of the consumption of pork, blood, alcohol, carrion and meat that has not been slaughtered according to Islamic prescriptions. Meat is the most strictly regulated food. The animal (of a permitted species) must be slaughtered by a sane adult Muslim by cutting the throat quickly with a sharp knife. The name of Allah must be invoked while cutting. The question whether stunning is allowed remains an issue of debate, both within and beyond the Muslim community. The rules for foods that are not explicitly prohibited by the *Quran* may be interpreted differently by various scholars.[413]

Jewish dietary laws (*kashrut*) determine which foods are fit for consumption by observant Jews (*kosher*). It is a complex and extensive system with many detailed prescriptions concerning the production, preparation and consumption of food. The prescriptions are laid down in Jewish biblical and rabbinical sources. Kosher laws deal predominantly with three issues: prohibited foods (e.g. pork, shellfish and rabbit), prescriptions for religious slaughter (*shechita*) and the prohibition on preparing and consuming dairy products and meat together. In addition, there are numerous prescriptions dealing with special issues such as wine and grape juice, cooking equipment and Passover. Ruminants and fowl must be slaughtered by a specially trained religious slaughterer (*shochet*) using a special knife. Prior to the slaughter, the *shochet* makes a blessing. The animal is not stunned. Slaughtered animals are inspected for visible defects by rabbinically trained inspectors, particularly the lungs. Red meat and poultry have to be soaked and salted to remove all the blood. Any ingredients derived from animal sources are generally prohibited because of the difficulty of obtaining them from kosher animals. The prohibition of mixing milk and meat requires that the processing and handling of all materials and products fall into one of three categories: meat, dairy or neutral (*pareve*). To assure the complete separation of milk and meat, all equipment must belong to a specific category. After eating meat, one has to wait 3 to 6 hours before eating dairy. There is some disagreement over what constitutes kosher between the Orthodox, Conservative and Reform Jewish schools of thought.[414]

It cannot be visibly determined whether food is halal or kosher (as with other credence quality attributes relating to organic food and Fair Trade products). So, how does a consumer know which food is halal or kosher? There are basically three options:

- buying from someone of known reputation (e.g. an Islamic butcher);
- asking a religious leader which foods are permitted; or
- buying foods with a halal or kosher label.

In traditional societies, a combination of the first two options is often applied. People living in a religious community that runs all political, economic and communal matters internally rely on religious leaders and food suppliers of known

[413] For an overview of Islamic dietary laws and their interpretations see Regenstein, J.M., Chaudry, M.M. and Regenstein, C.E., 2003. The kosher and halal food laws. Comprehensive Reviews in Food Science and Food Safety 2(3): 111-127; Bonne, K. and Verbeke, W., 2008. Religious values informing halal meat production and the control and delivery of halal credence quality. Agriculture and Human Values 25: 35-47; Milne, E.L., 2007. Protecting Islam's garden from the wilderness: halal fraud statutes and the first amendment. Journal of Food Law and Policy 2: 61-83.

[414] See Regenstein *et al.* (2003), Rosenthal (1997), Sigman (2004), Hodkin (2005), Milne (2007) and Popovsky (2010) for an overview of Jewish dietary laws and their interpretations.

reputation. In bygone days, this applied to many European Jewish communities, for whom the chief rabbi was often the final authority in *kashrut* supervision.[415] In the 1930s in the Netherlands, only a few food manufacturers were under rabbinical supervision as most foods were prepared in the home.[416]

Due to the increase in industrially manufactured foods and the growing geographical distance between production and consumption (internationalisation of the food market), reliance on local suppliers and religious leaders is often no longer sufficient. Traditional local arrangements are also disrupted by migration. Nowadays, consumers who seek kosher or halal foods are dependent on a label or trademark that identifies a product as kosher or halal. The consumer has to trust the source and message of the communication. These developments have resulted in a large number of kosher certified products in US supermarkets and a growing number of halal labelled or certified products in Western European supermarkets.

The growth of halal certification in the Netherlands fits into a general pattern of growing third-party certification and other regulatory arrangements involving a mix of private and public actors. Food safety regulation currently involves a large number of public and private organisations with complementary, overlapping or competing roles.[417] The relations between public and private actors in food regulation are varied and complex and form an interesting field of study. In some cases, private regulation is largely independent from public regulation (such as the Marine Stewardship Council label for sustainable fish); in other cases, private regulation is encouraged or enforced by governmental actors (such as many industrial hygiene codes).[418]

Most private regulatory arrangements are nonetheless deeply intertwined with governmental and intergovernmental regulatory structures.[419] Food products and producers are subject to multiple regulatory arrangements. At least some of them will be public in most cases. In this respect, the halal/kosher certification system in the Netherlands appears to be rather exceptional, since governmental and intergovernmental regulators are largely absent. Governmental regulation does not encompass kosher and halal labelling and certification. It only includes prescriptions for and oversight of (religious) slaughter and general food regulations. To the general issues involved in the relation between private and public actors, the case of halal and kosher food regulation adds freedom of religion and the responsibilities of the state vis-à-vis the autonomy of religious communities.

---

[415] Epstein and Gang (2002); Sigman (2004: 523).
[416] Information from the Chief Rabbinate of Holland provided to the author (9 July 2010).
[417] Fuchs *et al.* (2011), Havinga (2006), Levi-Faur (2010), Marsden *et al.* (2010) and Van Waarden (2006).
[418] Havinga (2006).
[419] Meidinger (2009: 234).

Comparison with the United States is particularly interesting. Like the Netherlands, the United States is not an Islamic or Jewish country. Unlike the Netherlands, however, the United States is characterised by an extensive kosher certification industry and an important role for legislators, governmental enforcement agencies and courts in kosher labelling.

In this chapter, I will analyse the division of roles between state actors, the food industry, certification agencies and religious authorities in regulatory arrangements connected with halal and kosher foods in the Netherlands and the United States.

12.3 Kosher certification in the Netherlands

The Second World War decimated the Jewish community in the Netherlands. A large proportion of the current community are non-observant or observe *kashrut* partially by abstaining from pork and shellfish or not drinking milk with a meat meal. According to a rough estimate of the Chief Rabbinate of Holland, only 300-400 households keep a kosher kitchen nowadays.[420] Kosher food has not been the subject of public debate in recent years. However, slaughter without prior stunning, including slaughter in accordance with Jewish dietary laws, has come up for discussion.

Since 1945, the Chief Rabbinate of Holland has been the Dutch kosher certifying agency for ingredients, semi-finished products and end products.[421] Kosher shops and restaurants are under the supervision of local rabbinates such as the Rabbinate of Amsterdam or the Rabbinate of The Hague.

Kosher certification is particularly important for exporting foods and ingredients to Israel and the United States (as the local Dutch need for kosher products is too small to legitimise certification). The Chief Rabbinate of Holland issues kosher certificates for a single product or several products or a certificate for the whole production process. The procedure starts with a food producer applying for kosher certification. After receiving the application a supervisor visits the location to judge whether the plant or production can be certified kosher.

The Jewish dietary norms and requirements for certification do not appear in a written document of the Chief Rabbinate of Holland. The Chief Rabbinate applies Jewish laws as laid down in the Old Testament (notably Leviticus and Deuteronomy) and rabbinical directives and interpretations. After rabbinical approval of the ingredients and the equipment, the Chief Rabbinate regularly supervises the production site. Products and plants under the supervision of the Chief Rabbinate of Holland may use a seal of approval (*hechsher*) stating 'onder toezicht van het

[420] Information from the Chief Rabbinate of Holland provided to the author (9 July 2010).
[421] See: http://www.kosherholland.nl.

Opperrabbinaat voor Nederland' ('under supervision of the Chief Rabbinate of Holland') around Hebrew text. This seal is a protected hallmark. The producer pays an hourly rate for the supervision and a fee for the certificates. The frequency of supervision visits depends on the hazards involved in the particular product and plant. Thus, a Jewish butcher is supervised on a daily basis, while a kosher plant that only produces kosher foods receives a year letter. Most plants produce both kosher and non-kosher foods; these plants are visited more frequently (e.g. monthly) and every product needs a kosher certificate.

Every year, the Chief Rabbinate publishes a *kashrut* list containing products generally available in Dutch supermarkets that are permitted for Jews to eat. The listed products are not produced under the supervision of the Chief Rabbinate and do not have a kosher certificate. Instead, the Chief Rabbinate investigated the product and decided that it is permitted for Jews (though with a lower kosher standard).[422] A list of forbidden E-numbers for food additives is also included. The list is published to assist Jews living far from kosher shops. In the Netherlands and other European countries, the principal method of rabbinical approval is via a kosher list published by local or national Jewish authorities, and Jewish consumers rely heavily on these authorities, such as the Chief Rabbinate.[423]

12.4 Halal certification in the Netherlands[424]

The domestic market for halal foods in the Netherlands is more sizeable than the kosher market. In the Netherlands, the influx of migrants has resulted in a growing number of Muslims. It is estimated that in 2006 about 5% of the population in the Netherlands was Muslim (837,000 persons).[425] This has only recently become visible in supermarkets, shops, hospital and corporate cafeterias, where halal products are being introduced. In the media and in the Dutch parliament, some persons and organisations have objected to this development as unwelcome Islamisation.[426] Religious slaughter has also been criticised from the perspective of animal welfare.[427]

Van Waarden and Van Dalen distinguish between the 'official' and 'international' halal market and the local 'uncle and auntie' market. The official market includes large-scale exporting companies, large supermarket chains and certification agencies. The 'uncle and auntie' market is a local market based on trust in the

[422] Information from the Chief Rabbinate of Holland provided to the author (9 July 2010).
[423] Bergeaud-Blackler *et al.* (2010: 27).
[424] The information on halal certification in the Netherlands is based mainly on Van Waarden and Van Dalen (2010).
[425] Van Herten and Otten (2007).
[426] E.g. Handelingen TK 17 (1013) of 29 October 2003. See also: http://scepticisme.prikpagina.nl/read.php?f=1056&i=241469; http://www.elsevier.nl/opinie/reacties_op_commentaar.
[427] See: http://www.dierenbescherming.nl/downloads/docs/offerfeest_2006.doc.

local butcher and grocery store of the same social and ethnic group. The domestic halal market in the Netherlands is still dominated by this local market.[428]

Unlike kosher certification, there is not one single halal certifier in the Netherlands. There are about 30-40 different halal certificates. These include larger, official certifying bodies (such as the Halal Feed and Food Inspection Authority, Halal Quality Control, Halal Correct and the Halal Audit Company), small - often individual - certifiers (imams), self-certifiers (businesses that label their brand or shop as halal, such as Mekkafoods) and the international certification bodies (such as JAKIM, IFANCA and IHI Alliance). This chapter does not consider the international certification bodies. Many certifiers operate under the supervision of or are recognised by an Islamic authority such as Majlis Al Ifta, the Association of Dutch imams, JAKIM Jabatan Kemajuan Islam Malaysia (Department of Islamic Development Malaysia),[429] Majelis Ulama Indonesia MUI (Indonesia Council of Ulama), the Islamic Board for Fatwa and Research of the Islamic University of Rotterdam, the Al Azhar University of Cairo or an imam. In the case of some certifiers, it is unclear whether they are under religious supervision or recognised by an Islamic authority.

Dutch halal certifying bodies are not recognised by the Dutch Council of Accreditation and most of them would not qualify for accreditation because a written document containing all requirements for certification is not available. Some halal hallmarks are legally protected by civil law, as is unauthorised use of the Halal Feed and Food Inspection Authority logo, which is protected by international copyright law.[430]

12.5 Religious slaughter in the Netherlands

Specific requirements for religious slaughter are included in both Jewish and Islamic dietary law. Orthodox Jewish communities and some Islamic communities do not accept that animals are stunned before slaughtering. In the Netherlands, slaughtering animals without prior stunning is prohibited, as in all European Union countries.[431] Since the adoption of the first Dutch laws prohibiting slaughtering without prior stunning in 1922, an exception has been made for Jewish slaughter.

[428] Van Waarden and Van Dalen (2010); Bonne and Verbeke (2008).

[429] There are two Dutch bodies on the list of approved foreign halal certification bodies of MUI: Halal Feed and Food Inspection Authority and Total Quality Halal Correct. See: http://www.mui.or.id. These halal certifiers are also on the list of JAKIM. The Control Office of Halal Slaughtering also appears on the JAKIM list. See: http://www.jurnalhalal.com/2010/04/halal-bodies-recognized-by-jakim.html.

[430] See: http://www.halal.nl.

[431] Oosterwijk (1999); Havinga (2008); Kijlstra and Lambooij (2008); Ferrari and Bottoni (2010). Council Directive 93/119/EC of 22 December 1993 on the protection of animals at the time of slaughter and killing, replaced by Council Regulation (EC) No. 1099/2009 of 24 September 2009 on the protection of animals at the time of killing, which will enter into force in 2013.

Since 1996, a similar exception has been made for Islamic slaughter.[432] Religious slaughter is legally defined as slaughter of animals without prior stunning taking place according to Jewish or Islamic rite. Slaughter according to Islamic or Jewish rite can only take place in authorised slaughterhouses after notifying the Dutch Food and Consumer Product Safety Authority (VWA) in advance.[433] More detailed requirements are laid down in a special Regulation on Religious Slaughter. These requirements include the avoidance of suffering, instructions for the handling and restraining of animals and slaughter techniques.[434] Veterinarians of the VWA supervise religious slaughter, and non-compliance with the above-mentioned requirements may result in a warning, a fine or the stoppage of slaughtering. This supervision does not include compliance with religious laws.

With its specific provisions for religious slaughter, Dutch law implicitly assumes that religious slaughter equals slaughter without prior stunning. In the Netherlands, religious slaughter is disputed. The current wave of criticism comes from three sides. First of all, animal rights organisations object to the inhumane and painful treatment of animals. Some of these organisations want an official ban on religious slaughter, while others are trying to convince Islamic and Jewish organisations to accept some form of reversible stunning. Secondly, some right wing politicians and political organisations object to the growing Islamisation of Dutch society. These Islamophobic critics perceive religious slaughter as a clear sign of the intrusion of Islamic norms and the unwanted permissiveness of the left-wing elite. Finally, veterinarian organisations in the Netherlands and Europe advocate obligatory stunning prior to slaughter. They argue that scientists agree that slaughter without prior stunning causes unnecessary pain and suffering for the animals. The Federation of Veterinarians of Europe is of the opinion that the practice of slaughtering animals without prior stunning is unacceptable under any circumstances.[435] Dutch veterinarians point to an ethical dilemma for veterinarians who have to supervise religious slaughter. They see religious beliefs as dynamic and as allowing change in order to improve animal welfare.[436] As long as slaughter without prior stunning is allowed under national or European legislation, the veterinarians recommend stipulating specific minimum requirements.

[432] Gezondheids- en welzijnswet voor dieren III-44 (Animal Health and Welfare Act of 24 September 1992). Many other EU countries also allow slaughter without prior stunning for religious reasons (e.g. Germany, UK, Italy and Belgium). However, some countries do not allow slaughter without prior stunning (e.g. Sweden, Norway and New Zealand). Denmark, Finland and Austria do allow the killing of unstunned animals but require immediate post-cut stunning (Kijlstra and Lambooij 2008: 5-6; Ferrari and Bottoni 2010: 10). Until 1975, the Dutch government was not prepared to make an exception for Islamic slaughter, as an explicit prohibition on stunning is not found in the *Quran* (Oosterwijk 1999: 111-112).

[433] Until 2006, a declaration by the religious authority regarding the number of animals was required.

[434] Besluit Ritueel Slachten 1996.

[435] Federation of Veterinarians in Europe, 'Slaughtering of animals without prior stunning', FVE position paper FVE/02/104 (2002). Available at: http://www.fve.org.

[436] KNMvD-standpunt over het onbedwelmd slachten van dieren, 15 March 2008.

At the end of 2007, a motion in favour of a prohibition on slaughter without prior stunning was rejected by the Dutch parliament. Almost half the votes (68 of the 150 votes) were in favour of such a prohibition. Freedom of religion was the most common reason for opposing the motion. A pending private member's bill prescribing prior stunning in ritual slaughtering has gained a majority in Dutch Parliament in April 2011.[437]

The coming regulation of the European Parliament and of the Council on the provision of food information to consumers includes a provision that meat and meat products derived from animals that have not been stunned prior to slaughter (i.e. have been ritually slaughtered) should be labelled as such ('Meat from slaughter without stunning'). This amendment was adopted by the European Parliament at first reading on 16 June 2010 (326 votes in favour, 270 against and 68 abstentions).[438] The objective of the provision of food information is to provide a basis for informed choices and safe use of food. This provision was not adopted by Coreper and the Council.[439] In December 2010 Dutch Parliament accepted a similar resolution urging the government to provide for obligatory labelling of halal meat to enable free choice for consumers to avoid ritually slaughtered meat.[440]

12.6 Regulation of kosher food in the United States

In the United States, kosher certification and supervision is quite different from the situation in the Netherlands. The domestic kosher market is extensive. In the Northeast of the United States, nearly half of the products on supermarket shelves are certified kosher. In the United States, kosher food is also bought by many non-Jewish consumers, because they believe it to be healthier, natural and higher-quality food.[441]

The process of kosher supervision is very similar to kosher certification in the Netherlands. A food manufacturer initiates the supervision and certification process

[437] Bill from MP Thieme (Political Party for Animal Rights, Partij voor de Dieren) Kamerstukken II 2009/10, 31 571. The majority was gained after the Labour party decided to support the bill in April 2011: http://nu.pvda.nl/berichten/2011/04/PvdA-steunt-voorstel-voor-verplichte-verdoving-bij-rituele-slacht.html#.
[438] See: http://www.europarl.europa.eu/sides/getDoc.do?type=TA&reference=P7-TA-2010-0222&language=EN&ring=A7-2010-0109.
[439] European Parliament, MEPs set out clearer and more consistent food labelling rules, Press release, 16 June 2010. Available at: http://www.europarl.europa.eu/en/pressroom/content/20100615IPR76127; http://halalfocus.net/2010/12/08/world-halal-forum-europe-approves-eus-rejection-of-amendment-205/. Jewish organisations lobbied against this obligatory labelling: 'Shechita fears if European law changes' (http://www.thejc.com/news/uk-news/32930/shechita-fears-if-european-law-changes); http://halalfocus.net/2010/12/07/uk-a-cautious-welcome-from-shechita-uk-to-eu-council%E2%80%99s-rejection-of-205/); 'Kosher bosses welcome EU labelling decision' (http:// www.meatinfo.co.uk/news/fullstory.php/aid/11958/Kosher_bosses_welcome_EU_labelling_decision.html).
[440] Resolution of MP Graus (PVV), Kamerstukken II 2010/11 32 500 XIII, nr. 111. http://www.tweedekamer.nl/images/07-12-2010_tcm118-215256.pdf, last visited 17 February 2011.
[441] Sullivan (1993: 201), Sigman (2004: 537, 544-545).

(mostly in response to an appeal from consumers or a buyer). The certifier investigates the product, the production process and location based on a contract between certifier and manufacturer. The kosher supervision agency (KSA) will pay a qualified inspector to make continual visits. Sometimes a representative of the KSA is required to be present to monitor during production (e.g. in the case of matzo for Passover).

Unlike in the Netherlands, there are many competing KSAs in the United States. Four KSAs are estimated to certify 90% of kosher products. The largest KSAs are often non-profit organisations, such as the Kashruth Division of the Union of Orthodox Jewish Congregations of America (OU). The organisations and individuals who supervise and certify kosher food are all Jewish.[442] There are over 300 registered kosher symbols used by KSAs in the United States.[443] The legal status of these certification marks is that of a protected trademark. Kosher supervision agencies can be divided into three broad categories: the large organisations that dominate supervision of larger food companies, individual rabbis with standards beyond the normative Orthodox standard and individual rabbis who are more 'lenient' (e.g. Conservative rabbis).[444] Most KSAs do not work with a written document defining general standards for kosher certification.[445]

An even more important difference with the Netherlands is the involvement of state law and state enforcement in the United States. Many states have specific laws governing kosher food. Kosher food is a very attractive market, which creates a strong temptation to pass off non-kosher food as kosher. Federal and state governments have enacted laws to protect consumers from this fraud. In 1922, the state of New York passed the first state-wide kosher fraud law to protect consumers from non-kosher food sold as kosher. Many other states followed and issued a kosher fraud statute.[446] The Orthodox Union was the main force behind the campaign to enact kosher fraud statutes.[447] Most kosher fraud statutes operate in a similar fashion. They generally prohibit the advertisement or sale of food labelled 'kosher' unless it conforms to state-defined food preparation and handling requirements. In these laws, kosher is defined as 'prepared or processed in accordance with orthodox Hebrew religious requirements' or similar.[448] Some of these statutes are part of the state criminal code, while others are part of state codes on public health, food regulation or commerce and trade. The penalty for

[442] Sigman (2004: 536).
[443] Sigman (2004: 525). Some of the most important KSAs are: the Union of Orthodox Jewish Congregations of America (OU, est. 1924), the Organized Kashrus Laboratories (OK, est. 1935), the Star-K Kosher Certification (est. 1947) and the KOF-K Kosher supervision (est. 1968).
[444] Regenstein *et al.* (2003: 125).
[445] Sigman (2004: 531-532).
[446] Rosenthal (1997: 951, note 1) lists 23 kosher fraud statutes.
[447] This was before the OU entered the kosher supervision and certification business. Sigman (2004: 552).
[448] Gutman (1999: 2369); Sigman (2004: 553).

violating these laws can be fines or even imprisonment.[449] Some statutes vest the power to inspect compliance with the law in the attorney general, a commission or a special agency. States such as New York and New Jersey established a Bureau of Kosher Enforcement and employed rabbis to enforce compliance.[450] In 2002, the Kosher Law Enforcement Division of the New York Department of Agriculture and Markets conducted 7,500 inspections in New York State to assure consumers that food products offered for sale as kosher were indeed kosher.[451] The situation seems to be quite different now, as the division is facing cuts of more than 95% to its budget for staff and kosher food inspections.[452]

Based on his analysis of reputation-based non-legal sanctions, private law remedies and consumer protection laws, Sigman concludes: 'There is no evidence that state kosher fraud enforcement plays a significant role in preventing wilful kosher fraud; nor is there evidence that enforcement addresses the problems facing kosher consumers today.'[453]

Kosher fraud statutes have been challenged in court for being unconstitutional. At first, the courts upheld the kosher statutes. But from 1992 onwards several courts invalidated kosher statutes for creating excessive state entanglement between church and state and advancing and inhibiting religion.[454] Most of these court cases are initiated by kosher establishments or kosher certifiers, often after a state inspector had found violations whereas a supervising rabbi or kosher certifier claims that everything is in compliance with the Jewish laws of *kashrut*. Two elements have led the courts to decide that kosher fraud statutes are unconstitutional under the First Amendment: kosher is defined according to Orthodox standards and the relevant state or local governments employed a rabbi as state-appointed official. The court found that the statute required the state to take an official position on the interpretation of Jewish dietary laws and advance the Orthodox definition of kosher. In response to this decision, several states changed their laws into a kosher disclosure statute (e.g. New Jersey, New York and Georgia). For example, the new 2010 Georgia Kosher Food Consumer Protection Act requires that 'a person who

---

[449] Sigman (2004: 554); Gutman (1999: 2369, note 144).

[450] Sigman (2004: 554). Division of Kosher Law Enforcement, New York State Department of Agriculture and Markets. See: http://www.agmkt.state.ny.us/kosher.

[451] See: http://www.agmkt.state.ny.us/KO/KOHome.html.

[452] See: http://www.theyeshivaworld.com/news/General+News/65637/Do-Not-Cripple-NYS-Kosher-Law-Enforcement.html. New York State Kosher-Enforcement Division wiped clean. Available at http://newyork.grubstreet.com/2011/01/new_york_state_cuts_every_last.html; Restore kosher division, new N.Y. Gov. Cuomo urged (3 January 2011). Available at: http://jta.org/news/article-print/2011/01/03/2742388/jewish_groups_call_on_ny_gov.

[453] Sigman (2004: 601).

[454] *Ran Dav's County Kosher Inc* v. *State*, 608 A.2d 1353 (N.J. 1992), *Barghout* v. *Mayor of Baltimore*, 66 F.3d 1337, 1342-46 (4th Cir. 1995), *Commack Self-Service Kosher Meats, Inc.* v. *Weiss*, 2002 U.S. App. LEXIS 9576 (2nd Cir. June 21, 2002).
Ciesla (2010); Levenson (2001); Milne (2007); Popovsky (2010); Rosenthal (1997).

makes a representation regarding kosher food shall prominently and conspicuously display on the premises on which the food is sold, in a location readily visible to the consumer, a completed kosher food disclosure statement...'.[455] A kosher food disclosure statement shall state in the affirmative or negative whether the person operates under rabbinical or other kosher supervision, the name and address of the supervising rabbi, agency or other person and the frequency with which the supervising person visits the establishment. The statement shall state whether the person sells or serves only food represented as kosher, or both kosher and non-kosher food, and whether meat, dairy and *pareve* food is sold or served. The Georgia law covers many other issues on which information has to be disclosed, such as rabbinical or kosher supervision in the slaughterhouse, glatt kosher meat and the use of separate work areas and utensils for kosher and non-kosher food and for kosher meat, kosher dairy and kosher *pareve* food. In the case of violation of this law, the administrator or the court may issue a cease and desist order or a civil penalty.

New York State's Kosher Law Protection Act 2004 requires producers and distributors of kosher food to 'have registered with the department the name, current address and telephone number of the person certifying the food as kosher'.[456] Special requirements are included for disclosure of information on the soaking and salting of kosher meat.

Only recently, some states also enacted a similar law for halal food.[457] These laws often define halal as 'prepared under and maintained in strict compliance with the laws and customs of the Islamic religion' or 'in accordance with Islamic religious requirements'.[458] 'Despite the widespread disagreement among and within Islamic 'schools of thought' over halal food, various individual states in the United States have attempted to define, by legislative edict, this inherently religious term. The stated purpose behind such legislative definitions of *halal* is to prevent the fraudulent representation of food as being *halal*. The constitutionality of these government-enacted definitions of *halal* is uncertain.'[459]

12.7 Religious slaughter in the United States

Religious slaughter is an important part of kosher and halal requirements and needs permanent supervision by certifying agencies or religious authorities. The

[455] See: http://www.legis.ga.gov/legis/2009_10/pdf/hb1345.pdf.
[456] See: http://public.leginfo.state.ny.us.
[457] E.g. the Halal Food Consumer Protection Act New Jersey (2000). Available at: http://www.njleg.state.nj.us/2000/Bills/a2000/1919_i1.htm. Also California (2005), Illinois (2005), Michigan (2005), Minnesota (2005) and Texas (2005). See Milne (2007: 63, note 9); Regenstein *et al.* (2003: 127).
[458] Milne (2007: 71).
[459] Milne (2007: 63).

Humane Methods of Slaughter Act (HMSA) stipulates that the slaughter of animals should be humane. In the Act, two methods are found to be humane:

1. when the animal is rendered insensible to pain by a gunshot or electrical, chemical or other means; or
2. when the slaughter takes place in accordance with the ritual requirements of the Jewish faith or any other religious faith.

Furthermore, the US Code contains the following clause: 'nothing in this chapter shall be construed to prohibit, abridge, or in any way hinder the religious freedom of any person or group...'.[460] The inspectorate of the Department of Agriculture (USDA) has only a minimal role in monitoring ritual slaughter. It is required to request written verification of slaughter methods from the religious official who has the authority over the enforcement of religious dietary requirements and to verify that animals are handled in a humane manner prior to the slaughter.[461]

Hodkin argues that if an animal is not stunned prior to slaughter and the kosher slaughter fails, the Humane Methods of Slaughter Act is violated. Hodkin advocates that USDA inspection personnel are well versed in the requirements of Jewish law to guarantee compliance with Humane Methods of Slaughter Act. She reached this conclusion after analysing a scandal about animals treated cruelly in a kosher slaughterhouse (video clips on YouTube and elsewhere generated protest).[462]

12.8 Comparative conclusions

In both the Netherlands and the United States, slaughter without prior stunning in accordance with religious requirements is permitted. The US regulations exclude religious slaughter from legal requirements by including a provision in the Humane Methods of Slaughter Act stating that slaughter in accordance with ritual requirements of a religious faith is humane. In the Netherlands, religious slaughter is subjected to more legal requirements, and compliance with these requirements is monitored and enforced by the Dutch Food and Consumer Product Safety Authority. These requirements do not extend to halal or kosher claims.

In the Netherlands, the certification of halal and kosher foods is left entirely to private actors. Halal and kosher certification is not regulated by public law,

---

[460] See: http://assembler.law.cornell.edu/uscode/html/uscode07/usc_sec_07_00001902----000-.html. See also Milne (2007).

[461] Hodkin (2005: 146).

[462] The animal rights group PETA (People for the Ethical Treatment of Animals) has filed a complaint against Agriprocessors (one of the largest kosher slaughterhouses in the world) for abusing animals. See: http://www.peta.org/Automation/AlertItem.asp?id=1192; http://www.goveg.com/kosher.asp. PETA complaint against kosher slaughterhouse. Available at: http://www.jewfaq.org/peta.htm; Statement of rabbis and certifying agencies on recent publicity on kosher slaughter. Available at: http://www.ou.org/other/5765/shichita2-65.htm. Agriprocessors is certified as kosher by the Orthodox Union. See also Gross (2005).

and governmental agencies are not involved in monitoring and enforcing halal and kosher regulations. Halal certification in the Netherlands is dominated by commercial actors, verified by religious authorities. Kosher certification is executed mainly by the Chief Rabbinate of Holland, a religious actor.

In the United States, religious authorities dominate the kosher and halal certification industry. State laws and state enforcement agencies are in place to protect consumers of kosher or halal foods from misrepresentation. The current laws focus on public disclosure of information and trademark protection. Relics of the previous legitimation of Orthodox Jewish standards by state law and institutions can still be found. It is not clear how the new kosher disclosure acts are being enforced. The enforcement of these acts is the subject of new law suits.[463]

In the United States, which is a liberal market economy, one would expect to find minimal governmental interference in the kosher and halal industry. In the Netherlands, which is a corporatist welfare state, one would expect to find a high level of state involvement in the regulation of the kosher and halal industry.

The regulation of religious slaughter is in line with these expectations. In the United States, one finds loose regulation with only minimal requirements providing freedom for (religious) variation. In the Netherlands, one finds more detailed regulations with state inspectors and state veterinarians controlling compliance in slaughterhouses.

However, a comparison of halal and kosher certification in the two countries reveals a different pattern. Contrary to expectations, the regulation of kosher certification is more state-centred in the United States than in the Netherlands. Even after the constitutional challenges to the former kosher fraud statutes, many US states have laws to protect the halal or kosher logo. Several states also have a special kosher enforcement agency to inspect compliance with legal requirements. Conflicts over the interpretation and enforcement of kosher laws have resulted in a substantial amount of case law. Although the role of governmental institutions in regulating the kosher industry has been limited in the past decades, governmental agencies still play a significant role.

By contrast, halal and kosher certification has been left entirely to commercial and religious organisations in the Dutch corporatist welfare state. 'Halal' and 'kosher' are not legally defined and protected designations. State authorities in

[463] Butchers file another lawsuit challenging NY State's kosher certification authority. Long Island Business News, 4 March 2005. Available at: http://www.allbusiness.com/legal/1027321-1.html. New York – conservative rabbis fight orthodox kosher state laws, 30 July 2008. Available at: http://www.vosizneias.com/post/read/18724/2008/07/30/new-york-conservative-rabbis-fight-orthodox-kosher-state-laws/print/with-images.

the Netherlands do not regard this as their task and avoid getting mixed up in religious matters.

This comparative conclusion raises questions concerning the protection of consumers in the Netherlands and the relationship between state and religion in the United States.

In the Netherlands, public law does not protect consumers from misrepresentation or fraud involving food sold as kosher or halal. Does this result in many stories of deception of Islamic and Jewish consumers? How do kosher and halal certificates succeed in establishing credibility without involvement of the state?

Focus groups of Jewish consumers in Amsterdam and five other European cities revealed that Jewish consumers rely on rabbinical supervision. Most participants said they themselves trusted all kosher certificates, but others only trust known stringent supervisors.[464] However, halal consumers in Amsterdam and five other cities questioned the reliability of halal food labels and certificates. They believed that halal labels should be authenticated by trustworthy religious institutions and preferred more traditional and personal networks of supply such as butcher shops. The focus group participants in the Netherlands believed that their food supply chain was trustworthy.[465]

In 1999, a Muslim woman (of Pakistani origin) bought and consumed a halal veal snack (croquette) in Amsterdam. After she found out that the meat was from animals that had not been religiously slaughtered, she claimed damages against the manager of the snack bar. The manager, a Muslim woman of Moroccan origin, claimed the snack was halal, because it contained no pork. She had put stickers with the word 'halal' in Arabic on products not containing pork at the request of some Moroccan youths. The judge in the summary proceedings declined to decide what 'halal' meat is. The chain of snack bars did order the manager to remove the 'halal' stickers.[466] This case illustrates the different interpretations of the concept of 'halal' ranging from 'contains no pork' to 'contains no pork and animals are religiously slaughtered without prior stunning and citing the name of Allah'. It also illustrates that it can be hard for an observant Muslim consumer to know what food is allowed and what is not.

Recently, complaints have been reported about 'unauthorised' halal certificates and about organisations in the Netherlands and Germany issuing 'fake halal

[464] Bergeaud-Blackler *et al.* (2010: 26-28).

[465] Bergeaud-Blackler *et al.* (2010: 46-48). Similar observations in Bock and Wiersum (2003: 22), Bijzondere ontmoetingen over vlees (2002: 7-9) and Bonne and Verbeke (2006).

[466] Rechter buigt zich over religieus gehalte van kroket. Het Parool, 1 September 1999; Geen straf voor foute kroketten. Het Parool, 11 September 1999; Halal-producenten wensen keurmerk voor rein voedsel. Trouw, 29 April 2000.

certificates'.[467] Van Waarden and Van Dalen found that all but one of their respondents from halal certifiers in the Netherlands agreed that the Dutch government should play a key role in the development of a national halal certificate in the Netherlands.[468] Only the largest halal certifier did not agree, as they already regard themselves as the supreme Dutch halal authority. Most of the 11 firms that produced or sold halal foods that we interviewed, indicated to be in favour of more government involvement.[469] The Islamic community in the Netherlands is said to have a need for a general halal hallmark, among other things because many products sold as halal are suspected of not really being halal.[470]

The Dutch Food and Consumer Product Safety Authority (VWA) is reluctant to become involved in religious matters. They are not involved in controlling whether foods are halal (or kosher) but only monitor and enforce legal requirements (hygiene, mandatory labelling and food safety). The Dutch Ministry of Agriculture has adopted the recommendation of a consumer platform to strive for a single halal certificate in the Netherlands and to define 'halal' in law.[471]

The newly established European Association of Halal Certifiers (AHC-Europe) aims to bring order and unity to the halal food sector in Europe. One of the founders has stated that governments should take the necessary measures to force certifiers to operate in line with the rules defined by AHC-Europe.[472] Currently, the Netherlands normalisation institute NEN explores the need of the market to be engaged in the development of a European Union halal standard.[473]

Van Waarden and Van Dalen conclude that governmental cooperation seems to be unavoidable and that a formal registration of the term 'halal' is required.[474]

---

[467] De Volkskrant, 27 November 2009; see: http://halalfocus.net/2009/10/04/new-organization-to-sort-out-rotten-apples-in-european-halal-food-market; Seada Nourhussen, Halal-label kan ook op 'onrein' toetje zitten. De Volkskrant, October 6 2003; Halal vlees blijkt vaak 'onrein'. Het Parool, 16 February 2005; Vlees van varkens in kosjere en halal kip. Het Parool, 15 June 2003; J. Siebelink, 2007. Herrie om halal. Reportage moslims weten niet wat ze eten. See: http://www.halalpagina.nl/index.php?pagina=halal2.

[468] Van Waarden and Van Dalen (2010).

[469] Havinga and Gerards (2011: 16-17). However, respondents from firms that produced kosher products were not in favour of government involvement.

[470] Smits and Van den Berg (2003: 32).

[471] Smits and Van den Berg (2003), confirmed by telephone on 19 July 2010. No concrete measures have been taken so far. The Ministry of Agriculture expects European regulation on halal certification (or on certification more generally).

[472] See: http://halalfocus.net/2009/10/04/new-organization-to-sort-out-rotten-apples-in-european-halal-food-market.

[473] The norm will be developed from the Austrian halal norm ONR 14200 'Halal food – Requirements for the food chain', NEN organiseert halal-norm bijeenkomst. Available at: http://www.evmi.nl/nieuws/marketing-sales/9389/nen-organiseert-halal-norm-bijeenkomst.html; http://www.nen.nl/web/Normen-ontwikklen/Nieuwelopende-normtrajecten/Commissie-in-oprichting.htm.

[474] Van Waarden and Van Dalen (2010).

The *kashrut* expert rabbi of the Chief Rabbinate of Holland recalls two cases of fraud in which a *hechsher* (seal of approval) was used without the product being supervised. The first case, involving canned mushrooms that carried the rabbinical hallmark but were produced without rabbinical supervision, was detected by the Israeli authorities. The Chief Rabbinate initiated summary proceedings before a civil court which found it in favour, while the manufacturer had to pay compensation. In the other case, which was settled, a pastry manufacturer agreed to pay the amount claimed by the lawyer. The Chief Rabbinate spokesman noted that it would have been impossible to take legal action against a person who wrongfully describes his product or site as kosher: 'In the United States, when I say my product or premises are kosher, this should be on solid grounds: which rabbi declared it kosher, is this rabbi recognised and so on. The United States has a high penalty in such cases. Not in the Netherlands. Even calling a ham sandwich kosher is not an offence under Dutch law.'[475]

The involvement of state law, state enforcement officers and the court system in halal and kosher certification of food is a delicate issue, since it relates to the separation of state and religion and freedom of religion. The principle of the separation of state and religion implies that government should be neutral towards religious matters and not biased in favour of or against a particular faith. Derogating from a general legal provision implies that the situations and actions to which this exemption applies or does not apply need to be defined. Does the exemption only apply to religious slaughter according to Jewish rites or also to slaughter in accordance with Islamic law or other religions? An exemption is based on particular religious requirements. Does it allow slaughter without prior stunning in recognised slaughterhouses or are all Muslims allowed to slaughter an animal for the Feast of the Sacrifice? Dutch law implicitly assumes religious slaughter to be slaughter without prior stunning. In the United States, as Hodkin points out, state inspectors even have to decide whether slaughter is in compliance with detailed religious requirements when they want to enforce humane slaughter methods in a kosher slaughterhouse.[476] This type of decision is difficult for governments, because it means that they have to intervene in religious matters or to rely on some religious authority to make the decision. This is particularly hard when there is no consensus within the religious community or when it generates resistance in the rest of the society. There is no univocal interpretation of 'kosher' within the Jewish community in the United States, and there is no univocal interpretation of 'halal' in the Islamic community in either the Netherlands or the United States. So far, Dutch law does not lay down what constitutes 'halal' or 'kosher' food. Many US states used to have laws that define kosher food according to Orthodox Jewish standards. This has resulted in entanglement between state and religion.

---

[475] Information from the Chief Rabbinate of Holland provided to the author (9 July 2010).
[476] Hodkin (2005).

12.9 Explaining the different position of the government

How to explain the more state-centred regulation in the US compared to the Netherlands? Our findings resemble the findings of Boström and Klintman, who compare standardisation of organic food in the United States and Sweden.[477] The organic food scene in the United States is characterised by many different regulatory schemes, which poses problems for consumers, producers, retailers and importers. The federal US government controls organic food standardisation, framing organic food as a marketing label. In Sweden, a well-reputed non-governmental organisation (KRAV) is allowed to audit organic production and ensure that EU regulations are being followed. Organic food is framed as an eco-label in Sweden. For Boström and Klintman, these different patterns of standardisation reflect traditional political, organisational and regulatory characteristics in the two countries. The open, consensus-building political culture in Sweden makes state and non-state actors willing to communicate, negotiate and search for pragmatic solutions. The political culture in the United States is more polarised, political authorities and organic actors are antagonistic, the general level of trust in the federal government is low and the government is willing to regulate (regulatory culture). In addition, the national organisational structure influences practices and debates. In Sweden, KRAV has an inclusive form of organisation that brings together all interest groups (with members from environmental NGOs, organic farmers' organisations and organic food manufacturers). Such an inclusive organisational platform is lacking in the United States, which leads to polarised debates. Finally, the regulatory arrangement itself triggers conflict in the US case. The centralisation of the standardisation process leaves no space for an organic movement to set stricter standards of its own, because the federal government sets minimum and maximum requirements. In contrast, KRAV has gained legitimate status within the EU regulatory framework but at the same time also has its own stricter rules.

Can these factors also explain the different patterns in kosher and halal certification in the Netherlands and the United States?

First of all, political culture seems to be important. The debates and law suits on kosher laws and kosher certification in the US case show antagonistic relations between state, religious and commercial actors and also within the Jewish community. As in the case of organic food, governments are also willing to regulate kosher food. Even after the kosher fraud laws were found to be unconstitutional, the existing laws were replaced by new laws and enforcement agencies stayed in place. Consumer rights are particularly important in US political culture, and framing the issue as consumer protection has contributed to successful lobbying for kosher laws.

---

[477] Boström and Klintman (2006).

The traditional consensual political culture in the Netherlands makes the Dutch government reluctant to regulate an industry without consensus among all interested parties, as in the case of halal certification.

Comparative studies on environmental regulation show that the US government and the US public do not trust industry to comply with regulations without strict enforcement, whereas in the Netherlands and other European countries the government and the public do generally trust industry to comply with regulations and act responsibly.[478] In this context, industrial self-regulation is supported widely in the Netherlands. However, this does not imply that government takes no part in the regulation of industry. Thus, the Dutch government and the European Union are involved in the certification of organic food, with EU law defining what may be labelled as 'organic'. In addition, the Dutch Food and Consumer Product Safety Authority (VWA) inspects fast food outlets that claim to use healthy frying fat, although this is not a legal requirement but just a voluntary private regulation.[479] At the same time, however, the VWA does not inspect the deception of consumers by means of fake-halal certificates due to the lack of a legal requirement to do so. The difference between these two cases seems to be that the use of liquid frying fat is framed as a health issue, whereas halal is framed as a religious issue.

Secondly, national organisational structure also seems to be important in the case of halal and kosher certification. Kosher certification in the Netherlands is dominated by the Chief Rabbinate, which apparently manages to unite all Jewish voices (possible differences of opinion are not voiced externally). The situation in the Netherlands after the Second World War probably facilitated this (only a small Jewish population was left, united against a hostile environment). However, this is not the inclusive form of organisation of KRAV. In the United States, which has experienced Jewish immigration from different countries and lacks the centuries of rabbinical tradition found in Europe, the battles over supervision of Jewish dietary practice and over what is and is not kosher were particularly intense.[480]

In the world of halal certification in the Netherlands, a dominant or central organisation is lacking and many competing organisations strive for a share of the market. This is indicative of a situation in which migrants from many different countries with different food and religious traditions are not united in a single association. Initiatives to establish a national halal hallmark have failed so far.

[478] Vogel (1986); Kagan (1990); Verweij (2000).

[479] The 'Verantwoord Frituren' (Responsible Frying) campaign of the Dutch Catering Industry Association and the Public Information Office for Margarine, Fat and Oil. The VWA controls the use of liquid frying fat to protect consumers from deception. See: http://www.vwa.nl/onderwerpen/levensmiddelen-food/dossier/frituurvet/wat-is-er-geregeld.

[480] Epstein and Gang (2002).

In the United States Orthodox Jewish organisations have lobbied strongly for the establishment of kosher fraud laws. These laws and their public enforcement are particularly strong in US states with a large and powerful Jewish community (e.g. New Jersey and New York). Only recently, Muslim organisations have also lobbied for halal laws in the United States. To the best of my knowledge, the Jewish community in the Netherlands has not campaigned strongly for state regulation. Muslims in the Netherlands are not a powerful political group, being new migrants with internally mixed opinions. It is only recently that some Muslim organisations have advocated a leading role for the government in halal certification. Kosher food was never perceived as a social problem, and no claims were made on the government to take action. In contrast, halal food is currently associated with several 'problems', such as illegal slaughter, animal welfare in religious slaughter and unreliable halal certificates. Religious slaughter has successfully been framed as an animal welfare issue. At the same time, the political climate in the Netherlands is not in favour of Muslim immigrants, and animal rights groups have gained a strong position.

The position of religious groups and authorities in the state may also be important. Kosher certification arrangements fit into this pattern. The United States is a more religious country, while the Netherlands is more secular. The traditional system of pillarisation in the Netherlands (which has declined in recent years) implied that every faith had its own service organisations (with public funding) in such fields as broadcasting, hospitals and schools. These pillars were semi-autonomous, and the government was reluctant to interfere in their activities.

Goldstein concludes that a general principle that applied for centuries in the United States was that courts should avoid deciding religious questions. Since 1944, however, this principle has expanded into a seemingly absolute prohibition.[481] This shift may have contributed to the court decisions disabling kosher fraud laws as unconstitutional.

Thirdly, the regulatory arrangements around kosher food in the United States have reinforced the powerful position of Orthodox rabbis, excluding rabbis of non-orthodox groups from rabbinical supervision.

In conclusion, the different patterns in the regulation of kosher and halal foods in the United States and the Netherlands can be partly explained by a different division between the state and religion, the powerful Jewish political lobby in the United States in favour of state regulation, framing kosher and halal labelling as a consumer rights issue in the US and as a religious issue in the Netherlands, and the high level of trust in industry and self-regulation in the Netherlands.

[481] Goldstein (2004: 316).

References

Bergeaud-Blackler, F., A. Evans and A. Zivotofsky, 2010. Final report – consumer and consumption issues – halal and kosher focus groups results. Available at: http://www.dialrel.eu/publications.

Bijzondere ontmoetingen over vlees, 2002. Impressies consumentenpanels. Available at: http://www.minlnv.nl/portal/page?_pageid=116,1640321&_dad=portal&_schema=PORTAL&p_file_id=13830.

Bock, B., and J. Wiersum, 2003. Trust in food and the need for more information. The relationship between food attitude, trust in food safety and consumption behaviour. Reports on the Dutch focus-groups held in Wageningen. Florence University Press, Florence, Italy.

Boström, M., and M. Klintman, 2006. State-centered versus nonstate-driven organic food standardization: a comparison of the US and Sweden. Agriculture and Human Values 23: 163-180.

Bonne, K., and W. Verbeke, 2006. Muslim consumer's motivations towards meat consumption in Belgium: qualitative exploratory insights from means-end chain analysis. Anthropology of Food 5.

Bonne, K., and W. Verbeke, 2008. Religious values informing halal meat production and the control and delivery of halal credence quality. Agriculture and Human Values 25: 35-47.

Ciesla, M., 2010. New York kosher food labeling laws violate the establishment clause. Available at: http://org.law.rutgers.edu/publications/law-religion/new_devs/RJLR_ND_61.pdf.

EU, 1993. Council Directive 93/119/EC of 22 December 1993 on the protection of animals at the time of slaughter or killing. Official Journal of the European Union L 340, 31/12/1993: 21-34.

EU, 2009. Council Regulation (EC) No 1099/2009 of 24 September 2009 on the protection of animals at the time of killing. Official Journal of the European Union L 303, 18/11/2009: 1-30.

Epstein, G.S., and I. N. Gang, 2002. The political economy of kosher wars. Departmental Working Papers 200227, Rutgers University, Department of Economics, Rutgers, NJ, USA. Available at: http://ideas.repec.org/e/pep1.html.

Ferrari, S., and R. Bottoni, 2010. Legislation regarding religious slaughter in the EU, candidate and associated countries. Available at: http://www.dialrel.eu.

Fuchs, D., A, Kalfagianni and T. Havinga, 2011. Actors in private food governance: the legitimacy of retail standards and multistakeholder initiatives with civil society participation. Agriculture and Human Values 28(3): 353-367.

Goldstein, J.A., 2004. Is there a religious question doctrine? Judicial authority to examine religious practices and beliefs. Bepress Legal Series Working Paper 316. Available at: http://law.bepress.com/expresso/eps/316.

Gross, A., 2005. When kosher isn't kosher. Tikkun Magazine 20(2): 52-55.

Gutman, B.N., 1999. Ethical eating: applying the kosher food regulatory regime to organic food. Yale Law Journal 108: 2351-2384.

Havinga, T., 2006. Private regulation of food safety by supermarkets. Law and Policy 28(4): 515-533.

Havinga, T., 2008. Ritueel slachten. Spanning tussen religieuze tolerantie en dierenbescherming. In: A. Böcker, T. Havinga, P. Minderhoud, H. van de Put, L. de Groot-van Leeuwen, B. de Hart, A. Jettinghoff and K. Zwaan (eds.), Migratierecht en rechtssociologie, gebundeld in Kees' studies. Wolf Legal Publishers, Nijmegen, the Netherlands, pp. 211-220.

Havinga, T. and C. Gerards, 2011. Halal and koosjercertificering in Nederland. Een verkennend onderzoek naar deregulering van halal en koosjer voedsel in Nederland. Insitutuut voor rechtssociologie, Nijmegen, the Netherlands.

Hodkin, M., 2005. When ritual slaughter isn't kosher: an examination of Shechita and the Humane Methods of Slaughter Act. Journal of Animal Law 1(1): 129-150.

Judd, R., 2003. The politics of beef: animal advocacy and the kosher butchering debates in Germany. Jewish Social Studies 10(1): 117-150.

Kagan, R.A., 1990. How much does law matter? Labor law, competition, and waterfront labor relations in Rotterdam and U.S. ports. Law and Society Review 24(1): 35-69.

Kijlstra, A., and B. Lambooij, 2008. Ritueel slachten en het welzijn van dieren. Een literatuurstudie. Animal Science Group Wageningen UR rapport 161, Wageningen, the Netherlands.

Koolmees, P.A., 1997. Symbolen van openbare hygiëne. Gemeentelijke slachthuizen in Nederland 1795-1940. Erasmus Publishing, Rotterdam, the Netherlands.

Levenson, B.M., 2001. Not so strictly kosher. In: B.M. Levenson (ed.), Habeas codfish: reflections on food and the law. The University of Wisconsin Press, Madison, WI, USA, pp. 184-198.

Levi-Faur, D., 2010. Regulation and regulatory governance. Jerusalem Papers on Regulation and Governance No. 1. Available at: http://regulation.huji.ac.il.

Marsden, T., A. Flynn and M. Harrison, 2010. The new regulation and governance of food: beyond the food crisis? Routledge, New York, NY, USA.

Meidinger, E., 2009. Private import safety regulation and transnational new governance. In: C. Coglianese, A.M. Finkel and D. Zaring (eds.), Import safety: regulatory governance in the global economy. University of Pennsylvania Press, Philadelphia, PA, USA, pp. 233-253.

Milne, E.L., 2007. Protecting Islam's garden from the wilderness: halal fraud statutes and the first amendment. Journal of Food Law and Policy 2: 61-83.

Oosterwijk, H.G.M., 1999. Beleidsimplementatie tussen regels en religie. De Rijksdienst voor de keuring van Vee en Vlees en het toezicht op ritueel slachten tijdens het offerfeest. Regulering en markten.

Popovsky, M., 2010. The constitutional complexity of kosher food laws. Columbia Journal of Law and Social Problems 44: 75-107.

Raes, K., 1998. Godsdienstvrijheid en dierenleed. Slachten door middel van de halssnede tussen levensbeschouwelijke tolerantie en ethische verantwoording. Ethiek & Maatschappij 1: 91-104.

Rath, J., R. Penninx, K. Groenendijk and A. Meijer, 1996. Nederland en zijn Islam. Een ontzuilende samenleving reageert op het ontstaan van een geloofsgemeenschap. Het Spinhuis, Amsterdam, the Netherlands.

Regenstein, J.M., M.M. Chaudry and C.E. Regenstein, 2003. The kosher and halal food laws. Comprehensive Reviews in Food Science and Food Safety 2(3): 111-127.

Rosenthal, S.F., 1997. Food for thought: kosher fraud laws and the religious clauses of the first amendment. George Washington Law Review 65: 951-1013.

Sigman, S.M., 2004. Kosher without law: the role of nonlegal sanctions in overcoming fraud within the kosher food industry. Florida State University Law Review 31: 509-601.

Smits, M.J.W. and J. van den Berg, 2003. Diversiteitsbeleid:(h)erkennen van meerstemmigheid. LEI, The Hague, the Netherlands.

Sullivan, C.B., 1993. Are kosher food laws constitutionally kosher? Boston College Environmental Affairs Law Review 21: 201.

Verweij, M., 2000. Why is the river Rhine cleaner than the Great Lakes (Despite looser regulation)? Law and Society Review 34(4): 1007-1054.

Voedsel en Waren Autoriteit, 2007. Slachten volgens de Islamitische ritus tijdens het Offerfeest. Informatie voor slachthuizen. Available at: http://www.vwa.nl.

Vogel, D., 1986. National styles of regulation: environmental policy in Great Brittain and the United States. Cornell University Press, Ithaca, NY, USA.

Van Herten, M. and F. Otten, 2007. Naar een nieuwe schatting van het aantal islamieten in Nederland. Bevolkingstrends 3. Centraal Bureau voor de Statistiek, Heerlen, the Netherlands. Available at: http://www.cbs.nl/NR/rdonlyres/ACE89EBE-0785-4664-9973-A6A00A457A55/0/2007k3b15p48art.pdf.

Van Waarden, F., 2006. Taste, tradition, transactions, and trust: the public and private regulation of food. In: C. Ansell and D. Vogel (eds.), What's the beef? The contested governance of European Food Safety. MIT Press, Cambridge, MA, USA, pp. 35-59.

Van Waarden, F. and R. van Dalen, 2010. Hallmarking halal. The market for halal certificates: competitive private regulation. Paper presented at the Third Biennial Conference of the ECPR Standing Group on Regulation and Governance, Dublin 17-19 June, 2010.

13. Organic food

A private concept's take-over by government and the continued leading role of the private sector

Hanspeter Schmidt

13.1 Introduction

On a first level of analysis it is clear, that today public law, not private rules, regulate organic food labelling in major markets (such as the market in the EU, the US or in Japan). Statutory regulation by government appears comprehensive, however, analysis on a second level, of private commercial practices, demonstrates, that it is a function of the private sector, to draw the clear line between products which can and those which cannot be sold as 'organic'. It is the private and not the public sector, which developed rules, that force organic operators to avoid pesticide contamination of organic products by conventional agriculture.

This chapter presents two intertwined lines of argument. One is about the shifts in nature of organic norms between public and private. The second, making the other concrete, is about contamination of organic products with residues of pesticides and chemicals as perceived by public and private regulators.

13.2 'Bio' and 'Eco' and 'Regular'?

In July 2010 I asked the sales manager of a Dutch horticulture nursery what it means, when I read on his invoices to organic farmers in Germany the indication 'Bio' next to the botanical name of seedlings. He said: 'Here in Holland we have Normal, Bio, and we have Eco'. He continued to explain: 'Bio means no pesticides and a clean substrate. Eco is with SKAL certification'. So it became clear, that he had supplied organic farmers with seedlings, which were 'Bio' as he defined it. And not certified by the Stichting Skal (NL-BIO-01) in Zwolle, which is the sole institution in the Netherlands, admitted by the authorities to certify organic production.

This player in the 'Bio' market was not aware of the rules: today no 'Bio' indication is permitted in the European Union for the marketing of food, feed, seeds and seedlings not certified in organic controls as required by two EU Regulations: Regulation (EC) No 834/2007 and Regulation (EC) No 889/2008. Thus, this manager was mistaken to believe that there is a 'Bio' for his products outside of the statutory EU organic food certification. He was required to seek the certification by SKAL for the product and for his company before he used the term 'Bio' for the marketing of herbal seedlings.

13.3 Comprehensive protection of organic terminology

There is no distinction between 'Bio' and 'Eco', since they (as well as 'organic') are statutorily defined as synonyms for all of the languages[482] of the European Union.[483] The same applies to 'ökologische' (Landwirtschaft) and 'biologischer' Landbau as attributes to farming in German language: these terms are synonyms to 'organic' by statutory definition. However, not only these terms, but all indications suggesting the origin from organic agriculture, such as the trade mark 'Biobronch'[484] or 'Biogarde'[485], trigger the full statutory program of the two EU Regulations. They encompass performance requirements, both for farming and food processing, and they require third party certification.

13.4 The friendly take-over by government in the 1990s

The concept of 'organic food' had been the private sector's original creation. It had defined two elements as distinctive: 'organic' products originate from 'organic' farming. And they undergo not more than merely 'organic' food processing.

In the 1970s organic farmers founded growers' associations and the International Federation of Organic Agriculture Movements (IFOAM) as their transnational roof organisation. The IFOAM Basic Standards for Organic Agriculture[486] were published in 1980.

The Regulation (EEC) No 2092/91 established organic labelling rules as statutory law directly applicable in all EU Member States first for plant and in 2000 also for animal products[487]. In the United States the National Food Production Act of 1990[488] provided for the basis for a National Organic Program congruent in many ways to the European development. Thus 'organic food' as a private marketing profile was subject to a friendly, but comprehensive take-over by government.[489]

[482] European Court of Justice, 14.07.2005, C-107/04.
[483] EU, 2007. Council Regulation (EC) No 834/2007 of 28 June 2007 on organic production and labelling of organic products and repealing Regulation (EEC) No 2092/91. Official Journal of the European Union L 189, 20/7/2007, Article 23 (1).
[484] BVerfG, 30.01.2002, 1 BvR 1542/00, NJW 2002, 1486.
[485] Landgericht Leipzig, 20.04.2004, 1 HK O 7140/03, GRUR-RR 2004, 337.
[486] Current IFOAM Basic Standards for Organic Production and Processing (corrected version 2007) is available at: http://www.ifoam.org/about_ifoam/standards/norms/norm_documents_library/IBS_V3_20070817.pdf.
[487] EU, 1991. Council Regulation (EEC) No 2092/91 of 24 June 1991 on organic production of agricultural products and indications referring thereto on agricultural products and foodstuffs. Official Journal of the European Union L 198, 22/07/1991: 1-15. Repealed by Regulation (EC) No. 834/2007.
[488] US Government, 1990. Public law 101-624, Food, Agriculture, Conservation and Trade Act: Title 21 Organic Foods Production Act.
[489] A similar take-over by government occurred at the same time in the USA (Federal Organic Foods Production Act 1990).

13.5 Contaminants

The absence of traces of pesticides was not introduced as a parameter of statutory distinction of conventional and organic food production. The considerations of Regulation (EEC) No 2092/91 merely mentioned the possibility to later introduce 'provisions aimed at avoiding the presence of certain residues of synthetic chemicals from sources other than agriculture (environmental contamination) in the products obtained by such production methods'.

Provisions aimed at avoiding the presence of residues of synthetic chemicals from agricultural sources (contamination by drift from neighbouring conventional cultures) in organic products were not mentioned in 1991 nor were they introduced into the public law at a later date. Obligations to avoid agricultural pesticide contamination by conventional agriculture were never introduced: EU organic food law still does not require the minimisation of residues of synthetic chemicals from conventional agricultural sources (contamination by drift from neighbouring conventional cultures) into organic products.

13.6 2011: still the same concept

In 2007 the 1991 Regulation of the European Union was revised and split into two parts, but its concept and its language remained the same: 'organic' refers to the process of food production rather than product properties such as the absence of residues of agrochemicals. This splitting was to make more clear which parts of the organic rules were in the responsibility of the EU Commission (administration) and which in the realm of the EU Council (Member States). To make sense of the food regulations for producers, zigzag-reading is required.

Avoiding the presence of residues of synthetic chemicals from conventional agricultural sources has not been addressed in the revision process. The reason in 2007 was probably, as it had been in 1991, that the EU Rules on organic agriculture were not to be amended in any manner, that would necessitate organic farmers to force their conventional neighbours to avoid spray drift in the application of synthetic pesticides in their cultures. Avoidance of rules in the EU Organic Production Law with respect to the avoidance of toxins introduced into organic food products from conventional agriculture serves to protect the interests of conventional agriculture. Some may say, that this makes life of organic farmers easy: they are not responsible for traces of synthetic pesticides in their products, but their conventional neighbours are responsible for this. Traces of synthetic pesticides in organic food products caused by conventional agriculture do not impair, in accordance to the rules of EU Public Law, the selling as organic. The second level of analysis will show, that private commerce cannot live with such a half-complete system, since its application could damage consumers' trust in the integrity of organic food.

The public law definition of organic food versus conventional products emphasises aspects of the production process, as it had been the case in the private rules of the 1970s. Organic management measures are mentioned, for example the maintenance of soil fertility by the cultivation of legumes, green manures and multi-annual crop rotation. But the practical implementation of such active measures is a function of the farmer's skills, climatic and soil conditions in the region, and other rather 'soft' factors. There is no requirement to force conventional neighbours to avoid spray drift.

Also, any descriptive outline of organic management measures does not provide for a clear line of separation between organic and conventional food production. Practically efficient distinctiveness was achieved by the use of 'positive lists', that permit the use of a narrow catalogue of traditional plant protection substances, while all other are prohibited in organic agricultural production. The 'positive lists' remain the pivotal point of the EU law on organic food labelling. And it would blur the distinctiveness of organic agriculture to drop the concept of 'positive lists' for a system of ad-hoc decisions. It is the positive lists and their management by public statutory law, which ascertain that there remains nothing nebulous in the definition of 'organic farming' and its produce.

13.7 Positive lists for farming and processing

The organic farmers' associations imposed a full prohibition of plant protection products (synthetic pesticides) and (synthetic nitrogen) fertilisers combined with a scheme of selective authorisation by registration in a 'positive list'.

Only very few off-farm substances were registered for these purposes, such as copper compounds (Bordeaux and Burgundy mixture) to control fungi on grapes and stone meal to repel insects. The separating line between 'conventional' and 'organic' was thus drawn by general prohibition of agrochemical inputs and the cautious use of the positive listing mechanism. In the 2010 EU organic food law it is this scheme of general ban and selective admission, which draws the clear line.[490]

The 'positive list' determines organic food processing as well. The organic farmers' associations required 'organic' food to come not only from organic farmers but from 'organic processing' where most additives or processing aids permitted and used in mainstream food processing are not used. This dual concept is the model of the EU organic food law today. The distinction of conventional and 'organic' food products is not only defined by positive lists for agricultural inputs but by a

[490] EU, 2008. Commission Regulation (EC) No 889/2008 of 5 September 2008 laying down detailed rules for the implementation of Council Regulation (EC) No 834/2007 on organic production and labelling of organic products with regard to organic production, labelling and control. Official Journal of the European Union L 250, 18/9/2008, Annexes I, II, V to VII.

ban of additives and processing aids with a limited number permitted for organic processing by having been positive listed.[491] Thus only about 1/5 of food additives used in conventional food processing are allowed for organic food.

Today it is the prohibition of radiation, GMOs and the ban on non-agriculture inputs with the selective permission of a limited number by positive listing which draws the sharp line between conventional and organic food production and processing.

13.8 The friendliness of the take-over

The organic farmers asked in the 1980s for the take-over by government. Their products were challenged in Germany as misleading labelling.[492] Conventional competitors claimed, that the public perceives 'organic' as an indication of a 'pristine' condition and as a guarantee for the absence of all traces of pollution. At the time German courts applied the prohibition of the use of synonyms of 'naturrein' (*nature pure*) rather broadly and it was not clear how 'organic' label claims would fair in the German courts when challenged by conventional competitors.

The European IFOAM members asked the European Commission to use its competence for legislative initiative to define organic labelling requirements by statutory law. The interest on the Commission's side was to legally define an alternative form of agriculture as a basis for selective support by agricultural subsidies.

13.9 Take-over of norms, but not of controls

Take-over by government however stopped on the normative side. The set-up of organic certification was left to the Member States and they were given the choice of implementing organic certification by private bodies or public authorities or a mix of both. In 1991, most, with the exception of Denmark, opted for private bodies. While organic certification had been executed within the framework of the organic farmer associations, now independent organisational units were established.[493]

The question, whether organic certification is a governmental task or a private function has remained a rather controversial or blurred issue in some of the Member States. The European Court of Justice held organic certification not to be one of the activities which are connected, even occasionally, with the exercise

[491] In Annex VIII of Regulation (EC) No 889/2008.

[492] Max Forstmann, Sind Bezeichnungen wie 'Bioland', 'Biodyn', 'biologisch-dynamisch', 'auf Spritzmittelfreiheit geprüft', 'biologisch kontrolliert' unzulässige Angaben im Sinne des § 17 Abs. 1 Ziff. 4 oder § 17 Abs. 1 Ziff. 5 LMBG, Zeitschrift für das gesamte Lebensmittelrecht 1985, 16 ff.

[493] As prescribed by the '<WWP>General requirements for bodies operating product certification systems' (ISO/IEC Guide 65:1996 = EN 45011:1998), available at: http://www.beuth.de/langanzeige/DIN+EN+45011/3357759.html.

of official authority.[494] The Advocate General of the EU Court had argued, that the system put in place by the Regulation is one in which, essentially, the private inspection bodies operate a product certification system under the supervision of a public authority.[495] The German Federal Administrative Court claimed, that clearly, organic certification is a governmental function (Staatsaufgabe).[496]

Other German courts believe, that organic certification by private inspection bodies does not lead to a decision of a particular case (such as a German 'Verwaltungsakt'). Organic certification is here understood to have no binding quality. It is meant to be open to cancellation with no regard to the narrow circumstances prescribed by the national law on administrative procedures for the cancellation of administrative acts.[497] This legal opinion perceives organic certification as a non-binding expert assessment. This leads to uncertainty with respect to the status of a product as organic. All this is not in line with requirements of procedural fairness and due process. And thus the protection of property rights is not well balanced with the rightful target to remove products of doubtful origin from the organic market.

The issue, whether organic certification should become a part of the regular public food inspection was quite controversial, when Regulation (EC) No. 2092/91 was revised. The text of Regulation (EC) No. 834/2007 repeats now many sentences of the Food Inspection Regulation (EC) No. 882/2004. The idea was to convey, that the rules laid down in these sentences shall apply to organic certification, while the issue, whether organic certification shall be regarded as an integral part of the public food inspection in the Member States is left undecided on the level of the Union, to be decided by the Member States in their national legal order.

Germany has decided in its federal statute (Ökolandbaugesetz), that private organic certifiers are entrusted with the implementation of organic inspection. On the federal level this allocation of responsibility is, however, limited to implementation measures which are not 'administrative procedures' (Verwaltungsverfahren).

Administrative procedures are defined by German law as those, which lead to administrative decisions. Administrative decisions are those, which decide a particular case in the field of public law. Whether organic certification is an administrative act is thus still open to a quite controversial discussion.

In Germany it is the Federal Agency (Bundesanstalt für Landwirtschaft und Ernährung), which admits private organic certifiers to practice, but it is in the realm of the sixteen German States (Bundesländer) to decide, whether they treat private

[494] At that time Articles 45, 49, and 55 EC Treaty.
[495] Opinion of Advocate General Sharpston delivered on 12 July 2007 in Case C-393/05, Commission v. Austria, and Case C-404/05, Commission v. Germany.
[496] Bundesverwaltungericht, 13.06.2006, 3 BN 1/06.
[497] Oberverwaltungsgericht Lüneburg, 10.06.2008, 13 ME 80/08.

organic certifiers as assistants to the public administration (Verwaltungshelfer) or as bodies entrusted with official functions of the administration (Beliehene).

Twenty years after the friendly take-over of the normative standards of organic food products the relationship between public food inspection and private organic certification remains widely unclear. In Germany numerous cases are pending in the administrative court system, comprising issues such as whether the States may limit their liability for wrong-doings of private organic certifiers. Or whether government is restricted to a supervision of private organic certifiers which is bound by the law or, the alternative concept, whether government may supervise the organic certifiers guided by considerations of societal opportunity.

There is a comprehensive lack of clarity whether it is the civil or the administrative courts which control acts of private organic certifiers and consequently a lack of effective judicial review of decisions taken in the complex interwoven activities of private inspection bodies and supervising public authorities.

13.10 Toxins from non-regulated sources

The EU organic food law remains silent on drift from neighbouring conventional productions. And silent on the application of pesticides not as plant protection products (on the plants or their harvested products) but (as biocides) to control pests in storage facilities, containers, trucks or ships before they are used for organic produce, risking they may leave traces.

While Regulation (EC) No 889/2008 requires a management of the storage of products, that avoids contamination by products and/or substances not in compliance with the organic production rules,[498] there are no positive lists for the use of biocides for pest control in empty facilities.

When, for example, organic durum products (wheat) arrives in 2011 with traces of pirimiphos-methyl in Germany from Mediterranean production regions the seller is likely to justify this with drift or the permitted use to sanitise facilities before their use for organic goods. Consequently the seller is likely to argue, that any traces of pirimiphos-methyl lower than the threshold under public law for conventional durum, will not impair the possibility to sell the product as 'organic'.

13.11 BNN orientation values

It is the private regulations of organic associations which deal with these sources of contamination in requiring minimisation and suitable measures of quality improvement. The German Federal Association for Natural Foods and Products of

[498] Articles 35 (1), 63 (1)(c) of Regulation (EC) No 889/2009.

Traders and Processors (Bundesverband Naturkost Naturwaren – BNN) published orientation values as a practical means of dealing with traces of active toxic substances in organic products.[499]

The General Organic Buying Conditions of major EU organic food companies refer to the BNN Orientation Values (Orientierungswerte). These have first been published in 2001. They are indicators for the assessment of whether small amounts of synthetic plant protection agents or preservatives point to their prohibited active use. In practice BNN members pledge to trade only those goods as organic that do not exceed the BNN Orientation Values.

In the EU market organic food companies (BNN members or not, and not only those which process infant formulae products) use the BNN Orientation Values 'as a sample or model' for the contractually required product qualities (in the sense of Article 35 (2)(b) CISG 1980[500]).

The BNN Orientation Value for each pesticide agent is 0.010 mg/kg. For processed products that are diluted or concentrated during processing (by dehydration, extraction, pressing or other processes), the analytical result must be converted to reflect the original fresh product. If there is evidence of post-harvest contamination by harvest protection agents, or mixing with non-organic produce, conversion of results by calculation back to the original product is not permitted. The exceeding of a total of two pesticide agents is *not* allowed.

The BNN Orientation Values serve 'to distinguish between such trace quantities of residue resulting from contamination, and excessively high quantities of residue requiring investigation'. The BNN orientation value is perceived as 'a practical and helpful decision-making support mechanism'. Where the BNN orientation value quantities are exceeded, the requirement arises to investigate the origins of the residue and whether the statutory norms on organic production were complied with. This is explained not to 'impact the fundamental concept that organic foods and products are defined by their cultivation and not by their residual substance quantities'.

The BNN Orientation Values of 2001 developed in a decade to the gold standard of trade with organic commodities and thus closed a logical and practical gap in organic production requirements.

[499] See http://www.n-bnn.de/html/img/pool/Orientierungswert_EN_0906.pdf?sid=7c5ed2ee9ec528d1f0d8bee9324c44d6.

[500] United Nations Convention of Contracts for the International Sale of Goods.

13.12 Pesticide traces as misleading labelling

In 2003 a less differentiated approach of one of the advisory services in the General German food inspection ('Edelhäuser Proposition' by CVUA Stuttgart) suggested to consider organic label claims on products with detectable traces of plant protection products beyond 0.01 mg/kg as misleading labelling and therefore as per se prohibited. The organic grower associations in Germany considered this concept of disregard of particular circumstances a reopening of the 'battle of ideologies and cultures inherent in the completion' of organic and conventional agriculture as they had experienced it in the 1980s.[501] Today the positions have become quite close: the authorities agree with a case-to-case review of the causes of pesticide traces in organic food products. All sides agree, that organic products are rightfully expected to show very low traces. In particular traces much lower than those in conventional products.

Reports on the Organic Monitoring Program of Baden-Württemberg are published annually on the Internet. In 2008, as in preceding years, organic fruit and vegetable samples were found to differ significantly from conventional samples regarding residues of synthetically produced pesticides, both in frequency of their occurrence and the total amount found:

> 'In most of the organic samples no detectable residues of plant protection products were found. In the few cases where residues were detected, they mainly involved one active substance at trace level (below 0.01 mg/kg), thus, far below the usual concentrations, which come about in plant products after pesticide treatment. (...) The overall average amount of pesticides detected in the analysed fruit samples labelled as organically grown was 0.004 mg/kg. If those samples, where it was suspected that the product was not organically grown or a blend with conventionally grown fruit may have occurred, are omitted in the calculation, the overall average pesticide amount results to low 0.001 mg/kg. In contrast, conventionally grown fruits contained on the overall average 0.44 mg/kg pesticides'.[502]

A factor of 100 thus divides the average pesticide trace level in organic foods from conventional products. This is one of the aspects which make organic food attractive to consumers. Many of them doubt that testing of pesticide chemicals reliably accounts for hormone effects and other less obvious paths of damages to human health.

---

[501] See: http://www.bvl.bund.de/DE/07__DasBundesamt/05__Veranstaltungen/00__doks__downloads/symposium__2010__vortrag__edelhaeuser,templateId=raw,property=publicationFile.pdf/symposium_2010_vortrag_edelhaeuser.pdf.

[502] See: http://oekomonitoring.cvuas.de/english.html.

13.13 The statutory role of doubt

BNN Orientation Values are applied in the EU organic industry as pre-emptive expert witness assessments. When these values are exceeded, doubts with respect to the organic origin of the produce arise. This doubt triggers a mandatory review of the circumstances of the particular case. In the course of this review organic marketing is stopped. Here the private BNN values interact with the procedural rules of the statutory law in a quite particular manner.

Regulation (EC) No 889/2008 prescribes 'Measures in case of suspicion of infringements and irregularities'[503]: 'Where an operator ... suspects that a product is not in compliance with organic production rules, he shall ... withdraw from this product any reference to the organic production method ... . He may only put it ... on the market after elimination of that doubt, ... In case of such doubt, the operator shall immediately inform the control body or authority. The control authority or control body may require that the product cannot be placed on the market ... until it is satisfied, by the information received from the operator or from other sources, that the doubt has been eliminated'.

This statutory prescription of a mandatory self-suspension of the marketing of certified organic food products as organic in cases where any doubts arise, is triggered when the BNN Orientation Values are exceeded. The CVUA Reports on the considerable gap of pesticide residue levels between organic and conventional products render the BNN Orientation Values a convincing expert documentation of what residue patterns may and may not be expected in organic food products.

The assessment of a specific organic production may, however, demonstrate the causation of traces beyond the BNN values by particular circumstances such as the unavoidable impact due to the excessive use of agro-chemicals by conventional farmers in the region of origin.

13.14 Conclusion on the role of private organic food regulation

Organic Foods are a success story of private regulation. The past shows a practically full and comprehensive take-over of private rules by the government. In its origins this take-over was friendly. Presently some swift and decisive measures to protect the integrity of organic products are required. Any unregulated influx of toxins into organic foods contradicts consumers' rightful expectations. Where the public organic food law lacks stringent measures, these are in trade practice imposed by private rules. Such as the BNN Orientation Values.

[503] Article 91 (1) of Regulation (EC) No. 889/2008.

For the future the private organic sector is likely to regulate open issues, such as toxins from conventional agriculture, from nano materials, and from packaging.

Thus innovation in the protection of organic food integrity remains a pivotal function of the private sector.

References

EU, 1991. Council Regulation (EEC) No 2092/91 of 24 June 1991 on organic production of agricultural products and indications referring thereto on agricultural products and foodstuffs. Official Journal of the European Union L 198, 22/07/1991: 1-15.

EU, 2007. Council Regulation (EC) No 834/2007 of 28 June 2007 on organic production and labelling of organic products and repealing Regulation (EEC) No 2092/91. Official Journal of the European Union L 189, 20/7/2007, 1-23.

EU, 2008. Commission Regulation (EC) No 889/2008 of 5 September 2008 laying down detailed rules for the implementation of Council Regulation (EC) No 834/2007 on organic production and labelling of organic products with regard to organic production, labelling and control. Official Journal of the European Union L 250, 18/9/2008: 1-84.

Forstmann, M., 1985. Sind Bezeichnungen wie 'Bioland', 'Biodyn', 'biologisch-dynamisch', 'auf Spritzmittelfreiheit geprüft', 'biologisch kontrolliert' unzulässige Angaben im Sinne des § 17 Abs. 1 Ziff. 4 oder § 17 Abs. 1 Ziff. 5 LMBG, Zeitschrift für das gesamte Lebensmittelrecht 1985, 16 ff.

ISO, 1998. ISO/IEC Guide 65:1996 = EN 45011:1998. General requirements for bodies operating product certification systems. Available at: http://www.beuth.de/langanzeige/DIN+EN+45011/3357759.html.

UN, 1980. United Nations Convention of Contracts for the International Sale of Goods (CISG). Available at: http://www.uncitral.org/pdf/english/texts/sales/cisg/V1056997-CISG-e-book.pdf.

US Government, 1990. Public law 101-624, Food, Agriculture, Conservation and Trade Act.

14. Food online

Reconnaissance into a consumer protection no-man's land between food law and the Civil Code

Lomme van der Veer[504]

14.1 Introduction

The concept of distance contract, defined as a contract entered into by parties who did not physically come together for this purpose, has a long history. In 1744, Benjamin Franklin published a catalogue in colonial North America from which readers could order scientific books through mail; the first mail-order delivery firm was born.

Apparently, Franklin thought that an agreement that was entered into without the parties being physically near one another entailed risks for purchasers, and they had to be protected from the possibility that the supplier's anonymity would make it possible for him/her not to fulfil his/her obligations. Franklin included a warranty in the catalogue: 'Those persons who live remote, by sending their orders and money to B. Franklin may depend on the same justice as if present'. The catalogue constituted the origination of the first mail-order delivery firm as well as the initial rights for the benefit of distance buyers.

A few centuries later, the Pony Express had developed into an extremely efficient and refined global system of mail delivery. The paper catalogue, however, is on the wane because the internet has enabled sellers to present their goods in a shop along the digital highway. The purchase agreements entered into in these shops are distance contracts, the same as Franklin entered into. But the technology is new.

As always, legislators follow the technology. Legislators, both European and national, believe that, with the introduction of web shops, a new situation has emerged in legal traffic which deviates from what already existed to such an extent that modification of the law is required. So legislators apply imperative rules of law to intervene in the contractual relationship between distance sellers and buyers.

The new technology has resulted in the expansion of the world of distance buying to foodstuffs. In the shadow of the turbulence of food safety crises and related modernisation of public and private food laws, a silent modernisation of the food market is taking place. Suppliers of 'food online' create their very own private rules, which must steer a middle course between the Scylla of the new legislation

504 This chapter is written in the context of his PhD research with the Law and Governance Group at Wageningen University.

under private law for the *way of contracting* and the Caribdis of the public food law for the *nature of the product*. This chapter envisages mapping out the passage. The analysis is executed within the context of Dutch civil law.

In this book about private food law, emphasis lays on the way in which contracting parties mutually bind each other to private standards. This contribution considers whether, in the case of a distance contract, civil law legislation contributes to the food law objectives of food safety and food security. An ever-increasing amount of food is entering physical kitchens via digital supermarkets.

14.2 The distance contract, buying food online

Most considerations of the European Parliament and Council for Directive 97/7/EC[505] regarding distance contracts do not in any way consider the element that distinguishes transactions via the Internet from Franklin's 'mail-delivery transactions': digital technology. This is different in the following considerations regarding the provision of information and privacy. The nature of the technology plays a role here:

> (13) Whereas information disseminated by certain electronic technologies is often ephemeral in nature insofar as it is not received on a permanent medium; whereas the consumer must therefore receive written notice in good time of the information necessary for proper performance of the contract;
> (17) Whereas the principles set out in Articles 8 and 10 of the European Convention for the Protection of Human Rights and Fundamental Freedoms of 4 November 1950 apply; whereas the consumer's right to privacy, particularly as regards freedom from certain particularly intrusive means of communication, should be recognised; whereas specific limits on the use of such means should therefore be stipulated; whereas Member States should take appropriate measures to protect effectively those consumers, who do not wish to be contacted through certain means of communication, against such contacts, without prejudice to the particular safeguards available to the consumer under Community legislation concerning the protection of personal data and privacy;

The European Union is not the only entity with the desire to act in a regulatory manner; Member States also desire this. A motive for the Directive was found in the already existing and mutually differing national rules for distance contracts in the Member States. The Directive intends to remove the detrimental effects

[505] EU, 1997. Directive 97/7/EC of the European Parliament and of the Council of 20 May 1997 on the protection of consumers in respect of distance contracts - Statement by the Council and the Parliament re Article 6 (1) - Statement by the Commission re Article 3 (1), first indent. Official Journal of the European Union L 144, 4/6/1997: 19-27.

arising from this for the competition between businesses in the internal market by implementing a minimum of common rules on a Community level.

> (4) Whereas the introduction of new technologies is increasing the number of ways for consumers to obtain information about offers anywhere in the Community and to place orders; whereas some Member States have already taken different or diverging measures to protect consumers in respect of distance selling, which has had a detrimental effect on competition between businesses in the internal market; whereas it is therefore necessary to introduce at Community level a minimum set of common rules in this area;

In the explanatory memorandum to Dutch regulations for the 'Adjustment of Book 7 of the Civil Code to Directive 97/7/EG of the European Parliament and of the Council of the European Union of 20 May 1997 on the protection of consumers in respect of distance contracts', this is formulated as follows:

> Firstly, some Member States had taken various measures with a view to the protection of consumers in respect of distance selling, with detrimental effects on the competition between the businesses in the internal market, while for consumers cross-border distance selling could be one of the main tangible results of the completion of the internal market (considerations 3 and 4.). Secondly, the Council Resolution of 14 April 1975 (PbEG C 92) is mentioned where the need to protect consumers from demands for payment for unsolicited goods and from high-pressure selling methods (consideration 5) is highlighted. Furthermore, the use of technology for distance communication must not lead to a reduction in the information provided to the consumer or to the provision of ephemeral (not recorded on a permanent data carrier) information (considerations 11 and 13). It is also important for consumers to actually see the product or ascertain the nature of the service before entering into the contract (consideration 14).
> Finally it is pointed out that the consumer's right to protection of his/her privacy should be recognised, particularly as regards freedom from certain particularly intrusive means of communication (consideration 17).

Since 1 February 2001, the Dutch 'distance contracts' (Civil Code, book 7, title 1, part 9A. Articles 7:46a up to and including 46j)[506] is prevailing law.

It is thus the nature and manner in which contracting parties relate, based on which legislators judge that consumers require mandatory protection in addition to general contract law. It is not the object of the agreement, the merchandise, but instead the technology used in the offer and the technology used in the

[506] Civil Code (or Dutch: Burgerlijk Wetboek) will be abbreviated as CC in the text.

acceptance of the offer that requires modification. There are additional rules in the traffic with consumers on the digital highway.

We consider these additional rules as laid down by the Dutch legislator in part 9A of book 7 CC to be relevant in this scheme.

> **Article 46a**
> In this part, the following words have the following meanings:
> a. distance contracts: a contract for which, within the framework of a system organised by the seller or service provider for the purpose of distance selling or service provision, up to and including the conclusion of the contract, exclusive use is made of one or more technologies for distance communication;
> b. distance buying: the distance contract which is a consumer purchase;
> (...)
> e. technology for distance communication: a means that can be used without simultaneous personal presence of parties for entering the distance contract.
> (...)
> g. Directive 97/7/EG: by the European Parliament and the Council of the European Union of 20 May regarding the protection of consumers in respect of distance contracts (PbEG L 144);
> (...)

The legislator has designed things in such a way that 'distance purchasing' is a subset of the distance contract, where, in addition to the nature of the contract – it is after all a purchase agreement – characteristics of the parties involved have been laid down in the contract. After all, a consumer purchase involves a seller who acts within the execution of a profession or business and a buyer, a natural person, who does not act within the execution of a profession or business as parties.

> **Article 46b**
> 1. (...).
> 2. This part does not apply to distance purchasing:
> a. concluded by means of automatic vending machines or automated commercial premises;
> b. concluded at an auction.
> 3. Articles 46c-46^{e} and 46f, paragraph 1 do not apply to the distance purchasing of mainly foodstuffs supplied to the home of the buyers, to his/her residence or to his/her workplace by regular roundsmen.

The question of what exactly is meant by vending machines and automated commercial premises arises. Coffee, beverage and sweets vending machines – there are even fruit machines that supply actual fruit – must be considered as vending machines. In the case of an automated commercial premises, people probably think of 'a wall' in which people deposit money and pull out foodstuffs from behind a hatch. Although these machines can hardly be considered a 'technology for distance communication' as defined in Article 7:46a, a. CC, the legislator has apparently intended to do good by excluding this form of selling with technical tools.

It is also relevant to know whether it is only the baker and milkman who are meant by 'regular roundsmen', i.e. the suppliers who deliver with some frequency at set times. You may also see the occasional cheese and ice cream vendors in rural areas. Apparently, the legislator has considered the possibility of the buyer and seller not meeting near or in the vending car, but that instead the seller delivers orders to a private address based on a note or e-mail. The difference between the frequent and punctual roundsman to whom the rule of the distance purchase does not apply, and the roundsman who delivers at the buyer based on an order from a web shop or buyer, thus lies in the frequency and regularity of deliveries. This difference would be minimised if the buyer frequently orders from a web shop at set times, and the web shop delivers just as frequently and at set times. The intended distinction is that of the supplier who follows the usual route, contrary to a roundsman who comes by on the occasion of an order. This leads to the curious conclusion that if the sales contract is initiated by the seller, the protective measures of the distance purchase do not apply, and do apply in those cases in which the initiative is taken by the purchaser.

Article 46c deals with the provision of information to which the seller is obliged with distance purchases. Contrary to the conformity requirement to be discussed later, this is not mainly about information about the product. As regards the product, only the main characteristics of the item must be named. Usually, this involves information about prices, costs and the seller.

Article 46c

1. In good time prior to the conclusion of a distance purchase, the other party shall be provided with the following information using means adapted to distance communication technologies, of which the commercial purpose must explicitly be made clear:
   a. the identity of the seller, and, if the distance purchase requires (part) advance payment, the seller's address;
   b. the most important characteristics of the item;
   c. the price of the item, including any taxes;
   d. the costs of delivery, insofar as this applies;
   e. the manner of payment, delivery or execution of the distance purchase;
   f. the possible application of the option for dissolution in accordance with articles 46d, paragraph 1 and 46e;
   g. if the costs for the use of distance communication are calculated on any basis other than the basic rate: the amount of the applicable rate;
   h. the period for the acceptance of the offer, or the period for the fulfilment of the price;

i. where appropriate, in the case of a distance purchase which serves to permanent or periodic delivery of items: the minimum duration of the contract.

Pursuant to Article 46c, paragraph 1, the seller is obliged to provide the information meant 'in good time prior to entering into a distance purchase'. The legislator has thus not stipulated to make the information available to the purchaser 'prior to or while entering into the contract', in accordance with the system as formulated in Article 6:234, paragraph 1a CC regarding the applicability of general terms and conditions. This difference could be interpreted in wording, but it may also be that the difference is to be explained by causes less relevant for the meaning.

Paragraph 2 of Article 46c deals with the way and time at which information is to be provided in the execution insofar as this is not already the case based on paragraph 1.

2. In good time during the execution of the distance purchase and, as far as this does not concern items to be supplied to third parties, not later than the time of delivery, the buyer must be provided with the following information in a clear and comprehensible manner in writing, or insofar as the information under a and c-e are concerned, in another durable medium available and accessible to him/her, unless this information has already been provided prior to entering into the distance sale:
   a. the information meant in part a-f of paragraph 1;
   b. the requirements for exercising the right of dissolution in accordance with articles 46d, paragraph 1 and 46[e], paragraph 2;
   c. the visiting address of the place of business of the seller to which the buyer can submit complaints;
   d. insofar as applicable; information concerning the warranty and the services offered within the framework of the distance purchase;
   e. if the distance sale has a duration exceeding one year or an indefinite duration: the requirements for dissolution of the contract.

In addition to a general termination option for seven days after receiving an item, the arrangement for the distance sale has an extended period in which dissolution is an option if the provisions are not fulfilled as laid down in Article 46c.

Article 46d

1. During a period of seven working days after receiving an item, the buyer is entitled to dissolve the distance purchase without stating the reasons. If all requirements of Article 46c, paragraph 2 are not met, this term is three months. The first sentence applies accordingly as from the moment all requirements of Article 46c, paragraph 2 have been met within the period meant in the second sentence.

2. In the case of dissolution pursuant to paragraph 1, the seller cannot charge compensation from the buyer, except for the direct costs for returning the item.
3. In the case of a dissolution pursuant to paragraph 1, the buyer is entitled to the free refund of what he/she has paid to the seller. The reimbursement shall be carried out as soon as possible and in any case within thirty days after the dissolution.
4. Paragraphs 1-3 do not apply to the distance purchase:
   a. in the case of items of which the price is bound to fluctuations in the financial market over which the seller has no control;
   b. in the case of items that:
      1. have been produced according to the buyer's specifications;
      2. are clearly personal;
      3. cannot be returned by reason of their nature;
      4. are susceptible to rapid decay or ageing;
   c. of audio and video recordings and computer software if the buyer has broken the seal;
   d. of newspapers and magazines.

Despite the fact that the arrangement for distance purchasing applies unimpaired to the distance purchasing of foodstuffs, the exception as formulated in Article 46d, paragraph 4, under b, 3rd and 4th appears to result in that a violation of the provisions in Article 46c, paragraph 2 remains without any threat for dissolution for many foodstuffs. After all, it must be stated of many foodstuffs that, because of their nature, they cannot be returned, or that they may decay or age rapidly.

In early 2008, the European Commission published a proposal for a Directive regarding consumer rights.[507] In this Directive, the exceptions to the right to revoke are somewhat extended. Deliveries of wine if the delivery takes place after thirty days and agreements concluded in a sale by auction were added.

For the time being, it must be established that the legislator intends to provide consumers in distance selling with additional protection by ordering the provision of information which is not aimed at the product as much, but instead at the supplier and the costs involved in the contract; after all, Article 7:46c, paragraph 1b CC provides only the marginal order to state 'the most important characteristics of the item'. It must also be established that the violation of this limited obligation will regularly fail to lead to the option to avail of the dissolution options as formulated in Article 7:46d relating to the exceptions as presented in paragraph 4.

[507] EU, 2008. Proposal for a Directive of the European Parliament and Council regarding consumer rights, Brussels, 8 October 2008, COM (2008) 614 def. 2008/0196 (COD).

In relation to the fact that consumers are not in the physical vicinity of the article to be purchased in the distance purchasing, the legislator has provided him/her with ample opportunity without putting forward arguments to dissolve the contract during a period in which s/he has been in the physical vicinity of the product. This is impossible in the case of foodstuffs that can decay or age rapidly. The provision of information about products prior to the purchase is thus of significant importance in the case of foodstuffs.

14.3 Information and expectations about the product

Where foodstuffs are concerned, the European and national legislators attach sufficient importance to proper information about products that the provision of information on foodstuffs is regulated in a special way. Directive 2000/13/EC[508] orders Member States to harmonise legislation where labelling is concerned, which in the Netherlands was realised through the Food Labelling Decree. The Netherlands has entirely included the Directive into national legislation, so if the labelling of foodstuffs complies with the provisions of the Food Labelling Decree, these foodstuffs can be traded freely in other European Union Member States, provided that the indications and notifications on the packaging are stated in the language of the country where the foodstuff is marketed.

In distance contracts, the merchandise is presented in a digital display case. A brief study shows that foodstuffs are hardly ever displayed in such a way that the label is legible, and the accompanying text hardly ever provides the information prescribed by labelling law.

The buyer cannot examine the label prior to or during the purchase and is therefore unable to base his/her expectations on this. Subject of further research could be whether legislators of the directive and decree intended to have buyers take cognizance of the label prior to or during the *purchase*, or prior to or during *consumption*.[509] If the outcome is that the law orders that buyers must be able to take cognizance of the label prior to the purchase, almost every Internet foodstuff provider will be forced to modify their website. In addition, the texts on labels

[508] EU, 2000. Directive 2000/13/EC of the European Parliament and of the Council of 20 March 2000 on the approximation of the laws of the Member States relating to the labelling, presentation and advertising of foodstuffs. Official Journal of the European Union L 109, 6/5/2000: 29-42.
[509] A strong indication in the direction of the former – though postdating current food labelling law and not entirely unambiguous in its wording – is Article 8(1) of the General Food Law (Regulation (EC) 178/2002) setting as objective for food law to enable consumers to make informed choices ('Food law shall aim at the protection of the interests of consumers and shall provide a basis for consumers to make informed choices in relation to the foods they consume...'). EU, 2002. Regulation (EC) No 178/2002 of the European Parliament and of the Council of 28 January 2002 laying down the general principles and requirements of food law, establishing the European Food Safety Authority and laying down procedures in matters of food safety. Official Journal of the European Union L 31, 1/2/2002: 1-24.

must be legible in all of the languages of the Member States from which people can buy digital products in the web shop.

Questions that require elaboration within this framework not only see to the readability of labels, but also whether labels can be exemplary or that people need to be able to take note of the actual labels of the product that is to be delivered after purchase. The difference between both variants is expressed most tersely in the notification of the best-before date and use-by date.

It has been determined that, pursuant to Article 7:46c, paragraph 1 CC, the seller is obliged to make the information regarding 'the most important characteristics of the product'[510] known prior to the purchase. Assuming that the mandatory information on a label contains more than 'the most important characteristics of the product', the buyer of foodstuffs must first avail of the information legislators consider necessary after they have been purchased and delivered. As regards products which cannot be returned due to their nature or to the fact that they may decay or age rapidly, this means that the buyer can take cognizance of the properties of a product at a time when s/he can no longer return it based on the provisions of distance purchase. In fact, the seller can even limit the options for return by placing the best-before or use-by date presented close to the date of delivery.[511]

The legislator thinks that the impossibility of coming into physical contact with a product in the case of distance purchasing is to be compensated with additional options for dissolution. The legislator also thinks that the provision of information as regards foodstuffs must take place extremely precisely and carefully through mandatory statements on labels. For many foodstuffs, distance buyers have to do without the information on the label and without the options of dissolution as laid down in the distance contracts section in the Civil Code. None of the arrangements stated provide protection in the areas in which the legislator has considered that additional protection is required. Do distance buyers of foodstuffs have any instruments to make up for this?

14.4 Conformity

The item delivered must comply with the agreement. The conformity requirement is no more complicated than this. In fact, it simply is the application of the Latin adage *pacta sunt servanda* on purchase agreements. The question whether that which has been delivered complies with the agreement proves to require more elaboration. The Dutch legislator answers the question *a contrario*:

[510] Article 7:46c, paragraph 1 b CC.

[511] The seller who labels has a margin in choosing the date. Sellers of pre-labelled foodstuffs may choose to send the oldest stock first.

Article 7:17 CC

1. The supplied product must be in conformity with the agreement.
2. A product does not meet the agreement if, also in view of the nature of the product and the statements the seller has made about it, it does not have the properties the buyer could have expected based on the agreement. The buyer may expect the product to have the properties necessary for normal use and the presence of which he/she does not need to question, as well as the properties that are required for a particular use as provided for in the agreement.
3. A product that differs from that which has been agreed or a product of another type does not meet the agreement. The same applies if that which has been delivered differs in number, weight or measure from what has been agreed.
4. If a sample or model has been shown or given to the buyer, the product is to correspond with this unless it was provided merely as an indication without the product having to correspond with it.
5. The buyer cannot rely on the fact that the product does not meet the agreement if he/she was informed or reasonably could have been informed of this at the time of the conclusion of the agreement. The same applies if this is due to defects in or the unsuitability of raw materials originating from the buyer, unless the seller should have warned him/her about these defects or unsuitability.
6. (...)

This Dutch law that prevails since 1 May 2003 is the implementation of Directive 99/44/EC.[512] In this Directive, the requirement for conformity is expressed as follows:

Article 2 Directive 99/44

1. The seller must deliver goods to the consumer which are in conformity with the contract of sale.
2. Consumer goods are presumed to be in conformity with the contract if they:
   a. comply with the description given by the seller and possess the qualities of the goods which the seller has held out to the consumer as a sample or model;
   b. are fit for any particular purpose for which the consumer requires them and which he made known to the seller at the time of conclusion of the contract and which the seller has accepted;

[512] EU, 1999. Directive 1999/44/EC of the European Parliament and of the Council of 25 May 1999 on certain aspects of the sale of consumer goods and associated guarantees. Official Journal of the European Union L 171, 7/7/1999: 12-16.

c. are fit for the purposes for which goods of the same type are normally used;
d. show the quality and performance which are normal in goods of the same type and which the consumer can reasonably expect, given the nature of the goods and taking into account any public statements on the specific characteristics of the goods made about them by the seller, the producer or his representative, particularly in advertising or on labelling.

3. There shall be deemed not to be a lack of conformity for the purposes of this Article if, at the time the contract was concluded, the consumer was aware, or could not reasonably be unaware of, the lack of conformity, or if the lack of conformity has its origin in materials supplied by the consumer.
4. The seller shall not be bound by public statements, as referred to in paragraph 2(d) if he:
   - shows that he was not, and could not reasonably have been, aware of the statement in question;
   - shows that by the time of conclusion of the contract the statement had been corrected; or
   - shows that the decision to buy the consumer goods could not have been influenced by the statement.
5. Any lack of conformity resulting from incorrect installation of the consumer goods shall be deemed to be equivalent to lack of conformity of the goods if installation forms part of the contract of sale of the goods and the goods were installed by the seller or under his responsibility. This shall apply equally if the product, intended to be installed by the consumer, is installed by the consumer and the incorrect installation is due to a shortcoming in the installation instructions.

In his article, entitled 'De koopregeling in het richtlijnvoorstel consumentenrecht'[513] (The purchase scheme in the consumer law Directive proposal), Hijma observes a difference in meaning between both arrangements as a result of a free translation of the Dutch legislator: One more peculiarity can be observed. Article 2, paragraph 3 of Directive 99/44 determines that there is no non-conformity if, at the time of entering the contract, the consumer 'was aware, or could not reasonably be unaware of, the lack of conformity'. In the implementation, the Dutch legislator has translated this text in a peculiar manner: 'was aware or reasonably could be aware' (Article7:17, paragraph 5 CC).

[513] Hijma, J., 2009. De koopregeling in het richtlijnvoorstel consumentenrecht. In: Hesselink, M.W. and Loos, M.B.M. (eds.), Het voorstel voor een Europese richtlijn consumentenrechten: een Nederlands perspectief. Boom Juridische uitgevers, The Hague, the Netherlands, p. 171.

A less free translation, however, would have resulted in the Dutch text being as introverted as the English text: there is no non-conformity if the buyer was aware or should have been aware of the lack of conformity. Hijma states that, as evidenced by the explanatory memorandum, the Dutch 'implementation text' does not intend to result in a buyer's obligation to examine. The formula merely prevents the buyer from relying on the fact that the defect was unknown to him/her, while it is virtually impossible that it would have escaped him/her.

In the Dutch version of the applicable text, the Vienna Sales Convention[514] is translated just as roughly. In the English text, Article 35, paragraph 3 reads as follows:

> The seller is not liable under subparagrpaphs (a) to (d) of the preceding paragraph for any lack of conformity of the goods if at the time of the conclusion of the contract the buyer knew of could not have been unaware of such lack of conformity.

The German text is just as clever:

> Der Verkäufer haftet nach Absatz 2 Buchstabe a) bis d) nicht für eine Vertragswidrigkeit der Ware, wenn der Käufer bei Vertragsabschluss diese Vertragswidrigkeit kannte oder darüber nicht in Unkenntnis sein konnte.

The Dutch text reads:

> De verkoper is niet ingevolge het in het voorgaande lid onder a)-d) bepaalde aansprakelijk voor het niet-beantwoorden van de zaken aan de overeenkomst, indien de koper op het tijdstip van het sluiten van de overeenkomst wist of had behoren te weten dat de zaken niet aan de overeenkomst beantwoorden.

The core issue translates as 'knew or ought to have known'.

On 8 October 2008, the European Commission published the proposal for a Consumer Rights Directive.[515] The proposal envisages the adoption of a Directive that is to form a recast of the European provisions regarding consumer rights, jointly specified with the term 'consumer acquis'.

The proposal includes a conformity arrangement that provides more room for the statement that there is some obligation to have something examined or investigated on the part of the buyer. Article 24, paragraph 3 reads:

[514] United Nations Convention on contracts for the international sales of goods (CISG).
[515] EU, 2008. Proposal for a Directive of the European Parliament and Council regarding consumer rights, Brussels, 8 October 2008, COM (2008) 614 def. 2008/0196 (COD).

> There shall be no lack of conformity for the purposes of this article if, at the time the contract was concluded, the consumer was aware, or should reasonably have been aware of, the lack of conformity, or if the lack of conformity has its origin in materials supplied by the consumer.

So in fact, with Article 7:17, paragraph 5, the Dutch legislator comes closer to the proposal for the Directive than the current Directive.

14.5 Conformity requirement and distance contracts

It is conceivable that the Dutch or European legislator would be aware of the fact that consumers who become involved in distance purchasing develop expectations about products along routes other than by actually seeing and possibly holding the product. After all, the expectations to be fulfilled for the distance buyer in connection with the conformity requirement are constituted by presentations of the product in the digital shelf. It would have been possible for the legislator to have acted as a controller in the presentation regarding the provision of information.

The legislator decided another road, the road of 'unmotivated' dissolution. After all, based on the article, the distance buyer can dissolve the agreement without stating reasons for seven days, which can be extended to three months, after receiving the product. The buyer thus has the time to decide whether his/her expectations about the product are met. But it is this instrument a foodstuff buyer will regularly have to do without, pursuant to Article 7:46 d, paragraph 4.

The distance buyer also has to do without the information the label provides. A label can adjust the expectations of foodstuffs. As a result of the lack of corrective action of the label, the seller is fully obliged to comply with that which the buyer may expect from the seller in view of the nature of the product and the statements of the seller.

The above described difference in meaning between Article 7:17, paragraph 5 CC, which indicates some level of obligation to have something examined or investigated; and the more reticent Article 2, paragraph 3 Directive 99/44 may have a considerable impact to the question whether the label modifies the buyer's expectations as such that deliveries are always conform if the label provides accurate information. In distance purchasing, the label will play an opposite role; after all, buyers cannot read the label until the product has been bought and delivered, and are not able to determine whether the product is in conformity with the expectations until it has been delivered. So the label cannot contribute to the formation of expectations of the product, but instead to the assessment that the product does *not* meet the expectations.

What is the set of instruments available to distance buyers of perishable goods if the label forces him/her to adjust expectations? Actually, the buyer can avail of the entire set of instruments for non-conformity.

Based on Article 7:21 CC, buyers can demand delivery of that which is missing, repair or replacement. For a vegan who reads on the label that his/her vegetarian hamburger contains products of animal origin, these options for demand are no consolation, as is the case with a buyer of cheeseburgers who reads that there is no real cheese in the cheeseburger. Article 7:22 CC outlines other powers:

> **Article 7:22 CC**
> 1. If, in the event of a consumer sale agreement, that which has been delivered does not comply with the agreement, the buyer also has the right to:
>   a. dissolve the agreement, unless the deviation of that which has been agreed does not justify this dissolution and its consequences in view of its minor importance;
>   b. reduce the price in proportion to the degree of deviation of that which has been agreed.
> 2. The rights meant in paragraph 1 arise only if repair and replacement are impossible or cannot be expected from the seller, or if the seller has failed to perform an obligation as meant in article 21, paragraph 3.
> 3. Insofar as this part does not deviate from this, the provisions of part 5 of title 5 of book 6 regarding dissolution of an agreement apply to the authority as meant in paragraph 1 b.
> 4. The buyer may exercise the rights and authorities stated in paragraph 1 and article 20 and 21 without prejudice to any other right and claim.

Now that it appears from Article 7:6 CC that Article 7:22 has also a mandatory character, the distance seller will not be able to rule out the option to dissolve in the case of non-conformity, even if the foodstuffs are subject to decay or ageing.

A superficial investigation into what distance sellers of foodstuffs stipulate for themselves provides a curious result. Let us take a little excursion to practice. In which cases do the Albert Heijn, Etos en Gall & Gall webshops rule out dissolution in their General Terms and Conditions (Textbox 14.1)?

As many other distance sellers, Albert.nl derives its General Terms and Conditions to the set of terms and conditions developed by the Dutch Homeshopping Organisation in conjunction with the Consumers' Association within the framework

Textbox 14.1. Albert.nl webshop: general terms and conditions.

1. Web shop General Terms and Conditions
These General Terms and Conditions of the Nederlandse Thuiswinkel Organisatie (Dutch Homeshopping Organisation) were established in consultation with the Consumers' Association within the framework of the Coördinatiegroep Zelfreguleringsoverleg (CZ; Self-Regulation Coordination Group) of the Social and Economic Council and came into effect as of 1 January 2009.

Article 2 – Identity of the entrepreneur
- Albert Heijn bv trading under the name/names: Albert
- Business and visiting address: Provincialeweg 11 1506 HA Zaandam
- Telephone number: +31 (0)800-2352523
- Accessibility: from Monday through Friday from 8:30 am until 10:30 pm and Saturday from 8.00 am until 2.30 pm
- E-mail address: info@albert.nl
- Chamber of Commerce number: 35012085
- VAT identification number: nl002330884b01

(...)

Article 5 – The agreement
1. Subject to the provisions of paragraph 4, the agreement is concluded at the time of the consumer's acceptance of the offer and the fulfilment of the conditions set for this purpose.
2. If the consumer has accepted the offer electronically, the entrepreneur will immediately confirm receipt of the acceptance of the offer electronically. As long as the receipt of this acceptance has not been confirmed, the consumer may dissolve the agreement.
3. If the agreement is concluded electronically, the entrepreneur will take appropriate technical and organisational measures to safeguard the electronic transfer of data and ensures a safe web environment. If the consumer has the option of paying electronically, the entrepreneur shall observe appropriate safety measures.
4. Within legal frameworks, the entrepreneur may investigate whether the consumer is able to meet his/her payment obligations and whether all facts and factors that matter in a responsible conclusion of the distance contract. If the entrepreneur has valid grounds not to conclude the agreement based on this investigation, he/she is entitled to refuse an order or application stating reasons or to attach special conditions to the execution.
5. The entrepreneur shall include the following information for the product or service to the consumer, in writing or in such a way that the consumer can save it on a permanent data carrier in an easily accessible manner:
   - the visiting address of the business location of the entrepreneur where the consumer can lodge complaints;
   - the conditions under which and the way in which the consumer may invoke the right to revoke, or a clear notification of being refused the right to revoke;
   - the information about existing service after sale and warranties;

- the information included in Article 4, paragraph 3 of these terms and conditions, unless the entrepreneur has already provided the consumer with this information prior to the execution of the agreement;
- the requirements for dissolution of the agreement if the agreement has a duration of over one year or if the duration is indefinite.

6. If the entrepreneur has committed him/herself to supply a series of products or services, the provision in the above paragraph applies to the first delivery only.

Article 6 – The right to revoke on delivery of products

1. When purchasing products, the consumer has the option to dissolve the agreement without stating reasons for a period of 14 days. This period comes into effect on the day after receiving the product by or on behalf of the consumer.
2. During this period, the consumer shall handle the product and its packaging with due care. He/she shall unwrap or use the product only to such an extent in so far as this is required to be able to assess whether he/she wishes to keep the product. If he/she makes use of the right to revoke, he/she shall return the product and any accessories – if reasonably possible – in the original state and packaging to the entrepreneur, in accordance with the reasonable and clear instructions by the entrepreneur.

Article 7 – Costs in the event of revocation

1. If the consumer exercises the right to revoke, the costs for return shipment at the most will be at his/her expense.
2. If the consumer has paid an amount, the entrepreneur shall repay this amount as soon as possible and no later than 30 days after the return shipment or revocation.

Article 8 – Exclusion of the right to revoke

1. If the consumer does not have a right to revoke, this can only be excluded by the entrepreneur if the entrepreneur has clearly stated this in the offer, in any case in good time prior to the conclusion of the agreement.
2. Exclusion of the right to revoke is possible only for products:

- that have been produced by the entrepreneur in accordance with the consumer's specifications;
- which have a distinct personal nature;
- that cannot be returned because of their nature;
- that can decay or age rapidly;
- of which the price is bound to fluctuations in the financial market over which the entrepreneur has no control;
- for single issues of newspapers and magazines;
- for audio and video recordings and computer software of which the consumer has broken the seal.

(...)

of the Self-Regulation Coordination Group of the Social and Economic Council.[516] As do many other suppliers, Albert.nl has, as appears from the formulation of its Terms and Conditions, *copied* the instructions of the Dutch Homeshopping Organisation, which in turn have been copied from Article 7:46d, paragraph 4 CC instead of *following* them.

Article 8, paragraph 1 is particularly confusing. Which right to revoke has been excluded if the consumer does not avail of a right to revoke? In Article 8, paragraph 2, Albert.nl does not express the cases in which they avail of the option to exclude the right to revoke. Similar to the legislator, they merely outline the cases in which they are entitled to exclude this right, and as such leave the buyer uncertain about their intentions.

14.6 Conclusions

In digital distance sale the provisions of Directive 2000/13/EG and the Food Labelling Decree miss their objective if the purpose is to inform consumers about foodstuffs before they buy them.

The stipulations in Directive 97/7/EG and part 9A of title 1, book 7 CC miss their objective insofar as these stipulations intend to compensate for the lack of physical contact with ample options for dissolution, now that these options for dissolution may be lacking in the case of foodstuffs.

Ultimately, online buyers of food might know their position is best protected not by a *lex specialis* of consumer law or food law, but by the *legi generali* of the law of obligations because these eventually provide an option for dissolution. After all, it is impossible for the distance seller to keep the consumer from the right of dissolution in the case of non-conformity, even if he/she has drafted the General Terms and Conditions correctly.

Now that the legislator fails to provide a proper arrangement, the expectation that private parties will provide the highly necessary regulation seems justified. But the initial attempts to this are as yet far from perfect.

[516] The Social and Economic Council is an advisory board to the government and the highest body within the corporatist structure of product boards and industrial boards. On this system, see also Chapter 9 by Maria Litjes. Members of the Social and Economic Council are appointed by the government, by trade unions and by employers' associations (each one third of the total number). See: http://www.SER.nl.

References

EU, 1997. Directive 97/7/EC of the European Parliament and of the Council of 20 May 1997 on the protection of consumers in respect of distance contracts - Statement by the Council and the Parliament re Article 6 (1) - Statement by the Commission re Article 3 (1), first indent. Official Journal of the European Union L 144, 4/6/1997: 19-27.

EU, 1999. Directive 1999/44/EC of the European Parliament and of the Council of 25 May 1999 on certain aspects of the sale of consumer goods and associated guarantees. Official Journal of the European Union L 171, 7/7/1999: 12-16.

EU, 2000. Directive 2000/13/EC of the European Parliament and of the Council of 20 March 2000 on the approximation of the laws of the Member States relating to the labelling, presentation and advertising of foodstuffs. Official Journal of the European Union L 109, 6/5/2000: 29-42.

EU, 2002. Regulation (EC) No 178/2002 of the European Parliament and of the Council of 28 January 2002 laying down the general principles and requirements of food law, establishing the European Food Safety Authority and laying down procedures in matters of food safety. Official Journal of the European Union L 31, 1/2/2002: 1-24.

EU, 2008. Proposal for a Directive of the European Parliament and Council regarding consumer rights, Brussels, 8 October 2008, COM (2008) 614 def. 2008/0196 (COD). Available at: http://eur-lex.europa.eu/LexUriServ/LexUriServ.do?uri=COM:2008:0614:FIN:EN:PDF.

Hijma, J., 2009. De koopregeling in het richtlijnvoorstel consumentenrecht. In: Hesselink, M.W. and Loos, M.B.M. (eds.), Het voorstel voor een Europese richtlijn consumentenrechten: een Nederlands perspectief. Boom Juridische uitgevers, The Hague, the Netherlands, pp. 167-182.

15. National public sector and private standards

Cases in the Netherlands

Irene Scholten-Verheijen

15.1 Public law and private standards

Sometimes public law refers to private standards. The question, whether this reference to private standards make them binding rules under public law, has been the subject of several – civil and administrative law – court cases, of which we will discuss some in this chapter. Furthermore we will go into a complaint before the Dutch Competition Authority regarding the (abuse of a) dominant economic position with regard to private standards.

15.1.1 Dutch civil law proceedings

A noteworthy case is the civil law case of a Dutch company, Knooble B.V., versus the Dutch State and the Dutch Normalisation Institute (Stichting Nederlands Normalisatie Instituut, NNI).[517] The NNI is a foundation that prepares the so-called NEN-standards. NEN is short for *NE*derlandse *N*orm (which means Dutch Standard). NEN-standards contain technical specifications and rules describing the performance criteria of a product, a process or a service. There are about 2,000 specific Dutch NEN-standards for all kinds of things, from utensils to machinery, fire protection and dangerous goods.[518] The NNI holds the copyright of the NEN-standards and provides them against payment of € 62 on average. It is prohibited to further disclose, reproduce or multiply the NEN-standards.

The Dutch Buildings Decree (*Bouwbesluit*) (an Administrative Order based on article 2 of the Dutch Housing Act, *Woningwet*) provides structural conditions with regard to safety, health, operability, energy efficiency and environment. In the Dutch Buildings Decree and the Dutch Buildings Regulation (*Regeling Bouwbesluit*) (a Ministerial Regulation based on the Buildings Decree) reference is made in this regard to specific NEN-standards. Knooble is a consultant in building issues. Knooble requested the Court to rule that despite the reference in the law, the NEN standards are not binding law, alternatively that they should be made available free of copyrights. The first question the Court had to answer was whether the performance criteria in these private standards become generally binding by the

[517] Knooble vs. State and NNI, CoJ The Hague, 31 December 2008 (LJN: BG8465). LJN is the number used to disclose case law (in Dutch language) at: http://www.rechtspraak.nl.
[518] See http://www.nen.nl and http://www.nni.nl.

reference to these standards in the Buildings Decree and the Buildings Regulation. The Court considers that the concept of a 'generally binding regulation' consists of three elements:

- is it generally applied;
- does it have external effect; and
- do the public authorities enacting the provision have legal authority?

The Court considers that by reference in public law the performance criteria in the NEN-standards obtain external effect, as if they were part of the law itself and therefore become binding. They apply – just like the Buildings Decree and Regulation themselves – to anyone who wishes to undertake any building activities. In order to obtain a building permit, one should comply with the Buildings Decree and the Buildings Regulation, and therefore with the performance criteria in the private standards referred to. By referring to the private standards, the legislator determines the content of the NEN-standards to be binding for everyone. The fact that the standards have been drafted by a private entity is of no concern. The public authorities enacting the provision have legal authority. The Court therefore answers the question, whether the performance criteria in private standards become generally binding by the reference to these standards, to the affirmative.

By consequence, the NEN-standards fall under the scope of the Dutch Constitution[519] and the Dutch Publication Act. Article 89, sections 3 and 4, of the Constitution determines that the law regulates the publication and the entering into force of Administrative Orders as well as other generally binding rules; they will not enter into force until they are properly published. Article 3 of the Publication Act provides that articles do not enter into force unless they are published in the Bulletin of Acts, Orders and Decrees (*Staatsblad*) (Acts, Administrative Orders and other Royal Decrees establishing generally binding rules) or in the Government Gazette (*Staatscourant*) (Ministerial Regulations establishing generally binding rules). Now that the standards have not been published – but are only available from the NNI upon payment – the Court rules that they cannot be considered to be binding as part of a public law. 'The law should be known and available for everyone'.

In short, the Court ruled that when legislation refers to private standards, these standards share in the legally binding character of the law. If, however, the requirements of publication (and public availability) have not been complied with, this (private part of the) legislation is void. In other words, the Court recognises the competence of the legislature to turn private standards into public law. To the exercise of this competence all requirements for public law apply.

---

[519] For a translation into English language see: http://www.rijksoverheid.nl/documenten-en-publicaties/publicaties-pb51/the-constitution-of-the-kingdom-of-the-netherlands-2008.html.

Both Knooble and the State/NNI filed for appeal with the Court of Appeal in The Hague. At the moment of submission of this chapter, the judgement was still pending.

15.1.2 Dutch administrative law proceedings

After Knooble, some administrative courts rendered a judgement regarding the same issues. The Court of Utrecht follows the Civil Court of The Hague in its judgement.[520] But the Court in 's-Hertogenbosch judges differently.[521] As in Knooble, the Court is of the opinion that the NEN-standards have binding effect, through the reference made to them in public law. But then the Court judges that the fact that the standards have not been published according to the Publication Act, does not deprive the standards of their binding effect. Article 3 of the Housing Act states that in the Administrative Order as meant in article 2 of the Housing Act, reference may be made to (parts of) standards and quality statements. The Court considers that Article 3 of the Housing Act provides an explicit legal basis for the reference to the NEN-standards in the Building Decree. The Court further considers that the Housing Act is an Act of Parliament and therefore is not inferior to the Publication Act. The Court is of the opinion that by reference alone in the Housing Act, the NEN-standards already have obtained the character of generally binding rules, and therefore the Publication Act does not apply. In so far as the Housing Act in that way interferes with Article 89 of the Constitution, the Court considers it is not authorised to assess Acts of Parliament (according to Article 120 Constitution). The Court comes to the conclusion that the NEN-standards should be considered binding as such.

The Court of Groningen, in conclusion, judges that the reference to the NEN-standards in the Building Decree cannot be seen as a statutory provision and therefore the Publication Act does not apply.[522] The Court considers that it is not uncommon to refer to generally applicable standards and directives; these are sufficiently available and therefore can be known to everyone. The reference to such standards is not in conflict with legal certainty. Given these considerations, the (summary trial) judge sees no reason to conclude that the NEN-standards should be considered non-binding.

Consensus seems far away. The case of 's-Hertogenbosch will be judged in appeal by the Council of State. It will be interesting to see how the (civil) Court of Appeal and the (administrative) Council of State will deal with this matter. It should come

[520] Portaal Vastgoed Ontwikkeling vs. Mayor and Aldermen of Nieuwegein, CoJ Utrecht, 6 July 2009, SBR 09/156 (LJN: BJ2496).
[521] City Crash vs. Mayor and Aldermen of 's-Hertogenbosch, CoJ 's-Hertogenbosch, 5 February 2010, AWB 08/1587 (LJN: BL3758).
[522] [Applicant] vs. Mayor and Aldermen of Groningen, CoJ Groningen, 29 July 2010, AWB 10/613 (LJN: BN2936).

as no surprise that the entanglement of private and public law, leads to puzzles and controversy.

15.1.3 Complaint before the Dutch Competition Authority

On August 31, 2010, the Dutch Competition Authority decided upon a complaint against the NNI.[523] The complaint was that the NNI abuses its dominant economic position by charging high rates for copies of the NEN-standards. The complainants are of the opinion that the NNI should provide these copies for free or against payment of reproduction costs only. Especially given the fact that these standards are being referred to in public law and legislation.

The NNI has been contractually appointed by the Dutch Government as the national normalisation institute; it therefore has a legal monopoly with regard to the formation and publication of the NEN-standards. The activities of the NNI may be considered economic activities. Subsequently, the Dutch Competition Authority establishes that the NNI most likely has a dominant economic position regarding the formation of the NEN-standards as well as regarding the publication and sale thereof. This, however, does not necessarily mean that the NNI abuses this position. The question is whether the NNI charges excessive high tariffs for those standards that are referred to in the law and legislation. Whether or not this is the case depends on the economic value of the formation and publication of NEN-standards in relation to the tariffs being charged. In order to establish this, the Dutch Competition Authority would need more time for investigation, which would take a huge effort on the part of the Dutch Competition Authority.

The Dutch Competition Authority also refers to a political discussion regarding the finance structure of the NNI. Parliament suggested that the NEN-standards should be available for free and that the NNI should be financed from public funds.[524] As the Dutch cabinet was under resignation, the decision was postponed until a new cabinet had been installed. The Dutch Competition Authority further refers to the pending civil and administrative legal proceedings, amongst others the pending case of Knooble versus the State/NNI.

Given the political discussion and the (pending) judicial cases, as well as the relatively low economic interest of complainants, the Dutch Competition Authority decided to give no priority to this complaint and to conduct no further investigation with regard to this matter.

[523] Rechtspraktijk BAWA c.s. / Kombiplast B.V. vs. Stichting Nederlands Normalisatie-instituut, Dutch Competition Authority, 31 August 2010, 6965/7.
[524] See: Kamerstukken II (documents of the second chamber of Parliament) 2008-2009, 28 325, nr. 109.

15.2 Public procurement and private standards

Private standards also play a role in public procurement. In this section two civil law cases will pass review regarding the use of private standards as criteria in a tender procedure. We will furthermore have a look at the European Commission's opinion regarding this matter.

15.2.1 Dutch civil law proceedings: Douwe Egberts

Here we will discuss two fairly similar civil law cases with Douwe Egberts playing the leading part.[525] The Province of Groningen respectively the municipalities of Den Helder and Alkmaar initiated an open European tender procedure – in accordance with Directive 2004/18/EC – for the supply, maintenance and service of hot drinks machines (coffee, tea, cocoa, hot water). The local governments are the contracting authorities according to the Decree on procurement rules for government tenders (*Besluit aanbestedingsregels voor overheidsopdrachten*), which applies to the public tenders. The tenders were to be assessed on grounds of exclusion criteria, selection criteria and award criteria. In the terms of reference the governments require that the coffee and tea should have a Fairtrade label, such as Max Havelaar or EKO label (this was established as a knock-out criterium). At least the following specific criteria should be fulfilled:

- coffee/tea/cacao will be purchased from small farmers co-operations;
- prices that aim at covering the costs of sustainable production (according to social and environmental standards);
- an additional Fairtrade premium based on the world market price;
- advance credit, enabling coffee farmers to do the necessary investments;
- longer term trade relationships.

Douwe Egberts holds the UTZ Certified label for coffee and tea. Upon request whether the UTZ Certified label would be considered equivalent to the Max Havelaar / EKO label, the answer was that a Fairtrade label was explicitly mentioned because of their specific objectives (sustainability, social and environmental standards – such as prices that aim at covering the costs of sustainable production, an additional Fairtrade premium, advance credit, longer term trade relationships, and decent working conditions for hired labour – all aiming at the highest rate of sustainability possible). Only other labels that would meet these objectives could be considered equivalent, and UTZ Certified would not be considered as such. Douwe Egberts was of the opinion that this was unlawful and started interim

[525] Douwe Egberts Coffee Systems Nederland B.V. vs. Province of Groningen (and intervening party Max Havelaar Foundation), CoJ Groningen, 23 November 2007, 97093/KG ZA 07-320 (LJN: BB8575); and Douwe Egberts Coffee Systems Nederland B.V. vs. Municipality of Den Helder/Municipality of Alkmaar (and intervening party Max Havelaar Foundation), CoJ Alkmaar, 18 March 2010, 117231 (LJN: BL7898).

injunction proceedings, requesting reconsideration of the specific sustainability criteria in the public tender.

The Courts reason that the requirements cannot be seen as technical specifications, describing the mandatory characteristics of a product or service. They can only be seen as additional conditions, relating to the way the assignment is performed. Under European and Dutch procurement law it is allowed to request such conditions. Moreover, the conditions at hand set by the local governments are acceptable, now that they are being subscribed by the European Parliament, the European Commission and the Dutch governments, according to the Courts.[526] Furthermore, the principle of equity was not breached, because there would still be a number of tenderers that could make a bid. In conclusion, Douwe Egberts did not succeed in substantiating that the UTZ Certified label, giving the results, is at least equal to the Fairtrade label.

15.2.2 Infraction procedure European Commission

The European Commission, however, holds another opinion. On 14 May 2009, the European Commission gave the Netherlands a notice of default in a similar case, not unlikely based upon a complaint from Douwe Egberts. The European Commission was of the opinion that the Netherlands infringed EU public procurement rules. The Province of Noord-Holland awarded a public contract for the supply and management of coffee machines through an open EU-wide tender procedure. In the Commission's view, the award procedure did not meet the requirements of EU public procurement rules, in particular those relating to technical specifications, selection criteria and award criteria.

In terms of technical specifications, the Province requested the tenderers to supply coffee and tea with one or two specific labels for organic and Fair Trade products: EKO and/or Max Havelaar. This is not allowed under the public procurement rules as it discriminates against certain tenderers. The Province stated that it would accept equivalent labels; however it did not specify substantive criteria

[526] The Court of Groningen refers to the Interpretative Communication of the Commission on the Community law applicable to public procurement and the possibilities for integrating social considerations into public procurement, Brussels, 15.10.2001, COM(2001) 566 final, p. 8; to the draft Resolution of the European Parliament on Fair Trade and Development, (2005/2245(INI)), p. 8; and to the Agenda of the Dutch cabinet 2015 regarding the realisation of the Millennium planning aims, The Hague, 29 June 2007, p. 10.
The Court of Alkmaar refers to the Communication from the Commission to the Council, the European Parliament and the European Economic and Social Committee, Contributing to Sustainable Development: The role of Fair Trade and nongovernmental trade-related sustainability assurance schemes, Brussels, 5.5.2009 COM(2009) 215 final, p. 6; and to the letter of the Dutch Ministers Cramer, Donner and Koenders to the Dutch Parliament with regard to the international social criteria of Sustainable Purchasing, 16 October 2009 (PDI 2009037807), p. 5.

that could clarify to tenderers when a label would be considered equivalent. This situation was not transparent for competing businesses.

With regard to the selection of tenderers, the Province required the tenderers, among other things, to indicate what they do to make the coffee market more sustainable and how they contribute to environmentally, socially and economically sound coffee production. However, the objective of such criteria is not to ensure that tenderers have the necessary technical and professional capabilities to perform the contract, as would be required under EU public procurement rules, but rather inform the contracting authority about the general business policy of the tenderers. Also, it was not clear how and according to which criteria the Province would assess the information submitted by the tenderers. This was detrimental to the transparency of the tender procedure.

Lastly, the Province also breached the rules on award criteria. It used a criterion under which additional points are granted to tenders that offer ingredients (sugar, milk) that have a specific label or a comparable label. In the Commission's view, a contracting authority cannot use such an award criterion, since a label as such is not a criterion suitable to identify the economically most advantageous offer. The Province did not specify any substantive criteria in this regard, which again is not transparent for tenderers.

The Dutch government does not agree. According to the Ministry, the Province did not exclusively request the Max Havelaar label; therefore the tender was open for other labels as well. The conditions of the Max Havelaar label are freely available and generally known for everyone. Moreover, the observed conditions regarding sustainability and corporate social responsibility should not be considered selection criteria. They are merely specific performance criteria, which authorities are allowed to request. And even if it were selection criteria, it would still be allowed, so the Ministry argues. In conclusion, the Ministry is of the opinion that neither the rules regarding the award criteria have been breached. A contracting authority – in this case: the Province of Noord-Holland – is allowed to choose a product that, because of certain quality features, may not be the best priced product per se. From this it follows that the Province of Noord-Holland was allowed to appreciate the tenders including the Max Havelaar or EKO label, or a comparable label, over other tenders.

The Commission still wasn't convinced and sent the Netherlands a reasoned opinion;[527] the Commission even announced to take the Netherlands to the

[527] EU press release, IP/09/1618, Brussels, 29 October 2009; Available at: http://europa.eu/rapid/pressReleasesAction.do?reference=IP/09/1618&format=HTML&aged=0&language=EN&guiLanguage=en.

European Court of Justice.[528] In the mean time the civil servants of the Province of Noord-Holland drank fairtrade coffee already as of 1 January 2009 from coffee machines from a company called Maas International. The legal proceedings came too late for the contract at issue. It expired on 31 December 2010...

The question, however, if public authorities in formulating a call for tender under EU procurement law are entitled to refer to private standards, remains highly relevant.

15.3 Public enforcement and private standards[529]

Private supervision in the form of audits and public supervision in the form of official controls exist next to each other, but they also interrelate. The EU framework for official controls is Regulation 882/2004 (the Official Controls Regulation).[530] The competent authority ensures that official controls are guaranteed in all stages of the food chain. Food businesses are responsible for meeting all the conditions in order to produce and supply safe food. Some conditions specifically state the standards a product should comply with. Others only relate to the safety of the food; the businesses may decide what is the best way to achieve this objective. On the one hand, these 'open standards' give the businesses the opportunity to find their own solutions and ways of working; on the other hand they have to convince the competent authority that what they do is the right thing and leads in the end to food that is safe. If not, the competent authority has the power to impose punitive or corrective measures (either through administrative or criminal law).

Regulation 882/2004 states that:
- 'official controls' must be performed by the competent authority;
- the competent authority may delegate specific tasks related to official controls to one or more control bodies;
- taking into account self control systems.

The official controls take place on the basis of risk analysis, taking into account specific risks, the way a company behaved in the past (does it normally comply with the law?), as well as the reliability of self control systems. And this is allowed under the Control Regulation. If and when a company or sector has a form of self-

[528] EU press release, IP/10/499, Brussels, 5 May 2010; Available at: http://europa.eu/rapid/pressReleasesAction.do?reference=IP/10/499&format=HTML&aged=0&language=EN&guiLanguage=en.

[529] This part of the chapter is based on: Beuger, H., 2010. Overheidstoezicht en voedselveiligheid, in: Voedselveiligheid: certificatie en overheidstoezicht. Praktijkgidsen Warenwet, Sdu, The Hague, the Netherlands.

[530] Regulation (EC) No 882/2004 of the European Parliament and of the Council of 29 April 2004 on official controls performed to ensure the verification of compliance with feed and food law, animal health and animal welfare rules. Official Journal of the European Union L 165, 30/4/2004, p. 1.

control that adequately guarantees food safety, the supervision can be performed differently by the competent authority.

We give you the example of the way the Dutch competent authority deals with private standards (the Food and Consumer Product Safety Authority, *Voedsel en Waren Autoriteit* VWA). The VWA performs the official controls, and where possible, takes into account private control systems. The controls of the VWA over the years therefore shifted more and more from product control to process and system control. The VWA takes into account and controls the requirement of HACCP. The VWA also takes into account other private schemes guaranteeing a well functioning food safety system. Therefore such a system should at least be:

- transparent;
- based on a standard;
- connected to independent third party auditing;
- sufficiently able to self-correct.

The VWA assesses self-control systems on the basis of these criteria and performs 'reality checks' in the businesses. It does not approve the systems, but it accepts them. In that way, the VWA keeps the opportunity of continually monitoring the system at hand as well as other systems. The VWA approach is based on 'legitimate trust', based on the results in practice. If and when it turns out that a private system contributes to the compliance with law by the businesses, the VWA will restrict its supervision to the minimum: (1) low frequency of company visits; (2) an incidental infringement of law will *not* be sanctioned with a punitive sanction immediately; businesses will have the opportunity to take corrective measures in order to be compliant. This will only be different when the company causes a dangerous situation for human or animal welfare.

In general, the Control Regulation in most member states resulted in intensified supervision. The VWA approach leads to a more remote supervision, reducing the burden the supervision causes for the businesses. This approach is based on (1) working risk based; (2) put responsibility of the businesses in the first place; (3) offer the businesses assistance in complying with the law (be more transparent); (4) take severe measures when things go wrong. The businesses are being divided by the VWA in three categories: red, orange and green; whereas green stands for most compliant and red for least compliant. The VWA supervision concentrates mostly on the red and orange businesses. The green businesses will be subject to a more remote supervision.

According to the VWA, the Netherlands is ahead of other countries on the point of remote supervision. Export interests however require that other parties in Europe and the rest of the world remain sufficiently confident regarding the quality of Dutch products and the supervision thereof. Obviously, there is a field of tension between (full) responsibility for businesses and substantial government supervision. It will always remain a challenge to find the right balance between the two.

15.4 Conclusions

Other chapters in this book show how the private sector uses private standards to express quality and safety wishes and to translate them into legally binding terms. The Dutch cases discussed in this chapter show that it is problematic to say the least if the public sector tries to use them in the same way. At the current stage of development, experience in the Netherlands does not provide firm answers. What it does do is placing important questions on the agenda: what happens to the private law character of private standards if reference is made to them in legislation? What happens if public authorities refer to private standards in calls for tender under EU procurement law? And finally, to what extent can enforcement bodies in setting their priorities rely on the performance of private quality management systems?

As regards the first question, Dutch courts express interpretations ranging from 'nothing out of the ordinary' to 'this turns them into full-fledged – but void – public law'. As regards the second question, in accepting that authorities in calls for tender express their wishes by reference to private standards Dutch courts position themselves in opposition to the European Commission who considers such practice contrary to the law. The last issue at first glance seems least controversial. The choice to turn enforcement attention away from effective private schemes in order to focus on sectors not applying such schemes seems to have the logic of risk based controls on its side.

Postscript

On November 16, 2010, the Court of Appeal in The Hague gave its judgement regarding the appeal in the Knooble-case.[531] The Court reversed the judgement of the Court of Justice in appeal. According to the Court of Appeal the reference to the NEN-standards in generally binding regulations does not make the standards generally binding themselves. In order to be generally binding – according to the law – the regulation must be laid down by or pursuant to an Act of Parliament. The NEN-standards cannot be considered as such, since they are private agreements established by representatives of organisations having an interest in the use by everyone of one general standard. The regulatory authority did not lay down the standards, it merely declared the NEN-standards applicable and therefore generally prevailing; setting the standards everyone at least or by equivalence should comply with. By consequence, according to the Court of Appeal the NEN-standards do not fall under the scope of the Dutch Constitution or the Dutch Publication Act.

[531] State vs. Knooble and Knooble vs. NNI, CoA The Hague, 16 November 2010 (LJN: BO4175).

The NEN-standards have been published by the NNI and are sufficiently known to the public; they can be examined at the NNI or acquired against payment. Knooble filed an appeal in cassation on 15 February 2011.

Contrary to the Court of Justice, the Court of Appeal does not recognise the competence of the legislature to turn private standards into public law. In the case of City Crash against the Mayor and Alderman of 's-Hertogenbosch the Council of State on 2 February 2011 denied the appeal. The Council of State judged that although the NEN-standards have binding effect, they cannot be considered generally binding rules within the meaning of Article 89 of the Dutch Constitution; therefore the Dutch Publication Act does not apply. The way the standards are made known (either against fair payment or by copy for inspection) is sufficient to safeguard that they can be consulted by everyone.[532]

In the mean time the Dutch Legislative Drafting Instructions (Aanwijzingen voor de regelgeving, Article 91a) were adjusted in such a way that in principle reference to private normalisation standards in legislation can only be done in a non-binding way (merely providing evidentiary presumption).[533]

If and when however a binding reference is made to private standards in legislation, the standards must be known and available to the public. The Dutch government on 30 June 2011 presented their intention to make some standards available free of charge.[534]

Furthermore, on November 30, 2010, Douwe Egberts announced that as of January 1, 2011, they will extend their sustainable product range with Fairtrade certified products for the whole-sale market.[535] However, there are still pending court cases where Douwe Egberts fights against the Fair Trade monopoly in public procurement.

References

Beuger, H., 2010. Overheidstoezicht en voedselveiligheid, in: Voedselveiligheid: certificatie en overheidstoezicht. Praktijkgidsen Warenwet, Sdu, The Hague, the Netherlands.

[532] City Crash vs. Mayor and Aldermen of 's-Hertogenbosch, Council of State, 2 February 2011, Case Number 201002804/1/H1. See: www.raadvanstate.nl.

[533] Staatscourant 2011, nr. 6743, 14 April 2001

[534] See: http://www.rijksoverheid.nl/documenten-en-publicaties/kamerstukken/2011/07/01/kabinetsreactie-op-project-kenbaarheid-van-normen-en-normalisatie.html.

[535] Press-release 30 November 2010, see: http://www.maxhavelaar.nl/nieuws/douwe-egberts-koffie-op-het-werk-nu-ook-fairtrade.

EU, 2001. Interpretative Communication of the Commission on the Community law applicable to public procurement and the possibilities for integrating social considerations into public procurement, Brussels, 15.10.2001, COM(2001) 566 final. Official Journal of the European Union C 333, 28/11/2001: 27-41.

EU, 2004. Regulation (EC) No 882/2004 of the European Parliament and of the Council of 29 April 2004 on official controls performed to ensure the verification of compliance with feed and food law, animal health and animal welfare rules. Official Journal of the European Union L 165, 30/4/2004: 1-141.

EU, 2006. European Parliament resolution on Fair Trade and development (2005/2245(INI)). Official Journal of the European Union C 303E, 13/12/2006: 865-870.

EU, 2009. Communication from the Commission to the Council, the European Parliament and the European Economic and Social Committee – Contributing to Sustainable Development: The role of Fair Trade and nongovernmental trade-related sustainability assurance schemes. Brussels, 5.5.2009 COM(2009) 215 final. Available at: http://eur-lex.europa.eu/LexUriServ/LexUriServ.do?uri=COM:2009:0215:FIN:EN:PDF

16. The outside of private food law

The case of braided private regulation in Dutch dairy viewed in the light of competition law

Maria Litjens, Bernd van der Meulen and Harry Bremmers[536]

16.1 Introduction

Regulation of food safety and food quality no longer is exclusively a public interest. Since the 1990s also entrepreneurs have taken responsibility by developing private regulation systems. Private systems in the domain of food production probably affect all or almost all businesses one way or another. They either adopt private rules to align the level of the quality of their products to their customers' demands, or, ultimately, they are excluded from the market. To counteract this last effect the Dutch Competition Authority (Nederlandse Mededingingsautoriteit; NMa) in 2000 took a decision that created a temporary hitch in the increase of private regulation. The NMa refused to grant exemption from the ban on cartels to a private regulation system of an association of dairy processors. In this quality arrangement virtually all dairy processors on the market in the Netherlands participated. It set rules for their suppliers, dairy farmers, and was considered likely to lead to exclusion of non-participating farmers because of lack of outlet. The NMa's decision seemed to limit the expansion of private regulation. Although private regulation more and more embeds in the food chain, no new similar NMa decisions followed. Have the private regulation systems become more adjusted to the requirements of competition policy or did the NMa change their policy or practices?

Key questions that can be asked are: Who makes private regulation systems, who should comply with them, how do systems relate to each other, what is their effect in the market and how do they relate to the first decision of the NMa regarding this topic? Did the competition policy change or did the competition authority in the Netherlands redefine her role regarding these initiatives of businesses?

The first objective of this chapter is to identify different private regulation systems. Once the regulatory private food law systems governing food safety and quality have been investigated, the competition policy and the positioning of the competition authority towards the new reality are addressed.

[536] This chapter elaborates on Litjens, M., 2009. Private regulation in the Dutch dairy chain. In: B. van der Meulen (ed.) Reconciling food law to competiveness, Wageningen Academic Publishers, Wageningen, the Netherlands, pp. 101-107.

The methodology which is applied is a case study. Data were gathered by means of interviews with key players in the feed and food domains in addition to a desktop study including literature, websites, decisions and other documents from the Dutch competition authority and case law. The topic covered is private food safety and quality regulation in the dairy supply chain in the Netherlands, including the feed chain supplying dairy farmers. This case study is based on an empirical inventory of actors and the systems these actors apply, giving special attention to the way different systems refer to and rely on each other, thus 'braiding' themselves into a web of connected private systems.

The remaining part of this chapter is structured as follows. First we will provide necessary background information on the concepts which are used and on some peculiarities of the governance of food safety and quality in the Netherlands and provide the structure of the analysis applied in this chapter (section 16.2). Then the main part of this chapter (section 16.3) takes the reader through the different stages of the feed and dairy chain by presenting the different elements of the entanglement of the systems. Subsequently, these are the processing businesses, their cooperation platforms and – last but not least – the private systems they have formulated. Next (section 16.4) all elements (private systems and their interconnections), are put together again to form a 'big picture' – literally. The big picture provides the foundation for addressing the repositioning of the public competition authority vis à vis the new private governance of quality assurance. This is done in section 16.5. This chapter concludes with summarising the findings with respect to the objective of this research in section 16.6.

16.2 Background

16.2.1 Nomenclature

In this chapter we use the label 'private regulation'. 'Regulation' refers to specific activities to standardise outcomes by rules. We prefer '*private* regulation' to label rules set by private actors over '*self*-regulation'. The latter label may invoke the image that the regulated party participates voluntarily in the system and has the initiative in the formulation of the rules or at least takes part in it. Participation may be voluntary as understood in contract law theory, because parties bind themselves through a meeting of minds resulting in explicit agreement. As will be shown below, it is not voluntary in economic reality, because saying 'no' is not a viable option. Also, generally regulated actors do not or only to a limited extent participate in setting the rules.

In analysing private regulation of food quality, including food safety and food hygiene[537] requirements, we can distinguish rules to be applied by the addressed businesses and additional rules necessary to make these rules work. The former we call 'standard', the latter 'scheme'. Standard and scheme taken together form a 'system'. A standard is a set of rules concerning topics such as product, process and business management of food companies. The standard may apply to quite a large number of regulated actors. A scheme provides rules regarding authority over the system, coordination and control with regard to its execution including topics such as certification and audit.

16.2.2 Private regulation and public hygiene law

Although the subject of this research is private regulation, some attention will be paid to public food law as well. From Hygiene Regulation 852/2004 European and national public authorities derive the power to acknowledge a private standard as a guide to good practice. Such recognition confirms the accuracy of the implementation of public rules on hygiene of foodstuffs in the private standard at issue.[538] Sometimes a public authority in its acknowledgement explicitly excludes certain rules in the standard that go beyond statutory requirements. Thus this exclusion expresses the view of public authorities on exclusively private rules in the standard. Rules of a scheme are not included in the acknowledgement. This does not mean, however, they are not relevant from a public law point of view.

Dutch public enforcement authorities shift emphasis from public controls on businesses and on products to supervision on private controls. In this context the rules of the scheme are at the centre of attention. While there is no formal acknowledgement as there is for standards, the acceptance of schemes as relevant in official controls, grants the some status under public law.

16.2.3 Public yet private product boards

An important role in the development of private regulation in the Dutch dairy sector is played by product boards. Product boards are a typical Dutch form of

[537] According to Article 2(1)(a) of Regulation 852/2004 'hygiene' means the measures and conditions necessary to control hazards and to ensure fitness for human consumption of a foodstuff taking into account its intended use. In this sense 'hygiene' equals 'food safety' viewed from business perspective. Regulation (EC) No 852/2004 of the European Parliament and of the Council of 29 April 2004 on the hygiene of foodstuffs. Official Journal of the European Union L 139, 30/4/2004: 1-54.

[538] Article 8 and 9 Regulation (EC) 852/2004 of the European Parliament and of the Council of 29 April 2004 on the hygiene of foodstuffs. Official Journal of the European Union L 139, 30/4/2004: 1-54.

government.[539] They are mandatory associations set up by Act of Parliament and for that reason enjoy the status of a public authority. It is compulsory for businesses dealing with products within the scope of a board to be registered and to contribute a fee.[540] The management of a product board consists of representatives of associations of employers and employees in the relevant branches. The management has the statutory power to issue regulations of a public law nature that are binding upon the members.[541] However, as associations of businesses, instead of exercising their regulatory powers, product boards can also act on a private law basis for the sake of the businesses in the branches within their scope.[542] This makes product boards public and private law hybrids.

16.2.4 Structure of analysis

All businesses engaged in production of any sorts need inputs. Outputs again are inputs for businesses downstream in the production chain. Upstream the chain stretches as far as businesses source inputs from other businesses. As we will see below, as far as private regulation is concerned, the dairy chain is closely connected to the animal feed chain from where dairy farmers source inputs. For this reason, in the analysis, we include the feed branches.

Products move downstream from producers of ingredients of animal feed to producers of compound feed, to dairy farmers, to dairy processors, to retail and finally to consumers. Private regulatory requirements move upstream: retail setting standards for processors, which in their turn make requirements to farmers, which put their demands on feed suppliers. In this sense we see two streams flowing in opposite directions. A product stream flowing from producers towards retailers. When we use the expressions 'upstream' and 'downstream' they refer to this product stream. In the second stream requirements flow from retailers towards producers. In the coming analysis of private regulation we follow this

[539] They are based on Article 134 of the Constitution of the Netherlands. In English translation (available at: http://www.rijksoverheid.nl/documenten-en-publicaties/publicaties-pb51/the-constitution-of-the-kingdom-of-the-netherlands-2008.html), this provision reads:
Article 134
1. Public bodies for the professions and trades and other public bodies may be established and dissolved by or pursuant to Act of Parliament.
2. The duties and organisation of such bodies, the composition and powers of their administrative organs and public access to their meetings shall be regulated by Act of Parliament. Legislative powers may be granted to their administrative organs by or pursuant to Act of Parliament.
3. (...).

[540] Product boards (productschappen) are vertically integrated public bodies for trades. Next to product boards exist industrial boards (bedrijfschappen) with a horizontal nature. In this chapter these are not addressed. On this system see also: http://www.ser.nl/sitecore/content/Internet/en/About_the_SER/Statutory_trade_organisation.aspx.

[541] On this issue, see also Chapter 7 by Otto Hospes.

[542] Article 71 Industrial Organisation Act (Wet op de bedrijfsorganisatie: WBO). Available at: http://wetten.overheid.nl/BWBR0002058/ [In Dutch].

flow of requirements moving upstream from the point of view of production, as expressed in Figure 16.1.

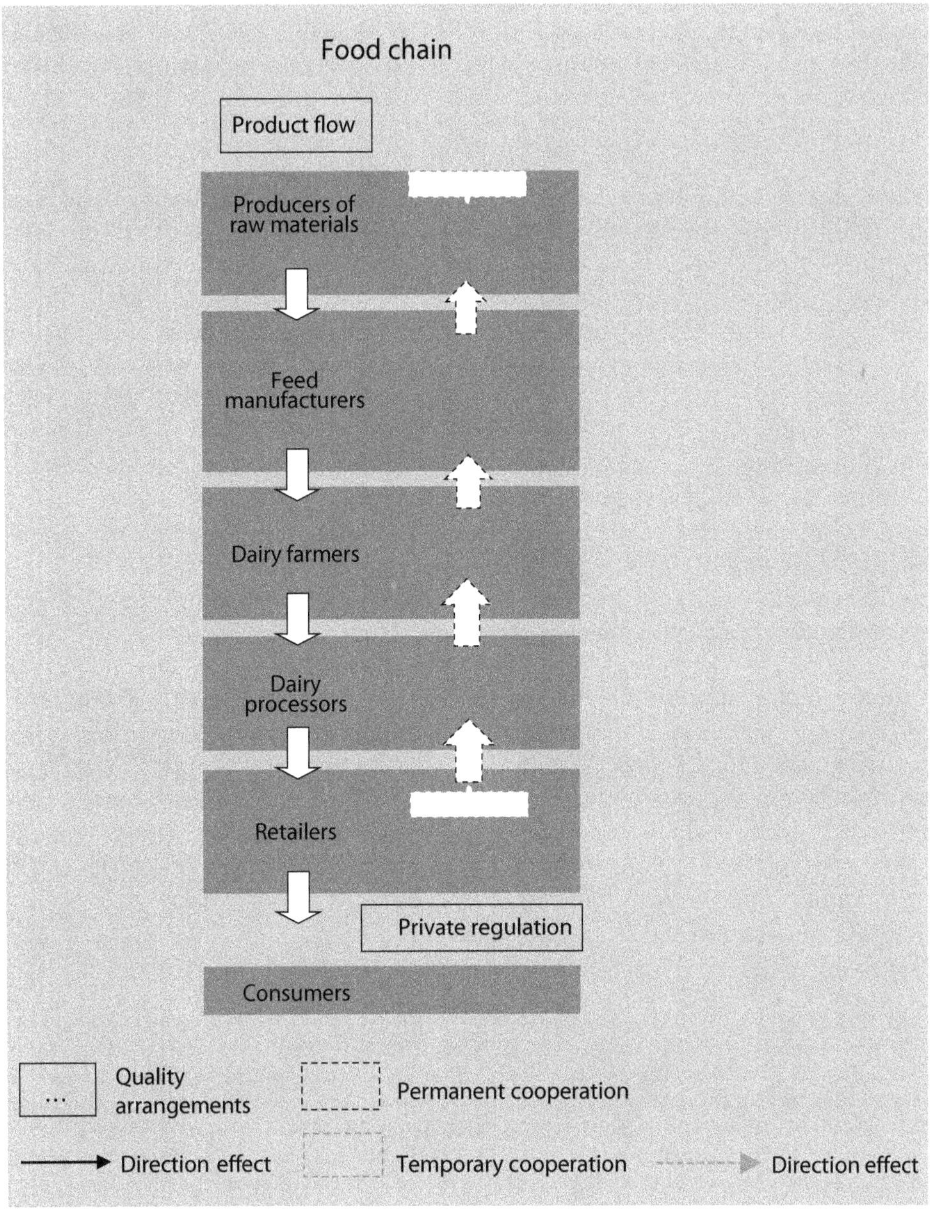

Figure 16.1. Structure of analysis of private regulation in the Dutch dairy chain.

16.3 Private regulation structured in the food chain

16.3.1 Sale of dairy products

In the Netherlands 99% of dairy sales to consumers take place in supermarkets.[543] The Netherlands is a small country with a high retail concentration. Retail chains cooperating in three retail purchase combinations account for more than 70% of all food sales.[544]

The Dutch association of retailers (Centraal Bureau Levensmiddelen: CBL) has drawn up a hygiene code for retailers. This code restates statutory requirements but does not add any requirements of its own. Therefore it is not stricter than the law. It is just a standard not connected to a scheme. Application is left to the discretion of the individual supermarket owners, no audits or certification takes place. No schemes exist which address retailers, neither at global nor at European level.

The so-called retailer standards set up by retailer organisations in several European countries do not address retailers, but provide retailers' requirements to be fulfilled by their suppliers. Nowadays all Dutch retailers require their suppliers of retailer-branded products to comply with the BRC or IFS[545] food safety standard.[546] Dutch retailers, as well as CBL, are member of GlobalGAP. GlobalGAP is a private regulation system with rules for the primary sector.

Dutch retailers support the Global Food Safety Initiative (GFSI). GFSI is an international cooperation of retailers. It has developed a guide on benchmarking food safety systems.[547] BRC, IFS and GlobalGAP systems as well a few other systems are recognised with a view to mutual exchangeability of these certification systems.[548] Dairy businesses have to apply GFSI benchmarked food safety standards set by retailer organisations to acquire access to the supermarket shelves in the Netherlands.

[543] See www.prodzuivel.nl/pz/productschap/publicaties/sjo/sjo05_engels/SJO_2005_H4.pdf, p. 42.
[544] Connor, J.M., 2003. The changing structure of global food markets: dimensions, effects, and policy implications. In: OECD Conference on Changing Dimensions of the Food Economy: Exploring the Policy Issues 6-7 February 2003, the Hague, Netherlands, pp. 3-4; Dobson, P.W., 2003. Buyer power in food retailing: the European experience. In: OECD Conference on Changing Dimensions of the Food Economy: Exploring the Policy Issues, 6-7 February 2003, The Hague, Netherlands, pp. 4-6.
[545] BRC = Global Standard of British Retail Consortium; IFS = International Food Standard of German and French retailer organisations.
[546] Havinga, T., 2006. Private regulation of food safety by supermarkets. Law and Policy 28(4): 515-533. As an example see: http://www.frieslandcampina.com/english/responsibility/foqus-for-quality.aspx.
[547] See http://www.mygfsi.com/information-resources/gfsiguidancedocumentsixthedition.html.
[548] For more detail see Chapter 3 by Bernd van der Meulen and Chapter 4 by Theo Appelhof.

16.3.2 Processed milk products

In the Netherlands 10 dairy businesses, one huge and nine relatively small, process virtually all milk.[549] This small number of businesses is the result of a long process of concentration. The businesses process raw milk into cheese (60%), butter (30%) and consumption milk (10%). Some 60% of the Dutch dairy production is for export, mainly to other EU-countries (Germany, France and Belgium). To comply with retail's requirements, Dutch dairy processors are certified for the BRC standard or standards recognised as equivalent under GFSI, such as IFS, FSSC 22000 or Dutch HACCP.

Processors in turn request their suppliers to comply with private standards. At the end of the last Millennium processors with a total market share of 98% in the Netherlands and so virtually all dairy farmers as their suppliers created the joint quality system KKM (Keten Kwaliteit Melk; chain quality milk) and according to the Netherlands' Competition Act as it stood at that moment, submitted an application for an exemption from the ban on cartels to the NMa. In 2000 the NMa refused to exempt this joint system. KKM-requirements went beyond statutory requirements. Dairy farmers producing according to legal requirements but not according to KKM should not be denied all market access.[550] The NMa judged the total exclusion of non-KKM milk which resulted from the agreement of the dairy processors to be incompatible with competition law. Then each processor went for its own private quality system. These processor systems are legally positioned as an element of the general terms and conditions of purchase or as a recognition scheme linked to an article in the cooperative society constitution.[551] All these systems have been constructed by the same certification organisation, the successor of the foundation KKM, which is currently named Qlip. The only exception was the quality system of Friesland Foods. After the merger of Friesland Foods and Campina in 2008 into FrieslandCampina the quality standard ('Foqus') replaced the previous standards of both businesses. Foqus is based on retailer schemes benchmarked by GFSI.[552] This merger resulted in one quality system for 80-85% of the market of Dutch raw milk. Foqus and KKM, the standard used by other processors,[553] are both monitored by Qlip (who owns the KKM system). The requirements of processors stretch further upstream into the feed chain. All Dutch dairy processors' quality standards for raw milk (Foqus and the individual KKM based systems) require dairy farmers to only purchase feed that has been GMP+ certified and meets some further requirements. The processors of meat in

[549] An insignificant quantity is processed on the farm and by some very small processors. See http://www.nzo.nl and http://www.prodzuivel.nl.

[550] NMa 14 March 2000, case 1237: Ontheffingsaanvraag Stichting Keten Kwaliteit Melk. Available at: http://www.nmanet.nl [in Dutch].

[551] See http://www.qlip.nl and http://www.frieslandcampina.com.

[552] See http://www.frieslandcampina.com/english/responsibility/foqus-for-quality.aspx.

[553] For example, http://www.deltamilk.nl, also http://www.Noorderlandmelk.nl.

the Netherlands apply the same strategy. Together these requirements virtually exclude non-GMP+ certified feed from the market for feeding cattle.

The requirements of processors regarding feed go beyond GMP+ certification. In 2005 Friesland Foods and Campina (at that moment still two independent processors in the dairy chain, together with >80% market share) and a processor in the meat chain (Vion, earlier named Sovion; a company with >80% of the Dutch slaughterhouse capacity) published their intention to require farmers to only use certified feed from suppliers with an adequate liability insurance. They obliged feed suppliers to have an insurance covering all losses caused by contaminated feed including consequential losses suffered in processing and sale such as recall costs.[554] At that time only six Dutch feed processors could fulfil these requirements. They had created a new foundation named TrusQ to collaborate in control of ingredients, to share data and to conclude a common insurance high enough to cover all possible damages. The costs of this insurance would be too high for each individual processor. In the text of its quality scheme, Qarant, Friesland Foods chose not to mention TrusQ by name, but it phrased the requirements for feed in such a way that only TrusQ companies would be able to comply. In reaction to TrusQ other feed processors set up a foundation called Safe Feed. In 2007 Campina wrote Safe Feed a letter communicating a joint approach with six other dairy processors. On top of the requirements at that moment met by Safe Feed, the dairy processors required quality guarantees, electronic monitoring of feed and checks of the private insurance of feed manufacturers.[555]

GMP+, TrusQ and Safe Feed will be discussed in more detail in the sections below.

16.3.3 Raw milk

About 18,000 dairy farmers in the Netherlands produce 12 billion kg of milk a year. To comply with the requirements made or passed on by the processors, virtually all of the farmers are certified for Foqus or KKM.

Primary producers are organised in the Dutch Organisation for Agriculture and Horticulture (Land- en Tuinbouw Organisatie: LTO)[556] and/or the Dutch Dairy Farmers Union (Nederlandse Melkveehouders Vakbond: NMV).[557] LTO in cooperation with TrusQ, Safe Feed and the slaughterhouse Vion have drafted a

[554] Sources: http://www.trusq.nl/nieuws_details.php?pID=22 (no longer available, consulted 2008) and http://www.frieslandfoods.com/content/zoeken/news.asp?id=2520 (no longer available, consulted 2008). This joint press release is included in a thesis by M. Willems, 2007. Productaansprakelijkheid in de agrarische sector. Thesis Juridische Hogeschool Avans-Fontys, 's Hertogenbosch, the Netherlands, Bijlage VIII. Available at: http://hbo-kennisbank.uvt.nl/cgi/av/show.cgi?fid=2419.
[555] Letter dated 30-03-2007 of Campina to Safe Feed (referentienr. 2007116/AS/avl) (unpublished).
[556] http://www.lto.nl.
[557] http://www.nmv.nu.

model agreement for use by farmers when purchasing feed. It complies with the requirements of FrieslandCampina (dairy) and Vion (meat), in that it requires GMP+ certification and additional insurance such as organised in the context of TrusQ and Safe Feed.[558] No additional requirements on behalf of the farmers seem to have been set by the farmer's organisation. The model agreement grants the dairy and meat processors, as the customers of the farmers, the right to access information relating to product quality and the insurance policy of the feed processor.

NMV does not occupy itself with food safety and quality but focuses on fair prices for its members.[559]

16.3.4 Compound feed

Among the most vulnerable products used on dairy farms are feeds not produced on the farm. Most feeds (around 75%), such as grass and maize, are produced on the farm. The remaining 25% is purchased feed. This feed is known as compound (or mixed) feed. Usually a compound feed comprises about twenty different ingredients.

Some 150 compound feed manufacturers supply the dairy farmers in the Netherlands. About half of the mixed feed used in the Netherlands originates from outside the EU, a quarter from other EU Member States and a quarter is of Dutch origin but may include imported ingredients.

Almost all Dutch compound feed producers are certified by the quality system GMP+. GMP+ has been developed by the Product Board Animal Feed (Productschap Diervoeder; PDV).[560] For some years GMP+ was a public law-based, but voluntary, regulation. The Product Board, however, changed it to a private law based system. Despite this private law character, due to its association with the Product Board GMP+ to some extent retained its public law image which contributed to its smooth implementation. Almost all mixed feed processors use GMP+. The branch association of feed manufacturers 'Nevedi', representing 96% of feed production in the Netherlands, requires in its code of conduct the use of GMP+.[561] Among

---

558 See http://www.gezond-ondernemen.nl/nl/252228I2-Basis_Inkoopvoorwaarden_Varkenshouderij.html. Also applicable for dairy farmers.

559 In 2006 NMV initiated the Dutch Dairymen Board (DDB); an association intended to negotiate jointly on behalf of the member dairy farmers with processors, because they consider current positions of parties to be such that the milk price is not determined by the market but by the processors (http://www.ddb.nu). DDB signed up to the European Milk Board (EMB) to reach a fair milk price of 40 cent/kg milk (http://www.europeanmilkboard.org).

560 http://www.pdv.nl.

561 Source: http://www.nevedi.com/uploads/Gedragscode%20Nevedi.pdf (no longer available, consulted 2008).

the requirements of milk and meat processors a GMP+ certificate has become a license to deliver for producers in the primary sector.

In 2009 a part of the associations of businesses involved in the Product Board set a next step in privatisation of GMP+. The public law disposition of the Product Board became too restrictive. The associations established a foundation: 'GMP+ International'. The ownership of the GMP+ system is transferred into the hands of this private organisation.[562]

The scope of GMP+ continuously expands. GMP+ started as a regulation for compound feed processors only, but now GMP+ comprises different standards to regulate all kind of activities performed by other branches in the feed chain, like production of raw materials, transport, storage, compound feed processing. This expansion, however, is not reflected in the composition of the board of GMP+ International. Not all business associations have a say in GMP+ International.

GMP+ has been brought in line with other standards like OVOCOM (Belgium), QS (Germany) and AIC (Great Britain). Certification for any of these systems is accepted as fulfilment of GMP+.

Together with the European Feed Manufacturers' Federation (FEFAC)[563] the owners of these systems established a joint association, the International Feed Safety Alliance (IFSA), with the aim – among other things – to construct a common standard IFIS (IFSA Feed Ingredient Standard) for compound feed and raw materials.[564] As is explained below, this attempt to turn these standards into a single standard did not meet with success.

In practice GMP+ certification alone was not considered to be sufficient assurance of a safe supply of raw materials. So, further collaboration in the mixed feed branch was undertaken, going beyond private regulation. Six compound feed producers created the foundation TrusQ. TrusQ is a co-operative alliance with a total market share in the Netherlands of 60 to 80%.[565] The aim of TrusQ is to ensure safe feed (and food), but also to provide insurance against safety incidents. A common insurance for the TrusQ-members covers damages related to legal liability to a maximum of €75 million.[566] TrusQ was closed to other Dutch manufacturers (foreign manufacturers, however, have always been welcome to join). This prompted the establishment of Safe Feed.[567]

[562] http://www.gmpplus.org.
[563] http://www.fefac.org.
[564] Source: http://www.ifsa-info.net (no longer available, consulted 2009).
[565] As stated in 2008 on http://www.trusq.nl.
[566] Source: http://www.trusq.nl/nieuws_details.php?pID=26 (no longer available, consulted 2008).
[567] Source: http://www.demolenaar.nl/nieuws/show-nieuws.asp?id=596 (no longer available, consulted 2008).

The two foundations, TrusQ and Safe Feed, go beyond audits and certification on the basis of GMP+. They also organise control of supply and suppliers. In addition they collect and exchange information to improve transparency in the chain.[568] Individual businesses contribute with capital and labour. The foundations monitor the input of materials for the benefit of the participants. Participants in these schemes are not allowed to purchase raw materials from non-complying suppliers.[569] Unlike TrusQ, Safe Feed does not organise insurance for its participants, but it requires participants to take care of insurance individually.[570] Such insurance should cover damages from €2 to €5 million for different types of risks.

The supplementary requirements to GMP+ are tuned to the demands of processors of primary products downstream in the product chain.[571] In June 2009 Safe Feed and TrusQ started to explore possibilities to cooperate.[572] It took a while before concrete results became visible. On 30 March 2011 a press release notified the world of the decision to merge as from the 1st of July 2011 and to create a new organisation under the name TRUST FEED. This new organisation will come to cover close to 100% of the market.

16.3.5 Raw feed materials

Raw materials for compound feed are the so-called single feed ingredients, by-products and additives. Single feed ingredients are soy, grain, maize, and tapioca. By-products come from the food industry. Additives have the form of single additives or of premixes (a combination of additives). Ingredients are interchangeable. Mixed feed has different composition depending on the requirements. There are no data available about quantities of specific materials used in compound feed.

The feed ingredients sector in Europe consists of a considerable number of producers and traders. Many of them are organised in associations on national, European and global level. More than a dozen of the European associations jointly formed the European Feed Ingredients Platform (EFIP).[573] EFIP is a voluntary platform encouraging the use of guides to good practice. It evaluates these private law sector guides on the implementation of public law requirements. A benchmark standard has been drawn up to describe the minimum requirements. One of the member organisations of EFIP, the foundation FAMI-QS, owns a sector guide for

[568] See http://www.safefeed.nl/data/Feed%20Safety%20Data%20Sheet%20EN.doc, including explanation [English text, but only available on the Dutch website].
[569] http://www.trusq.nl and http://www.pdv.nl/lmbinaries/08_leverancierbeoordeling_(roordink).pdf.
[570] http://www.safefeed.nl.
[571] http://www.safefeed.nl/desktopdefault.aspx?panelid=90&tabindex=2&tabid=175, and http://www.gezond-ondernemen.nl/nl/25222812-Basis_Inkoopvoorwaarden_Varkenshouderij.html, also applicable for dairy farmers.
[572] Source: Press release http://www.safefeed.nl: SAFE FEED and TRUSQ examine the possibilities of coöperation (no longer available).
[573] http://www.efip-ingredients.org.

feed additives and pre-mixtures and linked it to a certification scheme. Establishing FAMI-QS was considered the perfect way for a dozen of multinationals in Europe to have a voice in private rules addressing their products. FAMI-QS is now required for virtually all imports into the European Union. Participation in FAMI-QS means submitting to an obligation to use the guide of good practice and to contribute financially for controls and certification. The FAMI-QS-guide is approved by the European Commission and recognised as a Community Guide to Good Practice.[574] EFIP contributes successfully to the development of standardisation of other raw materials. The activities of the processors of feed ingredients in the EFIP platform outshone the IFSA initiative mentioned above. Their IFIS-standard was not put into operation.

The owners of the four national standards continued with their own standards. The owner of GMP+ has recognised FAMI-QS as equivalent. Certified products in one system are accepted in the other.

16.4 The big picture

The findings discussed above, can be graphically represented as shown in Figure 16.2. The starting point is Figure 16.1 (see section 16.2.4). The first column represents the feed and food production with a downstream flow of products. The other columns form an elaboration of the flow of private regulation and additional activities as explored in the former sections. The sequence in the discussion above is in the diagram pictured upwards, just like the right column in Figure 16.1. The second column in Figure 16.2 holds different private regulation systems. Collaboration based on private regulation and public-private activities are placed in separate columns.

The second column expresses private food safety arrangements established by single businesses, associations of businesses or vertically integrated organisations such as Product Boards. Beside the sector in which a private system is established, also the effects in other branches are shown. Associations of retailers establish common systems for suppliers and the primary sector. These regulations apply in a vertical direction (BRC, GlobalGAP). The use of the standards is voluntary, but economic dependency brings such a need to do so that choice is an illusion.

Dairy processors require corporate standards to be applied by their suppliers. The relation between regulator and regulated is in a vertical direction, upstream. The quality systems are either part of general conditions and terms of purchasing contracts and thus of the existing contractual relations or linked to the articles of association of a cooperative society (FrieslandCampina) and form in this way

[574] Press release January 2007: FAMI-QS recognized as Community Guide to Good Practice. Available at: http://www.fami-qs.org/Press%20Release%20EU%20Guide%201.pdf.

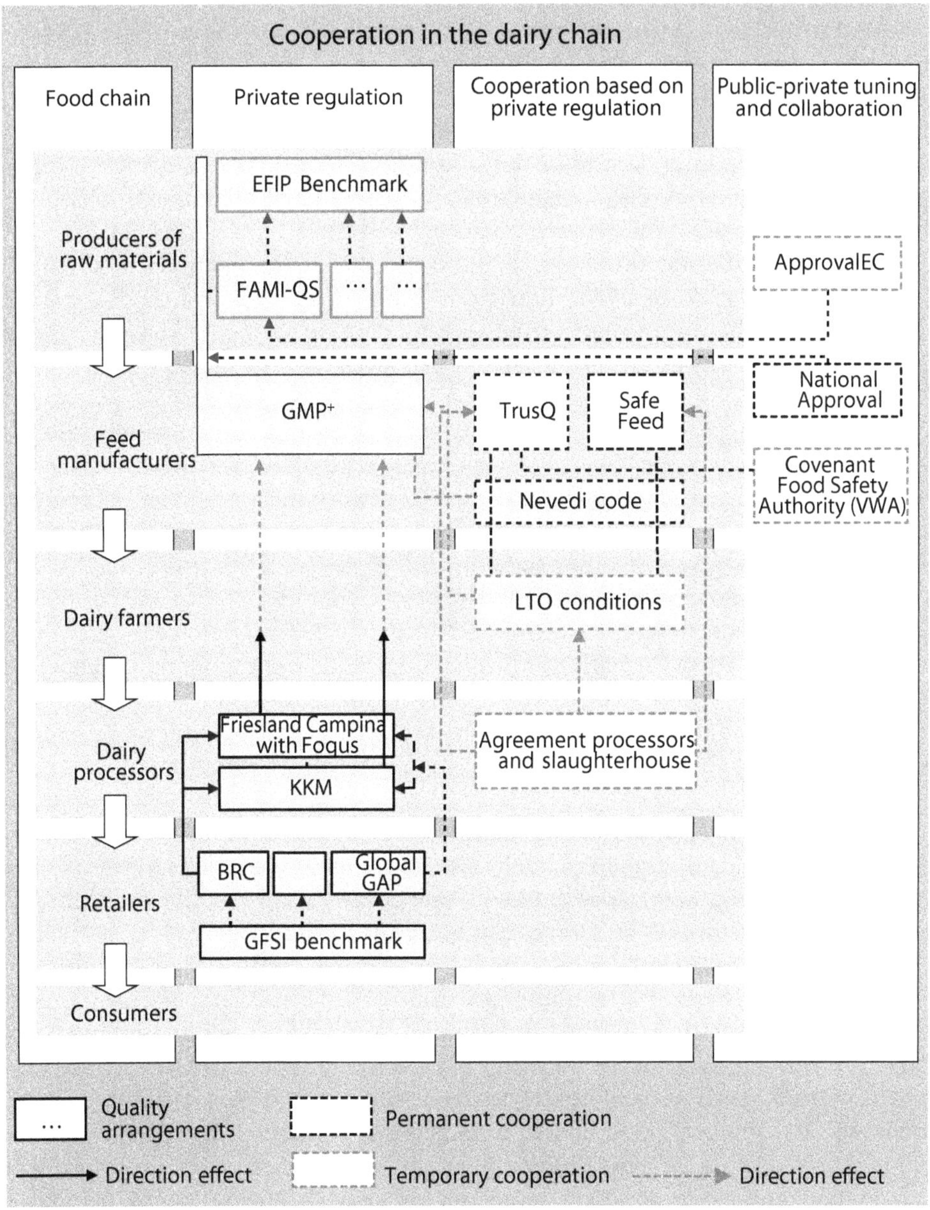

Figure 16.2. The external relations of private schemes in the Dutch dairy chain.

part of membership obligations. Either way they acquire a contractual binding character. These two legal constructions can also occur in combination.

Other businesses use some form of KKM created by the same independent foundation. Unsurprisingly, all KKM schemes are similar.[575] Foqus and the KKM-standards include the obligation for farmers to use GMP+ certified feed. Horizontally developed common standards and corporate standards have effects in vertical regulation where regulator and regulated are not the same. In other parts of the production chain standards are established by company networks, Product Boards or associations of companies. These regulations apply in horizontal as well as vertical direction, thus also imposing standards on suppliers (GMP+). The use of the standards is voluntary, but again strongly motivated by external drivers. Some quality arrangements have a chain character in that they set a number of standards for different sectors that are geared to each other (GMP+).

The third column expresses how on top of food quality arrangements stakeholders have joined forces on the basis of private law. Cooperation has different aspects. TrusQ and Safe Feed are established as foundations with input from and output for individual businesses. GMP+ provides an interesting case in that even an association not involved in setting the standard (Nevedi) applies it as a prerequisite for membership. By mutual agreement with associations of suppliers and customers LTO created a model for general terms and conditions to feed purchasing contracts including GMP+ and information requirements. Gentlemen's agreements between dairy processors and a slaughterhouse chain produced additional insurance requirements regarding feed purchased by farmers. These joined forces increase the necessity to submit to participation in a private regulation system.

Public-private cooperation is positioned in the fourth column. Private regulation does not operate in isolation from public law. Public recognition of quality standards as guides of good practice and public-private covenants create the possibility to reduce public controls. The food safety authority (Voedsel en Waren Autoriteit; VWA) entered into consultations with the standard owner (GMP+) and with the businesses cooperating in the implementation of the standard (covenant TrusQ-VWA). The objective was to come to recognition of the food safety arrangements but also to establish trust in their implementation. Such trust would justify a policy of the food safety authority to reduce public controls. Regulation 882/2004, in Article 3(1), requires food safety controls to be risk based. The (former) Dutch ministry of agriculture (now Economic affairs, Agriculture and Innovation) has formulated a policy on supervision of controls strongly advocating to include in the assessment of risk requiring controls the presence (or absence) of trustworthy private schemes.[576]

[575] Until 2007 all schemes were available online.

[576] See http://english.minlnv.nl/portal/page?_pageid=116,1640593&_dad=portal&_schema=PORTAL. On this topic see also Van der Meulen, B.M.J. and Freriks, A.A., 2006. Millefeuille. The emergence of a multi-layered controls system in the European food sector. Utrecht Law Review 2(1): 156-176.

The current case study shows that private systems do not stand alone. They are embedded in cooperation arrangements and connected to each other. The processes of linking food safety arrangements multiply the effect of these systems (dotted lines in the diagram). The connections have different styles.A quality system can impose an obligation on the regulated business operators to comply with another system as well, for example by using certified materials (all private systems of dairy processors addressing farmers hold the obligation to use GMP+ feed). A condition for participation in an association may do the same (Nevedi for GMP+ certification).

Mutual recognition of independent systems is a different style of connection. The owners of systems recognise each other's systems. Certification for one standard is accepted as proof of compliance with the requirements of another standard. This applies for example to FAMI-QS and GMP+, GMP+ with Ovocom, QS and AIC. This example shows that linking may have effects far beyond the production of feed and food in the Netherlands. Mutual recognition may have a similar effect in the market as one system at European or even worldwide level.

Benchmarking is a process whereby recognition of systems as equivalent is done not by the system owners among each other, but by a third party. Benchmarking may be done on the basis of an independent guide as yardstick.[577] The GFSI benchmark of retailers holds requirements for systems like BRC and GlobalGAP. The benchmark standard 'Feed Ingredient Standard for sector guides' of EFIP includes only legal requirements, which the sector guides of the member-organisations have to contain.[578] We even see that systems can be connected by simple agreement among businesses (Friesland Foods, Campina and Vion) to place obligations upon their suppliers.

For all practical purposes, it has become difficult - indeed next to impossible - to place feed and dairy products on the Dutch market without fulfilling private law requirements. Food quality and safety arrangements combined by all kinds of connection radiate the food quality requirements through the length and the breadth of the dairy chain.

16.5 Developments in competition law

As shown in the previous sections most systems and the connections between them are a result of cooperation between businesses. The empirical mapping provides a picture of such density, that it is appropriate to label the appearing

[577] The term 'benchmark' used in the dairy chain differs from the definition used in food quality management. See: Luning, P.A., Marcelis, W.J. and Jongen, W.M.F., 2002. Food quality management. A techno-managerial approach. Wageningen Academic Publishers, Wageningen, the Netherlands, p. 277.
[578] http://www.efip-ingredients.org.

private regulatory infrastructure as 'braided'. What we see is a web where private regulations are so closely 'knitted', 'entangled', 'felted' or however one wants to express it, that it seems to have become one interconnected structure.

Initially the NMa set limits, but what happened after that first decision? The developments in competition law and policy and their relevance to cooperation in the sphere of food quality are described in the following subsections.[579]

16.5.1 Competition law

Since 1998, Dutch competition law aligned itself to EU competition law. The Netherlands' Competition Act holds a prohibition on cartels in parallel to the first paragraph of Article 101 TFEU, the former Article 81 EC.[580]

Article 6(1) of the Competition Act reads:

> Agreements between undertakings, decisions by associations of undertakings and concerted practices of undertakings, which have the intention to or will result in hindrance, impediment or distortion of competition on the Dutch market or on a part thereof, are prohibited.

According to the third paragraph, of Article 101 TFEU (81(3) EC)[581] the prohibition on cartels may be declared inapplicable in the case of an agreement or concerted practice 'which contributes to improving the production or distribution of goods or to promoting technical or economic progress, while allowing consumers a fair share of the resulting benefit' if certain additional conditions are met. Initially, under Regulation 17/62[582], to become operational this exception needed a decision from the European Commission to exempt a certain agreement or practice on the basis of an application.

Dutch national law has a comparable rule initially resulting in an equal approach. Article 6(3) of the Competition Act reads:

> Section (1) shall not apply to agreements, decisions and concerted practices which contribute to the improvement of production or distribution, or to the

[579] On the details of competition law see Chapter 17 by Fabian Stancke.

[580] Mededingingswet. A brief explanation in English is available, see: http://www.nma.nl/en/legal_powers/dutch_competition_act/default.aspx.

[581] EU, 2010. Consolidated version of the Treaty on the functioning of the European Union (Lisbon). Official Journal of the European Union C 83, 3/3/2010: 47-199.

[582] EU, 1962. EEC Council Regulation No 17 of 6 February 1962 implementing Articles 85 and 86 of the Treaty. Offical Journal of the European Union 13/204 (last amended 2003 by Council Regulation (EC) No 1/2003 of 16 December 2002 on the implementation of the rules of competition laid down in articles 81 and 82 of the Treaty, Official Journal L 1/4/1/2003: 1) Official Journal of the European Union L 1: 1.

promotion of technical or economic progress, while allowing consumers a fair share of the resulting benefits, and which do not:

a. impose any restrictions on the undertakings concerned, ones that are not indispensable to the attainment of these objectives, or
b. afford such undertakings the possibility of eliminating competition in respect of a substantial part of the products and services in question.

Until 2004, businesses had to apply for an exemption for agreements with competition restrictions like the foundation KKM did for her quality arrangement in 2000.

16.5.2 Private systems grappling with competition law

Associations of dairy processors and farmers had initiated an integrated quality system for milk. This system was called KKM (Keten Kwaliteit Melk; chain quality milk). The processors concerned had a total market share of 98% of raw milk in the Netherlands. So virtually all dairy farmers were their suppliers and had to join this system. The NMa did not grant the exemption. KKM-requirements went beyond statutory requirements. Dairy farmers producing according to legal requirements but not according to KKM should not be denied all market access. The NMa judged the total exclusion of non-KKM milk which resulted from the agreement to be incompatible with the competition law ban on cartels. So at the turn of the Millennium, the deployment of a private regulatory infrastructure seemed to suffer a serious setback as it ran into trouble with competition law.

In 2002 the organisations involved in KKM made an attempt to bring KKM within the ambit of public law and thus outside the scope of competition law. They convinced the Dutch Dairy Board (Productschap Zuivel) to turn KKM from a private standard into a law-based regulation. For reasons not relevant in the current context, this regulation was struck down by the courts. The processors were back to square one.

After these setbacks to a collective quality system, dairy processors implemented individual private quality systems. Although these systems strongly resemble each other as well as the previous KKM and collectively they have probably the same excluding effect on non-qualified milk as KKM had, the NMa did not react as would have been expected on the basis of their previous decision. This decision included an explicit warning for businesses not to continue the forbidden agreement individually. However, in an informal opinion requested by Friesland Foods the NMa expressed no intention to look at the content of standards. In this informal opinion the NMa stated that individual processors, even with a dominant position, are permitted to set a quality standard for suppliers. Independent decisions of dairy processors fall outside the scope of the ban on cartels even if individual standards mutually bear resemblance.[583] This informal opinion did not repeal the former decision. Nonetheless, the informal opinion expresses more leeway for private standards. Just one private food safety system has been subject of investigation by the NMa since 2000.[584]

In December 2008 the two biggest dairy processors in the Netherlands, Friesland Foods and Campina, merged into FrieslandCampina. FrieslandCampina is now the third largest dairy processing company in the world. In the Netherlands it has a share of 80-85% of the dairy supply market. The European Commission authorised the merger. Neither the Commission nor the NMa raised any objections to the effect of a common quality standard (now Foqus), alone or together with KKM, on the market.

The informal opinions after the first decision did not throw light on the acceptability of private systems. Further on in the field of competition law private regulation submerged into silence.

16.5.3 Change in competition policy

A change in competition policy or law could provide an explanation. Initially cooperating businesses had to apply for an exemption. At EU level this changed with the coming into force of Regulation 1/2003[585] on 1 May 2004. From that date onwards the exception is considered to be directly applicable. It is up to the businesses concerned to decide if they meet the requirements for this exception. The role of the European Commission is limited to enforcement. The Dutch Competition Act has been changed in the same way. Until the 1st of August 2004 it employed a system of exemption from the ban on cartels granted by the NMa by decision on application. As from the 1st of August 2004 the exception applies directly. Article 6(4) of the Netherlands' Competition Act now reads:

[583] NMa 14 Januari 2005, case 4258, Informele zienswijze: borgingsysteem kwaliteit productie melk. Available at: http://www.nmanet.nl [in Dutch].

[584] NMa 25 October 2001, case 317, Algemene Voorwaarden overeenkomsten PVV/IKB Blanke Vleeskalveren 1997. NMa made one other informal opinion concerning agreements over anaesthetised castration of piglets. NMa 27 October 2008, case 6645, Informele zienswijze:verdoofd castreren van varkens. Available at: http://www.nmanet.nl [in Dutch].

This opinion has been analysed in: Litjens, M.E.G., 2009. Kleine ingreep. een onversneden prijsafspraak en ongesneden biggen. Markt en Mededinging 12(5): 161-165. This informal opinion only deals with the temporary price agreements included in the scheme. The effects of the quality agreements in the market were not taken into account.

[585] EU, 2003. Council regulation (EC) No 1/2003 of 16 December 2002 on the implementation of the rules on competition laid down in Articles 81 and 82 of the Treaty. Offical Journal of the European Union L 1, 4/1/2003:1.

> Any undertaking or association of undertakings invoking section (3)[586] shall provide proof that the conditions of that subsection are met.

From that point in time, businesses had to evaluate themselves whether they met the requirements. There is no obligation anymore to submit arrangements to the NMa.

From now on, the role of the NMa is limited to supervision and enforcement. This change in process does not necessitate a change in focus of the NMa. And indeed, the NMa-website shows that the authority keeps a close watch on the food and agriculture sectors of industry. Nevertheless, no private food safety systems have been subject of investigation by the NMa since 2004. We do not see how in the absence of a change in competition policy the existing silence can be explained. Is it a lack of interest or have the businesses convinced the NMa that initiatives have turned individual to the extent where they fall outside the scope of the ban on cartels? The last seems the most plausible explanation, but it is not easy to reconcile with empirical evidence. In the meantime the exclusionary effect from the market not only continues, but has dramatically increased as has been shown in the previous sections.

This sub-section shows that the competition authorities have adjusted their policies in two ways. The first one is that their entire role had to be redefined due to major shifts in European and Dutch competition law. It provided autonomy to businesses in evaluating competition effects of concerted action. However the NMa retained the task to enforce competition law. Thus the second change in policy is that the NMa has taken no further action with regard to private systems. However, the single safety and quality requirements as well as the integration in one overall system might be in conflict with basic competition rules. Additionally, no opinion is noticeable on the braiding of systems which all have similar or even the same characteristics. All in all, we see no other explanation of this anomaly in public response to private regulation than a lack of interest or insight in the 'system of systems', which we have called 'braided'.

16.6 Conclusions and discussion

The first objective of this chapter resulted in a description of the interconnected field of private food safety and quality regulation in the Dutch feed and dairy domains. Empirical findings show, that most private arrangements go beyond statutory requirements. In other words, they set stricter requirements than public law demands.[587] This case study shows that private food quality and safety arrangements usually do not primarily provide for obligations on the businesses

---

[586] Quoted in section 16.5.1.

[587] Information of expert stakeholders and investigation of quality systems.

setting them, but on one or more branches upstream. In this sense they are not 'self-regulation' but imposed regulation. The level of compliance which is required from businesses is further raised by the links between the arrangements.

Cooperatives may use their articles of association to impose requirements upon their members. In other cases economic dependence of suppliers may provide the customer the leverage to impose private standards. With the increase of the market share of collective private systems the possibility for suppliers not to participate decreases.

As the case study shows private systems do not stand alone. They are embedded in cooperation arrangements and connected to each other. System owners make systems exchangeable by mutual recognition and benchmark. In the feed and dairy domains this pattern unfolds both at national and international level.

Beside horizontal expansion by exchangeability, private systems become more and more linked in vertical direction in the chain. Although the connections have different styles, the overall effect is a more binding character of private rules. The linking of food safety and food quality arrangements increases buyers' power to steer the production process of suppliers. Finally private regulation enables buyers not only to regulate their suppliers' activities, but also those of players further upstream in the chain. The processes of linking food safety arrangements multiply the effect of the systems. Horizontal and vertical integration of private systems leads to a food chain that we labelled 'braided' in the subtitle to this chapter. If we look for a more 'legal' label, maybe an expression such as 'inter-regulated' could be considered.

In a world of inter-regulated food chains for most businesses non-participation will not be a viable option. In economic terms, businesses that do not participate in the private systems are excluded from the market. If we further take into account that many private standards are stricter than legislation, we are back at our point of departure. The exclusion that provided the main motive for the NMa not to grant an exemption from the ban on cartels for the KKM dairy quality system seems currently omnipresent.

The second objective was to describe changes in competition policy regarding these initiatives of businesses after the rejection of the application for exemption of one private regulation system: KKM. The exemption to the ban on cartels no longer requires an explicit decision on an application. This turns the issue from an administrative issue into an enforcement issue, but in our view this should not fundamentally alter the reasons for authority action. Thus NMa identified the problem in one single private system but thereafter the competition authority has not acted against other systems with no less exclusive effects in the market.

The NMa has no thought for the effect of 'braided' private regulation. Moreover, this observation leaves one to wonder if competition law based on the model of the EC Treaty (now Articles 101 and 102 TFEU with added merger control) is ready to deal with inter-regulated markets. It requires further research to answer this question.

The present developments need a more active involvement of the public competition authorities in the governance of private food safety and quality systems, while the repositioning of these has changed their role from active to reactive. Undesired effects of the present absence of authority involvement are exclusion of non-participants in several branches in the food chain and the increase of power positions of businesses mainly at the end-stages of the supply chain.

It appears that private food law not only has an *inside* worthy of investigation but also an *outside*. The inside consists of the details of individual systems – standards with schemes. The outside is the inter-relatedness of systems that turns private food law into a braided system of systems with an impact on businesses rivalling the impact of public food law.

Acknowledgements

Many thanks to Geronda Klop for helping design the figures.

References

Connor, J.M., 2003. The changing structure of global food markets: dimensions, effects, and policy implications. In: OECD Conference on Changing Dimensions of the Food Economy: Exploring the Policy Issues 6-7 February 2003, the Hague, the Netherlands, pp. 3-4.

Dobson, P.W., 2003. Buyer power in food retailing: the European experience. In: OECD Conference on Changing Dimensions of the Food Economy: Exploring the Policy Issues, 6-7 February 2003, The Hague, the Netherlands, pp. 4-6.

EU, 1962. EEC Council Regulation No 17 of 6 February 1962 implementing Articles 85 and 86 of the Treaty. Offical Journal of the European Union 13/204.

EU, 2003. Council regulation (EC) No 1/2003 of 16 December 2002 on the implementation of the rules on competition laid down in Articles 81 and 82 of the Treaty. Offical Journal of the European Union L 1, 4/1/2003:1.

EC, 2004. Regulation (EC) No 852/2004 of the European Parliament and of the Council of 29 April 2004 on the hygiene of foodstuffs. Official Journal of the European Union L 139, 30/4/2004: 1-54.

EU, 2010. Consolidated version of the Treaty on the functioning of the European Union (Lisbon). Official Journal of the European Union C 83, 3/3/2010: 47-199.

Havinga, T., 2006. Private regulation of food safety by supermarkets. Law and Policy 28(4): 515-533.

Litjens, M., 2009. Private regulation in the Dutch dairy chain. In: B. van der Meulen (ed.) Reconciling food law to competiveness, Wageningen Academic Publishers, Wageningen, the Netherlands, pp. 101-107.

Litjens, M.E.G., 2009. Kleine ingreep. een onversneden prijsafspraak en ongesneden biggen. Markt en Mededinging 12(5): 161-165.

Luning, P.A., Marcelis, W.J. and Jongen, W.M.F., 2002. Food quality management. A techno-managerial approach. Wageningen Academic Publishers, Wageningen, the Netherlands.

Van der Meulen, B.M.J. and Freriks, A.A., 2006. Millefeuille. The emergence of a multi-layered controls system in the European food sector. Utrecht Law Review 2(1): 156-176.

Willems, M., 2007. Productaansprakelijkheid in de agrarische sector. Thesis Juridische Hogeschool Avans-Fontys, 's Hertogenbosch, the Netherlands, Bijlage VIII. Available at: http://hbo-kennisbank.uvt.nl/cgi/av/show.cgi?fid=2419 [in Dutch].

17. The limit of private food law

Competition law in the food sector

Fabian Stancke[588]

17.1 Introduction

For a number of years, food producers and food retailers have experienced intense competition and margin pressure. Stagnating sales,[589] combined with the current structure of the market lead to an intense competition among retailers, a very low price level for groceries in Germany and corresponding pressure on margins in the food sector.[590] This margin pressure leads in turn to fierce competition among food producers. At the same time, competition authorities intensify their pressure on the industry by way of multiplying investigatory proceedings.[591] Ever since January 2010, when many premises in Germany were searched, the media suspect the food sector of having entered into illegal price fixing agreements.[592] As a consequence, market participants today face many uncertainties with regard to the legality of their business conduct. This uncertainty was reinforced by an informal letter[593] issued by the German Federal Competition Office (FCO) in April 2010 that was circulated among market participants and their business associations and in which the authority summarised its opinions regarding certain business practices. Representatives of the FCO have also announced the intention to intensify competition law enforcement in the food sector. As a consequence, many companies in the food sector now decline to discuss consumer promotion measures across the board. It can be expected that these uncertainties will at least remain until the current investigations have been closed. Against this background and in view of the risks that infringements of competition law entail, managers, employees and lawyers working in the food sector should know what types of conduct are problematic from a competition law point of view and what is still permissible.

---

[588] The author would like to thank Jakob Quirin, LL.M (Cambridge) for his contribution to this article. This is the English version of an article first published as Stancke, F., 2010. Das Kartellrecht der Lebensmittelbranche. Zeitschrift für das gesamte Lebensmittelrecht 2010: 543-565.

[589] Bundesvereinigung der Deutschen Ernährungsindustrie e.V., Kennzahlen der Ernährungsindustrie. Available at: http://www.bve-online.de.

[590] Siemes, J., Mager, G., Gerling, M. and Vogell, K., 2010. In: KPMG, Trends im Handel 2010, p. 51. Available at: http://www.kpmg.de/docs/trends_im_handel_2010_de.pdf.

[591] See the FCO's press release of 11/01/2010 regarding its investigation into the dairy sector; the press release of 14/01/2010 regarding the search of premises of retailers and brand producers in view of price fixing suspicions and the press release of 09/06/2010 regarding the imposition of fines against coffee roasters. Available at: http://www.bundeskartellamt.de.

[592] See Hasse, S., 2010. Kartellrecht für Privilegierte (commentary). LebensmittelZeitung, 16/04/2010, p. 2

[593] FCO, 2010. Vorläufige Bewertung von Verhaltensweisen in Verhandlungen zwischen Herstellern von Markenartikeln, Großhändlern und Einzelhandelsunternehmen zum Zwecke der Konkretisierung der Kooperationspflichten. Wirtschaft und Wettbewerb 2010: 786-791.

17.2 The requirements of competition law compliance

The risks connected to competition law infringements show that compliance is in the interest of companies and their management.[594] Even merely negligent infringements may seriously threaten the reputation of a company. Hefty fines of up to 10% of the company's total group turnover,[595] fining and imprisonment of employees,[596] D&O liability,[597] claims for damages and injunctive relief asserted by both clients and competitors[598] as well as the nullity of contracts[599] are further possible consequences of a competition law infringement. This last aspect may mean that already concluded contracts are worthless. Against this background, companies cooperating with others in the food sector should always carefully assess whether their conduct is in compliance with competition law.

17.3 Addressees of competition law in the food sector

All companies active in the food sector, whether they are producers, wholesalers or retailers, have to comply with competition law. The same holds true for their business associations, consultants and external researchers. However, agreements and concerted practices among companies that belong to the same corporate group do not usually fall within the ambit of competition law.[600]

Large food retailers are currently under particular attack. The most common allegation is that they abuse their buying power in order to demand unfairly low prices from their suppliers and to impose unfair terms on them. In addition, business associations and consultancies are increasingly becoming a focus of

[594] Competition authorities and courts generally assume that managers must ensure competition law compliance: Mitsch, W., 2006. In: Senge, L. (ed.) Karlsruher Kommentar zum OWiG, 3rd edition, C.H. Beck Verlag, Munchen, Germany, § 17, margin number ('mn.') 56 with further information.

[595] § 81 section 4 sentence 2 ARC; Article 23 section 2 of EU, 2003. Council Regulation (EC) No. 1/2003 of 16 December 2002 on the implementation of the rules on competition laid down in Articles 81 and 82 of the Treaty. Official Journal of the European Union L 1 of 4/1/2003: 1-25.

[596] § 81 section 1 in connection with § 9 OWiG; bid rigging is a crime in Germany which can be punished with imprisonment pursuant to §§ 298 und 263 German Criminal Code. In other countries, any kind of competition law infringement may constitute a criminal offence, e.g. in Austria, the United Kingdom and the United States.

[597] Directors and Officers Liability. See e.g. § 93 section 2 German Stock Companies Act (Aktiengesetz), § 43 section 2 Limited Liability Companies Act (GmbH-Gesetz).

[598] § 33 sections 1 and 3 Act against Restraints of Competition / ARC (Gesetz gegen Wettbewerbsbeschränkungen).

[599] Article 101 section 2 TFEU, § 1 ARC in connection with § 134 German Civil Code (Bürgerliches Gesetzbuch).

[600] Bechtold, R., 2010. In: Bechtold, R. (ed.) GWB, Kartellgesetz, Gesetz gegen Wettbewerbsbeschränkungen, 6th edition, C.H. Beck Verlag, Munchen, Germany, § 1, mn. 23; Schroeder, D., 2008. In: Wiedemann, G. (ed.) Handbuch des Kartellrechts, 2nd edition, C.H. Beck Verlag, Munchen, Germany, § 8, mn. 2 *et seq.* with further information; ECJ Case C-73/95 P Viho v Commission, ECR 1996: I-5457, mn. 16 *et seq.* and Zeitschrift für Wirtschaftsrecht 1997: 87; see also ECJ Case 15/74 Centrapharm v Sterling Drug, ECR 1974: 1147, mn. 41 and Neue Juristische Wochenschrift 1975: 516.

attention for competition authorities, mainly because they are suspected of playing important roles in cartels - for instance through recommendations on business conduct and information exchanges. A strict enforcement of competition law against such entities becomes possible through the German law of administrative offences which does not differentiate between the actual perpetrators of anticompetitive practices and persons aiding and abetting them.[601] The same holds true for European competition law as the General Court clarified in the test case *Treuhand v. Commission* in 2008.[602]

17.4 The restrictions on anticompetitive conduct

The conduct of companies in the food sector has to comply with both national (e.g. German) and European competition law, as the latter is directly applicable in the EU Member States.[603] The pertinent - and mostly congruent[604] - prohibitions of anticompetitive conduct can be found in § 1 ARC and in Article 101 section 1 of the Treaty on the Functioning of the European Union (TFEU)[605] which prohibits 'agreements between companies, decisions by associations of companies or concerted practices which have as their object or effect the prevention, restriction or distortion of competition'. The European courts have decided that companies must enjoy real autonomy in determining their course of action in the market.[606]

---

[601] Dannecker, G. and Biermann, J., 2007. In: Immenga, U. and Mestmäcker, E.J. (eds.) Wettbewerbsrecht Band 2: GWB. Kommentar zum Deutschen Kartellrecht, 4th edition, C.H. Beck Verlag, Munchen, Germany, Vor § 81, mn. 69; Vollmer, C., 2008. In: Hirsch, G. Montag, F. and Säcker, F.J. (eds.) Münchener Kommentar zum Europäischen und Deutschen Wettbewerbsrecht (Kartellrecht) Band 2: Gesetz gegen Wettbewerbsbeschränkungen: GWB, 1st edition, C.H. Beck Verlag, Munchen, Germany, § 81, mn. 43; Rengier, R., 2006. In: Senge, L. (ed.) Karlsruher Kommentar zum OWiG, 3rd edition, C.H. Beck Verlag, Munchen, Germany, § 14, mn. 4.

[602] CFI, Case T-99/04 Treuhand v Commission, ECR 2008, II-1505 and Europäisches Wirtschafts- und Steuerrecht 2008: 330.

[603] See (extensive) Böge, U. and Bardong, A., 2008. In: Hirsch, G. Montag, F. and Säcker, F.J. (eds.) Münchener Kommentar zum Europäischen und Deutschen Wettbewerbsrecht (Kartellrecht) Band 2: Gesetz gegen Wettbewerbsbeschränkungen: GWB, 1st edition, C.H. Beck Verlag, Munchen, Germany, § 22, mn. 1 *et seq.*; ECJ Case 127/73 BRT v SABAM, ECR 1974: 51, mn. 15/17 and Gewerblicher Rechtsschutz und Urheberrecht Internationaler Teil 1974: 342; Case 37/79 Marty v Estée Lauder, ECR 1980: 2481, mn. 13 and Neue Juristische Wochenschrift 1980: 2632; Case C-234/89 Delimitis v Henninger, ECR 1991: I-935, mn. 45 and Neue Juristische Wochenschrift 1991: 2204; Case C-453/99 Courage v Crehan, ECR 2001: I-6297, mn. 22 *et seq.* and Neue Juristische Wochenschrift 2002: 502.

[604] Karl, M. and Reichelt, D., 2005. Die Änderungen des Gesetzes gegen Wettbewerbsbeschränkungen durch die 7. GWB-Novelle. Der Betrieb 2005: 1436-1437; Stancke, F., 2005. Schadensregulierung und Kartellrecht. Versicherungrecht 2005: 1324.

[605] Until 30 November 2009: Article 81 section 1 EC. The Lisbon Treaty, that entered into force on 1 December 2009 amended the EU and EC - Treaty. The previous Article 81 section 1 EC became a new Article 101 section 1 TFEU.

[606] ECJ Case 43/73 *et al.* Société anonyme Générale Sucrière and others v Commission ECR 1975, 1663, mn. 174 and Wirtschaft und Wettbewerb 1976: 185 and Wirtschaft und Wettbewerb/Entscheidungssammlung EWG/MUV 347.

§ 1 ARC and Article 101 section 1 TFEU cover both horizontal and vertical restraints.[607] Agreements between competitors, e.g. between two food producers, may contain *horizontal* restraints; agreements between non-competitors, e.g. between food producers and food retailers, potentially contain *vertical* restraints. Agreements and concerted practices pursuant to which one party's leeway or discretion to negotiate with *third* parties on, e.g. prices, is restrained, possibly constitute infringements. Typically of concern are practices which restrain one of the parties to an agreement or third parties in determining important market features such as end-consumer prices, exclusivity rebates, sales, costs and marketing[608] or which harmonise costs.[609] The same holds true for information exchanges between competitors that concern these market features.[610] No restraints of competition exist, however, if the companies merely inform each other about objective circumstances that lie outside their individual influence and which concern the framework of competition such as general technical developments or new case-law.[611] Also, conduct will only be caught by § 1 ARC and Article 101 section 1 TFEU if it noticeably affects trade, i.e. if the restraint on competition has a noticeable effect.[612] And even if such noticeable effect is given, it is possible that the restraint on competition is exempted from the prohibition pursuant to a so-called block exemption regulation ('BER') or as a result of a case-specific substantive test on the basis of § 2 ARC / Article 101 section 3 TFEU. It should be borne in mind, however, that such an exemption will only be possible where

---

[607] Schwintowski, H.-P. and Klaue, S., 2005. Kartellrechtliche und gesellschaftsrechtliche Konsequenzen des Systems der Legalausnahme für die Kooperationspraxis der Unternehmen. Wirtschaft und Wettbewerb 2005: 370-378; draft bill on the 7th comprehensive amendment to the ARC, Bundestagsdrucksache 15/3640: 23 *et seq.*

[608] Bunte, H.-J., 2010. Einführung zum GWB. In: Langen, E. and Bunte, H.J. (eds.) Kommentar zum deutschen und europäischen Kartellrecht Bd. 1, 11th edition, Carl Heymanns Verlag, Koln, Germany, mn. 84.

[609] EU, 2011. Communication from the Commission. Guidelines on the applicability of Article 101 of the Treaty on the Functioning of the European Union to horizontal co-operation agreements (Horizontal Guidelines). Official Journal of the European Union C 11, 14/1/2011: 1-72. Available at: http://eur-lex.europa.eu/LexUriServ/LexUriServ.do?uri=OJ:C:2011:011:0001:0072:EN:PDF, mn. 27, 33 et seq.; Wagner-von Papp, F., 2004. Marktinformationsverfahren – Grenzen der Information im Wettbewerb: Die Herstellung praktischer Konkordanz zwischen legitimen Informationsbedürfnissen and Geheimwettbewerb. Wirtschaftsrecht und Wirtschaftspolitik: Vol. 191. Nomos Verlagsgesellschaft, Baden-Baden, Germany, p. 228.

[610] Stancke, F., Marktinformation, Benchmarking und Statistiken - Neue Anforderungen an Kartellrechts-Compliance. Betriebs-Berater 2009: 912 et seq. with further information; see also section 17.6.2 and 17.6.13.

[611] Federal Ministry of Economic Affairs, 1976. Kooperationsfibel - Zwischenbetriebliche Zusammenarbeit im Rahmen des Gesetzes über Wettbewerbsbeschränkungen, section II section 1.1; EU, 1968. Notice concerning agreements, decisions and concerted practices in the field of cooperation between enterprises. Official Journal of the European Union C 075, 29/07/1968: 3-6; Schumm, K., 2003. In: Schröter, H., Jakob, T. and Mederer, W. (eds.) Kommentar zum Europäischen Wettbewerbsrecht, 1st edition, Nomos Verlagsgesellschaft, Baden-Baden, Germany, Article 81 mn. 48.

[612] Established case-law since ECJ Case 5/69 Völk v Vervaecke, ECR 1969: 295, 302, mn. 7 and Neue Juristische Wochenschrift 1970: 399; Bechtold, R., 2010. In: Bechtold, R. (ed.) GWB, Kartellgesetz, Gesetz gegen Wettbewerbsbeschränkungen, 6th edition, C.H. Beck Verlag, Munchen, Germany, § 1, mn. 37 et seq.

the collusion has particularly positive effects which consumers must also profit from.[613]

17.5 The restrictions on non-collusive / unilateral conduct by market dominant companies

Competition law does not only restrict anticompetitive collusive conduct between market participants. It also restricts unilateral measures by companies that are dominant in their market or have a particularly strong market position, if these measures constitute an 'abuse' of that strong market position.[614] The abuse of a dominant market position is prohibited by both German and European competition law. The pertinent rules are §§ 19 *et seq.* ARC and Article 102 TFEU. Both provisions contain very similar rules,[615] but the German law tends to be stricter.[616]

The economic power of companies is usually constrained by competitors and the freedom of buyers to choose their supplier. Yet, some companies do not experience sufficient competitive pressure with the consequence that they have a particularly broad leeway to manoeuvre in their market(s). A company is dominant in this sense if, as a supplier or buyer of certain products, it is without competitors or does not experience significant competition or is particularly strong in a given market.[617] Theoretically, it is also possible that a number of companies are collectively dominant in a specific market.[618]

---

[613] Regarding the individual requirements see Kamann, H.-G. and Bergmann, E., 2003. Die neue EG-Kartellverfahrensverordnung - Auswirkungen auf die unternehmerische Vertragspraxis. Betriebs-Berater 2003: 1743, 1745 et seq.; Schwintowski, H.-P. and Klaue, 2005. Kartellrechtliche und gesellschaftsrechtliche Konsequenzen des Systems der Legalausnahme für die Kooperationspraxis der Unternehmen. Wirtschaft und Wettbewerb 2005: 370, 372 et seqq.; Roth, P. and Rose, V., 2008. Bellamy & Child: European Community Law of Competition, 6th edition, Oxford University press, Oxford, UK, 3.020 et seqq.; Bunte, H.-J., 2010. In: Langen, E. and Bunte, H.J. (eds.) Kommentar zum deutschen und europäischen Kartellrecht Bd. 2, 11th edition, Carl Heymanns Verlag, Koln, Germany, Art. 81 EG, mn. 146 et seqq.

[614] Wiedemann, G., 2008. In: Wiedemann, G. (ed.) Handbuch des Kartellrechts, 2nd edition, C.H. Beck Verlag, Munchen, Germany, § 23, mn. 16; de Bronett, G.K., 2008. In: Wiedemann, G. (ed.) Handbuch des Kartellrechts, 2nd edition, C.H. Beck Verlag, Munchen, Germany, § 22 mn. 7.

[615] See Bechtold, R., 2010. In: Bechtold, R. (ed.) GWB, Kartellgesetz, Gesetz gegen Wettbewerbsbeschränkungen, 6th edition, C.H. Beck Verlag, Munchen, Germany, § 20, mn. 115.

[616] Even where Article 102 TFEU and § 19 et seqq. ARC are parallely applicable, a national competition authority or court may apply the stricter national law, even if the conduct in question is permissible under European law: Article 3 section 1 sentence 2 Regulation 1/2003; Bechtold, R., 2010. In: Bechtold, R. (ed.) GWB, Kartellgesetz, Gesetz gegen Wettbewerbsbeschränkungen, 6th edition, C.H. Beck Verlag, Munchen, Germany, § 20, mn. 117.

[617] See in detail section 17.6.15.

[618] § 19 section 2 sentence 1 No. 2 ARC. See § 19 section 3 ARC for the presumptions of market dominance under German law; Bunte, H.-J., 2010. In: Langen, E. and Bunte, H.J. (eds.) Kommentar zum deutschen und europäischen Kartellrecht Bd. 2, 11th edition, Carl Heymanns Verlag, Koln, Germany, Article 82, mn. 52 et seq.

Examples of potentially abusive conduct by dominant companies are refusals to supply, discrimination between buyers or suppliers, loyalty and target rebates that are linked to the total volume of sales of a buyer, the imposition of unfair prices or terms, and tying practices.[619]

17.6 Groups of cases relevant under competition law in the food sector

In light of these preliminary observations, some groups of cases that are potentially relevant in the food sector will now be examined more closely. Quite naturally, this will not be a full list of relevant cases so that other forms of collusive and non-collusive / unilateral conduct should also be carefully assessed for their compliance with competition law in practice. Examples would be the sale of products over the internet and margin guarantees. Also, in view of the potential complexity of collusive and non-collusive conduct, only general legal considerations can be set out in this chapter.

17.6.1 Price fixing among competitors

Any direct or indirect agreement to fix prices among competitors, i.e. between producers of certain groceries or between wholesalers or retailers, which stops the parties from freely determining their buying or selling prices is problematic from a competition law point of view. In particular, the fixing of uniform sale prices,[620] be it with regard to gross price,[621] list price, net price, price components or rebates is prohibited.[622] Both collusion with regard to price increases and with regard to price decreases as well as fixing a point in time to change prices is problematic. Of relevance are not only agreements on fixed prices but also the fixing of a target or orientation price or of a price frame. Suggestions on how to calculate prices by e.g. business associations are, however, not prohibited if they merely give technical

[619] Wiedemann, G., 2008. In: Wiedemann, G. (ed.) Handbuch des Kartellrechts, 2nd edition, C.H. Beck Verlag, Munchen, Germany, § 23, mn. 41 et seqq., 51 et seq.; for details see section 17.6.15.

[620] Examples from German case law are Federal Court of Justice, Wirtschaft und Wettbewerb/ Entscheidungssammlung 40, 41 - Schulspeisung; Wirtschaft und Wettbewerb/Entscheidungssammlung 148, 150 - Freisinger Bäckermeister; Wirtschaft und Wettbewerb/Entscheidungssammlung DE-R 711, 717 - Ost-Fleisch; on the European level see e.g. EU, Wirtschaft und Wettbewerb/Entscheidungssammlung EV 917 and Wirtschaft und Wettbewerb 1982: 959 – Italian Flat Glass; Wirtschaft und Wettbewerb/ Entscheidungssammlung EV 1173 and Wirtschaft und Wettbewerb 1987: 263 – Roofing Felt (confirmed by ECJ Case C-246/86 Belasco *et al.* v Commission, ECR 1989: 2117 and Europäische Zeitschrift für Wirtschaftsrecht 1990: 323).

[621] See EU, Wirtschaft und Wettbewerb/Entscheidungssammlung EV 820 and Wirtschaft und Wettbewerb 1980: 770 - BP Kemi - DDSF.

[622] Bunte, H.-J., 2010. In: Langen, E. and Bunte, H.J. (eds.) Kommentar zum deutschen und europäischen Kartellrecht Bd. 2, 11th edition, Carl Heymanns Verlag, Koln, Germany, Art. 81 EG, Generelle Prinzipien, mn. 103.

help to find a price and if they cannot lead to uniform prices.[623] Price fixing cannot be justified by a reference to dumping practices by other companies,[624] alleged 'cut-throat' competition from other companies,[625] steep increases in commodity prices,[626] unfair trading practices by other companies[627] or over-capacity[628].[629]

17.6.2 Hub and spoke agreements

Prices cannot only be fixed directly between competitors but also indirectly through the mediation of a third party. An example is the wholesaler that distributes products of a number of suppliers and thus serves as a 'hub' to these suppliers that makes price fixing agreements possible.[630] Such mediated horizontal price fixing raises competitive concerns just as much as direct price fixing would.[631] That

---

623 Bunte, H.-J., 2010. In: Langen, E. and Bunte, H.J. (eds.) Kommentar zum deutschen und europäischen Kartellrecht Bd. 1, 11th edition, Carl Heymanns Verlag, Koln, Germany, § 1 GWB, mn. 182; for critical remarks see Regional Court Munich, Neue Juristische Wochenschrift 1985: 1906 and EU, Wirtschaft und Wettbewerb/Entscheidungssammlung EV 846 and Wirtschaft und Wettbewerb 1981: 572 – Re-rolled Steel, mn. 36; FCO, 2010. Vorläufige Bewertung von Verhaltensweisen in Verhandlungen zwischen Herstellern von Markenartikeln, Großhändlern und Einzelhandelsunternehmen zum Zwecke der Konkretisierung der Kooperationspflichten. Wirtschaft und Wettbewerb 2010: 786, 790. (786-791).

624 See e.g. EU, Wirtschaft und Wettbewerb/Entscheidungssammlung EV 1214 and Wirtschaft und Wettbewerb 1987: 430 - MELDOC; EU, Wirtschaft und Wettbewerb/Entscheidungssammlung EV 1101 and Wirtschaft und Wettbewerb 1986: 255 – Aluminium imports from Eastern Europe.

625 See e.g. EU, 1992. 92/204/EEC: Commission Decision of 5 February 1992 relating to a proceeding pursuant to Article 85 of the EEC Treaty (IV/31.572 and 32.571 - Building and construction industry in the Netherlands). Official Journal of the European Union *L 92, 7.4.1992: 1-30;* – SPO (confirmed by CFI Case T-29/92 SPO *et al.* v Commission, ECR 1995: II-289, 341 mn. 146).

626 See EU, 1994. 94/599/EC: Commission Decision of 27 July 1994 relating to a proceeding pursuant to Article 85 of the EC Treaty (IV/31.865 – PVC). Official Journal of the European Union *L 239, 14.9.1994: 14-35;* PVC II, (mainly confirmed by CFI, see e.g. joined Cases T-305/94 – T-335/94 Limburgse Vinyl Maatschappij, ECR 1999, II-931 and CFI, 1999. Umfang von Verteidigungsrechten. Wirtschaft und Wettbewerb 1999: 623-631; mainly confirmed by ECJ Case C-238/99 P Limburgse Vinyl Maatschappij *et al.* v Commission, ECR 2002: I-8375 and ECJ. Folgen der Nichtigerklärung einer Geldbußenentscheidung. Wirtschaft und Wettbewerb 2002: 1257-1266).

627 See ECJ joined Cases 43/82 and 63/82 - VBVB and VBBB v Commission, ECR 1984: 19 and Neue Juristische Wochenschrift 1985: 546.

628 See EU, Wirtschaft und Wettbewerb/Entscheidungssammlung EV 1165 and Wirtschaft und Wettbewerb 1987: 81 – Polypropylen; ECJ Case C-199/92 P Hüls v Commission, ECR 1999: I-4287 and ECJ, 1999. Bestätigung der Polypropylen-Entscheidung durch den Gerichtshof. Wirtschaft und Wettbewerb 1999: 902-907.

629 See Wägenbaur, B., 2009. In: Loewenheim, U., Meessen, K.M. and Riesenkampff, A. (eds.) Kartellrecht, 2nd edition, C.H. Beck Verlag, Munchen, Germany, Art. 81 EG, mn. 208.

630 The suppliers would correspondingly be the 'spokes'.

631 Higher Regional Court Düsseldorf, Wirtschaft und Wettbewerb/Entscheidungssammlung 4691, 4692 – Sternvertrag; Federal Court of Justice, Gewerblicher Rechtsschutz und Urheberrecht 2003: 633 - Ausrüstungsgegenstände für Feuerlöschzüge; see also FCO, Wirtschaft und Wettbewerb/ Entscheidungssammlung DE-V 1813 and FCO, 2010. Bußgeld wegen Beschränkung des Internetvertriebs. Wirtschaft und Wettbewerb 2010: 91-97; - CIBA Vision, mn. 45, 61 et seq.; see also FCO, 2010. Vorläufige Bewertung von Verhaltensweisen in Verhandlungen zwischen Herstellern von Markenartikeln, Großhändlern und Einzelhandelsunternehmen zum Zwecke der Konkretisierung der Kooperationspflichten. Wirtschaft und Wettbewerb 2010: 786, 788. (786-791).

'hub and spoke' situations are primarily problematic in view of their potential for horizontal collusion should not lead one to the assumption that the hub itself must not fear intervention by the competition authority. If, for instance, the wholesaler in the above example consciously facilitates the anti-competitive collusion by the suppliers, it will itself be held liable for a competition law infringement.[632] In order to avoid such risks parties should refrain from any disclosure of third party information on prices and conditions. It is also advisable to agree on confidentiality agreements prohibiting the disclosure of prices or price-relevant information to other market participants, in particular the other spokes.

17.6.3 Agreements on opening hours, certification marks and marketing

Non price-related agreements among competitors can also be problematic. For instance, agreements on opening hours directly regulate the conduct of parties in the market so that they are generally considered to have anti-competitive effects.[633] Agreements on certification marks[634] and other marketing agreements among competitors with regard to the sale, the distribution (e.g. through a common internet platform) or consumer promotion (e.g. through advertising associations[635]) can raise competitive concerns. Marketing agreements can, for instance, be problematic if they are merely a framework for agreements on prices or production quantities, if the parties to the agreement share markets among each other and/or if prohibitions on advertising or non-compete clauses are agreed on. Marketing agreements can also facilitate prohibited exchanges of information.[636] The European Commission has made critical comments on the harmonisation of distribution and marketing

---

[632] See also the relatively recent judgement by the Court of Appeal, Argos Ltd and Another v Office of Fair Trading [2006] EWCA Civ 1318, available at: http://www.bailii.org.

[633] This is the prevailing opinion in Germany: Kammergericht, Wirtschaft und Wettbewerb 1990: 945 – 'Ladenöffnungszeit'; FCO, Wirtschaft und Wettbewerb/Entscheidungssammlung 339 and Wirtschaft und Wettbewerb 1961: 305 – 'Sonnabendarbeitszeit'; Rehbinder, E., 1964. Betriebs-Berater 1964: 896; extensively Immenga, U., 1989. Grenzen des kartellrechtlichen Ausnahmebereichs Arbeitsmarkt. C.H. Beck Verlag, Munchen, Germany, p. 25, 39; Bunte, H.-J., 2010. In: Langen, E. and Bunte, H.J. (eds.) Kommentar zum deutschen und europäischen Kartellrecht Bd. 1, 11th edition, Carl Heymanns Verlag, Koln, Germany, § 1 GWB, mn. 193 with further information; Nordemann, J.B., 2009. In: Loewenheim, U., Meessen, K.M. and Riesenkampff, A. (eds.) Kartellrecht, 2nd edition, C.H. Beck Verlag, Munchen, Germany, § 1 GWB, mn. 135; with regard to collective agreements see Zimmer, D., 2007. In: Immenga, U. and Mestmäcker, E.J. (eds.) Wettbewerbsrecht Band 2: GWB. Kommentar zum Deutschen Kartellrecht, 4th edition, C.H. Beck Verlag, Munchen, Germany, § 1, mn. 262.

[634] Bunte, H.-J., 2010. In: Langen, E. and Bunte, H.J. (eds.) Kommentar zum deutschen und europäischen Kartellrecht Bd. 1, 11th edition, Carl Heymanns Verlag, Koln, Germany, § 1 GWB, mn. 186 with further information.

[635] Bunte, H.-J., 2010. In: Langen, E. and Bunte, H.J. (eds.) Kommentar zum deutschen und europäischen Kartellrecht Bd. 1, 11th edition, Carl Heymanns Verlag, Koln, Germany, § 1 GWB, mn. 189 with further information.

[636] Horizontal Guidelines, mn. 56, 233; on information exchanges between competitors see section 17.6.2. and 17.6.13.

costs – a possible effect of marketing co-operations.[637] Despite these objections marketing co-operations do not infringe competition law if they are objectively necessary for a company to access a market, e.g. through the cost saving effects that the co-operation generates. In addition, marketing co-operations can, under certain conditions, be exempted from the prohibition on collusive conduct by law. Whether or not a marketing co-operation is permissible must thus be determined in each and every individual case. It is generally possible to design them in ways that are non-objectionable.

17.6.4 Vertical price fixing agreements

The FCO currently focuses its activity on the direct or indirect fixing of resale prices[638] among suppliers and buyers / future sellers.[639] Quite recently, the FCO stated that one of the central elements of price competition in the food sector is that wholesalers and retailers can determine prices freely and that they carry the economic risk of their decisions.[640]

a. Agreements on fixed resale prices or minimum prices

Of particular concern is the oral or written fixing of resale prices or collusion with regard to minimum prices[641] irrespective of whether the resale price was explicitly agreed on or whether it is merely the consequence of the unilateral granting of advantages or exertion of pressure. Business meetings are typical occasions for such collusive practices. Even the mere mentioning of resale prices or minimum prices in ordering forms or other documents or the use of packaging with non-removable price imprints[642] is problematic. The mere fixing of maximum prices

[637] Horizontal Guidelines, mn. 233.

[638] Agreements on prices between a buyer / future seller and its supplier fall outside the pertinent prohibitions, if the buyer / future seller is a commercial agent, i.e. is simply meant to negotiate contracts on behalf of the principal and does not carry the economic risk, see Bechtold, R., 2010. In: Bechtold, R. (ed.) GWB, Kartellgesetz, Gesetz gegen Wettbewerbsbeschränkungen, 6th edition, C.H. Beck Verlag, Munchen, Germany, § 1, mn. 26.

[639] See FCO, 2010. Vorläufige Bewertung von Verhaltensweisen in Verhandlungen zwischen Herstellern von Markenartikeln, Großhändlern und Einzelhandelsunternehmen zum Zwecke der Konkretisierung der Kooperationspflichten. Wirtschaft und Wettbewerb 2010: 786, 787 et seq.

[640] See FCO, 2010. Vorläufige Bewertung von Verhaltensweisen in Verhandlungen zwischen Herstellern von Markenartikeln, Großhändlern und Einzelhandelsunternehmen zum Zwecke der Konkretisierung der Kooperationspflichten. Wirtschaft und Wettbewerb 2010: 786, 787.

[641] See FCO, 2010. Vorläufige Bewertung von Verhaltensweisen in Verhandlungen zwischen Herstellern von Markenartikeln, Großhändlern und Einzelhandelsunternehmen zum Zwecke der Konkretisierung der Kooperationspflichten. Wirtschaft und Wettbewerb 2010: 786, 787, 789.

[642] Regional Court Düsseldorf, Judgement of 18/3/2010, file number 14c O 234/09. Available at: http://www.justiz.nrw.de.

is not generally impermissible but can be exempted pursuant to § 2 ARC, Article 2, sections 1 and 4 letter a of the block exemption on vertical restraints.[643]

b. Recommended resale prices (RRPs)

RRPs do not necessarily infringe competition law.[644] Handing out a list with RRPs to a buyer that asks for it is unproblematic for instance. When handing over the list, the supplier may also explain the reasons for the price recommendations and the strategy that it pursues with regard to the positioning and marketing of the products. Price recommendations have a useful orientation function for consumers and can facilitate the calculation of prices for retailers. It therefore seems justified to assume that recommended resale prices generally do not noticeably affect competition.[645] The competition authorities do, however, keep a strict eye on RRPs that have the same effect as a fixed or minimum resale price because of the way in which they were suggested to a future seller. Such RRPs are considered to be hardcore restrictions pursuant to § 2 section 2 ARC, Article 4 letter a BER-vertical.[646] It should be noticed, however, that the question whether a resale price was agreed on or freely determined by the seller must be carefully assessed in each individual case.[647] The FCO has clarified that every attempt to influence the determination of prices by future sellers that goes beyond

[643] See EU, 2010. Commission Regulation (EU) No 330/2010 of 20 April 2010 on the application of Article 101(3) of the Treaty on the Functioning of the European Union to categories of vertical agreements and concerted practices. Official Journal of the European Union L 102, 23.4.2010: 1-7 ('BER-Vertical'); Kirchhain, S, 2008. Die Gestaltung von innerstaatlich wirkenden Vertriebsverträgen nach der 7. GWB-Novelle. Wirtschaft und Wettbewerb 2008: 167, 172 et seq.; Hildebrand, D., 2004. Wettbewerb in Recht und Praxis 2004: 470, 473; Bechtold, R., 2010. In: Bechtold, R. (ed.) GWB, Kartellgesetz, Gesetz gegen Wettbewerbsbeschränkungen, 6th edition, C.H. Beck Verlag, Munchen, Germany, § 1, mn. 58.

[644] FCO, Wirtschaft und Wettbewerb/Entscheidungssammlung DE-V 1813 and Wirtschaft und Wettbewerb 2010: 91 - CIBA Vision, mn. 43; Federal Supreme Court, Gewerblicher Rechtsschutz und Urheberrecht 2007: 603, 604 - Zulässige Verwendung einer Abkürzung für eine Preisempfehlung; CFI Case T-208/01 Volkswagen v Commission, ECR 2003: II-5141, mn. 33 et seqq. and Wirtschaft und Wettbewerb 2004: 203; ECJ Case C-74/04 P Volkswagen v Comission, ECR 2006: I-6585, mn. 35 et seq. and Wirtschaft und Wettbewerb 2006: 1082; Kirchhain, S., 2008. Wirtschaft und Wettbewerb 2008: 167, 175 et seq.

[645] See Baron, M., 2009. In: Loewenheim, U., Meessen, K.M. and Riesenkampff, A. (eds.) Kartellrecht, 2nd edition, C.H. Beck Verlag, Munchen, Germany, GVO-Vertikal, mn. 167; Veelken, W., 2007. In: Immenga, U. and Mestmäcker, E.J. (eds.) Wettbewerbsrecht Band 1: EG. Kommentar zum Europäischen Kartellrecht, 4th edition, C.H. Beck Verlag, Munchen, Germany, Vertikal-GVO, mn. 182; Bunte, H.-J., 2010. In: Langen, E. and Bunte, H.J. (eds.) Kommentar zum deutschen und europäischen Kartellrecht Bd. 2, 11th edition, Carl Heymanns Verlag, Koln, Germany, Art. 81 EG, Generelle Prinzipien, mn. 184.

[646] EU, 2010. Guidelines on Vertical Restraints. Official Journal of the European Union C 130: 1-46 (Vertical Guidelines), mn. 48, 226 et seq.; FCO, 2010. Vorläufige Bewertung von Verhaltensweisen in Verhandlungen zwischen Herstellern von Markenartikeln, Großhändlern und Einzelhandelsunternehmen zum Zwecke der Konkretisierung der Kooperationspflichten. Wirtschaft und Wettbewerb 2010: 786, 788.

[647] Kirchhoff, W., 2009. In: Wiedemann, G. (ed.) Handbuch des Kartellrechts, 2nd edition, C.H. Beck Verlag, Munchen, Germany, § 11, mn. 29.

non-binding recommendations is impermissible.[648] Any kind of contact with the future seller that makes RRPs a topic after the price was suggested, particularly with a view to the seller's previous 'price policy', puts a question mark behind the non-binding character of the RRP according to the somewhat questionable view of the FCO.[649] The creation of, or demand for, price comparison lists and/or receipt collections as well as the preparation of calculation schemes is looked at critically by the FCO. Somewhat beside the point seems to be the opinion of the FCO that advertising by the producer ('price cut, now only € 9.99') can unduly influence the free determination of prices by the distributor.[650] If distributors and retailers voluntarily participate in such campaigns they are not forced to lower their prices to the newly recommended price or to give rebates. According to the FCO, a particularly severe way of exerting pressure is the announcement of specific disadvantages in case of non-compliance with the RRP, e.g. the unilateral reduction of payments, delisting, worsening of terms, termination, deferment, suspension or limitation of the supplies or the downsizing of shelf space.[651] Other examples are limitations on gross margins,[652] prohibitions on selling at a loss[653] and obligations to return goods at net price to the distributor if the purpose of such provisions is to prevent retailers from pursuing a low price policy.[654] Finally, incentives that are meant to enforce compliance with RRPs, such as certain types of rebates,[655] kick-backs and margin-loss compensation raise the competition authority's concern.

17.6.5 Most-favoured-customer clauses

In view of the illegality of vertical price fixing the question arises how so-called most-favoured-customer clauses must be assessed. Such clauses oblige e.g. a food

[648] FCO, Wirtschaft und Wettbewerb/Entscheidungssammlung DE-V 1813 and FCO, 2010. Bußgeld wegen Beschränkung des Internetvertriebs. Wirtschaft und Wettbewerb 2010: 91-97; - CIBA Vision, mn. 44; FCO, 'Informationsblatt des Bundeskartellamts zu den Verwaltungsgrundsätzen bei UVP für Markenwaren', published in: Langen, E. and Bunte, H.J. (eds.), 2006. Kommentar zum deutschen und europäischen Kartellrecht Bd. 2, 9th edition, Carl Heymanns Verlag, Koln, Germany, p. 2858 et seq.

[649] FCO, 2010. Vorläufige Bewertung von Verhaltensweisen in Verhandlungen zwischen Herstellern von Markenartikeln, Großhändlern und Einzelhandelsunternehmen zum Zwecke der Konkretisierung der Kooperationspflichten. Wirtschaft und Wettbewerb 2010: 786, 788.

[650] Federal Supreme Court, Wirtschaft und Wettbewerb/Entscheidungssammlung 2256 - Herstellerpreiswerbung.

[651] FCO, 2010. Vorläufige Bewertung von Verhaltensweisen in Verhandlungen zwischen Herstellern von Markenartikeln, Großhändlern und Einzelhandelsunternehmen zum Zwecke der Konkretisierung der Kooperationspflichten. Wirtschaft und Wettbewerb 2010: 786, 789.

[652] Zimmer, D., 2007. In: Immenga, U. and Mestmäcker, E.J. (eds.) Wettbewerbsrecht Band 2: GWB. Kommentar zum Deutschen Kartellrecht, 4th edition, C.H. Beck Verlag, Munchen, Germany, § 1, mn. 250, 238.

[653] Federal Supreme Court, Wirtschaft und Wettbewerb/Entscheidungssammlung 1036 and Wirtschaft und Wettbewerb 1970: 80 - Lockvogel.

[654] Federal Supreme Court, Wirtschaft und Wettbewerb/Entscheidungssammlung 2479 and Wirtschaft und Wettbewerb 1990: 73 - Völkl.

[655] FCO, 2004. Tätigkeitsbericht 2003/2004, p. 161. Available at: http://www.bundeskartellamt.de.

producer to afford to all buyers the best prices and terms that it offers to any one buyer. The clauses show many variations, usually to the detriment of the supplier and in individual cases also to the detriment of the buyer. The FCO takes a particularly critical look at most-favoured-customer clauses that tend to harmonise prices among wholesalers or retailers.[656] This does not mean that most-favoured customer clauses are generally inadmissible. For instance, provisions to the detriment of the supplier are exempt from the prohibition pursuant to § 2 section 2 ARC, Article 2 BER-Vertical if the market share of each of the participating companies does not exceed 30%.

17.6.6 Price guarantees

Most-favoured-customer clauses must be differentiated from price guarantees. The point of the latter is to give buyers a claim against their suppliers if they can source the respective products from a competitor at a lower price. Despite their similarity with most-favoured-customer clauses, price guarantees do not raise competitive concerns. This is because a most-favoured-customer clause legally or factually obliges a company with regard to the future setting of prices and/or terms in relation to third parties, while price guarantees do not cause similar obligations: They merely oblige the guarantor towards the other contractual party.[657]

17.6.7 Listing fees

Advance payments to retailers in return for services and for listing groceries raise concerns both from a general competition law perspective and more specifically under the law of unfair competition. Of great importance are listing fees, so-called pay-to-stay fees and fees for access to a retailer's advertising campaign. Such agreements can develop 'pulling power' to the advantage of specific suppliers and can thus foreclose the market to the suppliers' competitors, in particular if the parties to the agreement hold market shares of more than 30%.[658] Under the German law of unfair competition payments with the intention to 'squeeze out' specific competitors from the market are problematic.[659] On the other hand, listing fees can also lead to an efficient allocation of shelf space for new products and other positive effects. Such agreements must therefore undergo an individual assessment in practice.

[656] FCO, 2010. Vorläufige Bewertung von Verhaltensweisen in Verhandlungen zwischen Herstellern von Markenartikeln, Großhändlern und Einzelhandelsunternehmen zum Zwecke der Konkretisierung der Kooperationspflichten. Wirtschaft und Wettbewerb 2010: 786, 790.

[657] Zimmer, D., 2007. In: Immenga, U. and Mestmäcker, E.J. (eds.) Wettbewerbsrecht Band 2: GWB. Kommentar zum Deutschen Kartellrecht, 4th edition, C.H. Beck Verlag, Munchen, Germany, § 1, mn. 401 et seq.

[658] See Vertical Guidelines, mn. 203 et seq.

[659] See section 17.6.16.

17.6.8 Selling at a loss

§ 20 section 4 sentence 2 ARC contains a provision that prohibits companies which are particularly strong compared to their small and medium sized competitors from offering goods or services below cost without an objective justification.[660] Even occasional selling at a loss is problematic.[661] In practice, difficulties are most often attached to the calculation of costs and prices.[662] For instance, the Higher Regional Court in Düsseldorf recently decided that contributions towards advertising costs may be taken into account when calculating costs for the purpose of the assessment whether goods were sold below costs.[663]

17.6.9 Category management

Category management has gained a foothold in Europe in recent years. It can be defined as the marketing of a whole group of goods at the point of sale by only one of the producers of the grouped goods. This 'category captain', chosen by the retailer, is responsible for designing the distribution of this group of goods at the retailer's point of sale. The support can be of varying intensity, from merely occasional advice up to the creation of a comprehensive marketing concept.[664] European Competition Authorities have only become aware of category management in

[660] See Alexander, C., 2010. Privatrechtliche Durchsetzung des Verbots von Verkäufen unter Einstandspreis. Wettbewerb in Recht und Praxis 2010: 727, 731 et seq. with regard to selling at a loss under the German law of unfair competition (§§ 3 sections 1, 4 Nr. 10 UWG).

[661] Loewenheim, U., 2009. In: Loewenheim, U., Meessen, K.M. and Riesenkampff, A. (eds.) Kartellrecht, 2nd edition, C.H. Beck Verlag, Munchen, Germany, § 20 GWB, mn. 146; See also Alexander, C., WRP 2010: 727, 728.

[662] Extensively: Markert, K., 2007. In: Immenga, U. and Mestmäcker, E.J. (eds.) Wettbewerbsrecht Band 2: GWB. Kommentar zum Deutschen Kartellrecht, 4th edition, C.H. Beck Verlag, Munchen, Germany, § 20, mn. 297 et seqq.; and also FCO, Bekanntmachung zur Anwendung des § 20 section 4 Satz 2 GWB. Available at: http://www.bundeskartellamt.de/wDeutsch/download/pdf/Merkblaetter_deutsch/Bekanntmachung_Einstandspreis.pdf.

[663] OLG Düsseldorf (VI-2 Kart 9/08 OWi). Schütze, J. and Beck, R.S., 2010. Drogerie Rossmann darf pauschale Werbekostenzuschüsse auf einzelne Produkte umlegen – Kein Verkauf unter Einstandspreis. Gewerblicher Rechtsschutz und Urheberrecht - Praxis: 208. See also: FCO, Sektorenuntersuchung Milch – Zwischenbericht, Dezember 2009: 95. Available at: http://www.bundeskartellamt.de.

[664] Loest, T., 2004. 'Category Captaincies' im Ordnungsrahmen des Kartellrechts. Wettbewerb in Recht und Praxis 2004: 454, 454 et seq.

recent years.[665] The concerns brought forward are of a rather complex nature. The main concern is that category management could serve as the basis for information exchanges that could then, in turn, serve as a basis for collusive conduct. In order to comply with its obligations, a 'category captain' usually needs information on the market behaviour of its competitors.[666] From a competition law perspective it is important to limit information flows to the degree that is objectively necessary for an effective category management. It is also advisable to make sure that the information which the 'category captain' receives is not freely accessible within its undertaking but protected by 'Chinese walls' from the curious look of others. Category management also creates the risk of an impermissible exchange of information on the level of retailers. Opportunities for such an exchange of information exist where competing retailers choose the same producer as their 'category captain'. Finally, an exchange of information between retailers and their 'category captain' can also be problematic if the retailer competes against the 'category captain' with a private brand.[667] Other pitfalls in connection with category management are: vertical price fixing,[668] an agreement between 'category captain' and retailer to the detriment of competitors of the 'category captain'[669] as well as the abuse of a possibly dominant position of the 'category captain'. While companies will not have a dominant position merely because they take over the category captaincy,[670] producers that are already dominant must be careful not to abuse their position as a category captain, for instance through attempts to convince 'their' retailers to delist competing products. Importantly, the 'category captain' may only give

[665] Apparently, the only case in which a European national competition authority considered category management is UK Office of Fair Trading, ME/1213/04 United Biscuits/Jacobs Bakery Limited, mn. 42 et seqq. (available at: http://oft.gov.uk/OFTwork/mergers/decisions/2004/united-biscuits). The FCO has, in its decision Pelikan/Herlitz (Wirtschaft und Wettbewerb/Entscheidungssammlung DE-V 1892 and FCO, 2010. Freigabe eines Zusammenschlusses in Schreibgerätemärkten. Wirtschaft und Wettbewerb 2010: 568-577) only pronounced its opinion on a similar method of managing products (see Besen, M. and Jorias, R., 2010. Kartellrechtliche Grenzen des Category Managements unter Berücksichtigung der neuen EU-Leitlinien für vertikale Beschränkungen. Betriebs-Berater 2010: 1099, 1100). The French Autorité de la Concurrence has been working on the topic since March 2010. On the European level should be noted: EU, 2005. Prior notification of a concentration (Case COMP/M.3732 – Procter & Gamble/Gillette). Official Journal of the European Union C 139, 08/06/2005: 35, mn. 134 et seq. in which the Commission came to a generally positive conclusion with regard to category management and the Vertical Guidelines, which dedicate mn. 209-213 to category management. Finally, it should be noted that the topic was discussed in the US already at the beginning of the decade; see Wiring, R., 2010. Sortimentsplanung im Supermarktregal – Welchen Spielraum lässt das Kartellrecht? Gewerblicher Rechtsschutz und Urheberrecht. Praxis 2010: 332.
[666] On the 'secrecy' of competition see section 17.6.13.
[667] See also section 17.4.13.
[668] See already section 17.4.4.
[669] See Besen, M. and Jorias, R., 2010. Kartellrechtliche Grenzen des Category Managements unter Berücksichtigung der neuen EU-Leitlinien für vertikale Beschränkungen. Betriebs-Berater 2010: 1099, 1101.
[670] Wiring, R., 2010. Sortimentsplanung im Supermarktregal – Welchen Spielraum lässt das Kartellrecht? Gewerblicher Rechtsschutz und Urheberrecht. Praxis 2010: 332.

non-binding recommendations to the retailer. The latter one must remain free to deviate from the course of action suggested by its 'category captain'.[671]

17.6.10 Exclusive supply/single branding obligations

Agreements on exclusivity are of natural relevance from a competition law point of view. Oftentimes, producers and retailers have an interest to agree on exclusive supply or single branding obligations. Such vertical restraints[672] of competition raise concerns because they can seal off the market to possible new entrants.[673] Exclusive supply obligations can e.g. make it difficult for non-participating producers to supply their products through certain distributors. On the wholesale or retail level, competition can be constrained if exclusive supply obligations lead to wholesalers and/or retailers being 'cut off' from the supply of certain goods. And finally, exclusive supply obligations can make it difficult for end consumers to gain access to products. Despite these problems, it is well accepted that agreements on exclusivity can, under certain circumstances, be permissible. Obligations to source less than 80% of total demand from a supplier are usually not problematic if each of the parties' aggregate market shares does not exceed 30%.[674] Exclusive supply obligations of short duration (about a year) do generally not cause problems.[675] Producers and suppliers with a strong (not necessarily dominant) position in their market that want to oblige their partners to source more than 80% of their demand over a longer period of time or distributors that intend to demand exclusive supply should be particularly careful when structuring their supply agreements.

[671] Loest, T., 2004. 'Category Captaincies' im Ordnungsrahmen des Kartellrechts. Wettbewerb in Recht und Praxis 2004: 454, 460; Besen, M. and Jorias, R., 2010. Kartellrechtliche Grenzen des Category Managements unter Berücksichtigung der neuen EU-Leitlinien für vertikale Beschränkungen. Betriebs-Berater 2010: 1099.

[672] In practice, such agreements take many different forms. Even merely factual exclusive supply obligations, e.g. on the basis of rebate systems can raise concerns.

[673] See Regional Court Mannheim, Wirtschaft und Wettbewerb/Entscheidungssammlung DE-R 298, 302 – Stromversorgung; EU, Wirtschaft und Wettbewerb/Entscheidungssammlung EV 827 and Wirtschaft und Wettbewerb 1980: 777 – Lab (confirmed by ECJ Case 61/80 Cooperatieve Stremsel- en Kleurselfabriek v European Commission, ECR 1981: 851 et seq.).

[674] See Article 1 section 1 lit. d) in connection with Article 2 section 1 BER-Vertical; Vertical Guidelines, mn. 66, 129; Vertical Guidelines, mn. 23, 87 et seq.; Holzmüller, T. and von Köckritz C., 2009. Zur Kartellrechtswidrigkeit langfristiger Bezugsbindungen und ihrer prozessualen Geltendmachung. Betriebs-Berater 2009: 1712, 1713.

[675] Holzmüller, T. and von Köckritz, C., 2009. Zur Kartellrechtswidrigkeit langfristiger Bezugsbindungen und ihrer prozessualen Geltendmachung. Betriebs-Berater 2009: 1712, 1713; Regional Court Dortmund, Wirtschaft und Wettbewerb/Entscheidungssammlung DE-R 1175, 1176 – Stadtwerke Lippstadt; CFI Case T-7/93 Langnese-Iglo v Commission, ECR 1995: II-0153, mn. 111; CFI Case T-9/93 Schöller v Commission, ECR 1999: II-1611, mn. 71 et seqq.; but see also Article 5 section 1 lit. a) BER-Vertical; Vertical Guidelines, mn. 133.

17.6.11 Purchasing cooperatives

The term 'purchasing cooperative' describes the many different forms of cooperation that companies enter into in order to manage their purchases.[676] Whether this cooperation takes the form of a simple agreement, a joint venture or a true cooperative is unimportant from a competition law perspective.[677] Purchasing cooperatives are often entered into in order to create or enhance buying power for the participating companies, often small and medium sized enterprises. In recent times, purchasing cooperatives are also increasingly used by big distributors in order to profit from economies of scale in the production of goods that are distributed under a private label.[678] Examples from practice are, the regional retailer cooperatives in Germany and international co-operations such as ALIDIS/Agenor. Since increased buying power can cause a reduction in prices on the selling market of the participating distributors, purchasing cooperatives are generally seen as a positive thing.[679] In individual cases purchasing cooperatives can, however, raise competitive concerns. This holds particularly true if the participating companies have a strong position in their selling markets and are thus able not to pass on the efficiencies gained to consumers.[680] Another concern is that competitors of the purchasing cooperative may be foreclosed from accessing important suppliers.[681] Finally, suppliers could be forced to reduce the number or quality of products offered, in view of the intense margin pressure created by purchasing cooperatives.[682]

Parties whose combined market share on both the purchasing and selling market does not exceed 15% usually do not have to fear an intervention by competition authorities when agreeing on a purchasing cooperative.[683] Companies which exceed these thresholds should carefully assess whether the intended cooperative is competition law compliant. Important criteria are whether the efficiencies

[676] Bunte, H.-J., 2010. In: Langen, E. and Bunte, H.J. (eds.) Kommentar zum deutschen und europäischen Kartellrecht Bd. 1, 11th edition, Carl Heymanns Verlag, Koln, Germany, § 1 GWB, mn. 159 et seqq.; Horizontal Guidelines, mn. 194 et seq.

[677] Schroeder, D., 2008. In: Wiedemann, G. (ed.) Handbuch des Kartellrechts, 2nd edition, C.H. Beck Verlag, Munchen, Germany, § 8, mn. 86, 107.

[678] EU, 2009. SEC(2009) 1449 Competition in the food supply chain, p. 9. Available at: http://ec.europa.eu/economy_finance/publications/publication16065_en.pdf.

[679] Horizontal Guidelines, mn. 194, 217; FCO, Tätigkeitsbericht 1995/1996: 72; with regard to farming cooperatives see ECJ Case C-250/92 Gøttrup-Klim, ECR 1994: I-5671, mn. 32 et seq. In this case, the ECJ even accepted the prohibition of a double membership in a purchasing cooperative and a minimum duration for the membership.

[680] Horizontal Guidelines, mn. 201. See FCO, Wirtschaft und Wettbewerb/Entscheidungssammlung DE-V 1607 and FCO, 2008. Kartellbehördlich angeordnete Cluster-Strategie. Wirtschaft und Wettbewerb 2008: 1119-1128; – EDEKA/Tengelmann. The pertinent passages are on p. 103 et seq. of the full decision that is available at: http://www.bundeskartellamt.de.

[681] Horizontal Guidelines, mn. 200, 203.

[682] Horizontal Guidelines, mn. 202.

[683] Horizontal Guidelines, mn. 208.

created through additional buying power are passed on when selling and whether the cooperative is accompanied by additional agreements (e.g. a market sharing agreement). An unfair use of the buying power gained through entering into the cooperative, e.g. through retroactive agreements that a supplier could not reasonably expect or through the arbitrary delay of payments to the supplier, should also be avoided.[684]

17.6.12 Subcontracting/supply agreements, e.g. for private label products

Supply agreements are mostly concluded among non-competing companies ('vertical' supply agreements) and are not objectionable if they are not accompanied by 'add-on' agreements such as agreements on exclusivity or resale prices.[685] However, it is also not uncommon that competing companies conclude supply agreements in order to save costs and to improve their offer of goods. A typical example is the supply of a distributor with products that are meant to be distributed under a private or 'white' label or the production for a brand-owner that does not own production facilities in the respective country. Supply agreements are problematic only if they contain non-compete obligations regarding future market activities[686] or if the supply agreement is a means to harmonise costs.[687] It is also of concern to competition authorities that supply agreements could serve as a 'bridge' for illegal exchanges of information.[688] In practice, companies should make sure that information is only exchanged to the extent that is necessary for the intended cooperation. Cost savings should also, at least in part, be passed on to consumers.[689] All in all it is possible to design supply agreements between competitors in unobjectionable ways.

17.6.13 Market information, benchmarking, business associations

Cooperation in the food sector for the purposes of market information and/or benchmarking gives companies an opportunity to determine their position in

[684] In practice, the successor to the British 'Supermarkets Code of Practice', that is the 'Groceries Supply Code of Practice' may be a helpful tool. Available at: http://www.competition-commission.gov.uk/inquiries/ref2006/grocery/pdf/revised_gscop_order.pdf. While the Code is only applicable in Great Britain and only for distributors with an annual turnover of more than 1 billion British pound, many of the Codes' guidelines contain helpful suggestions for approaching purchasing cooperatives.

[685] Whether such 'add-on' agreements would be permissible had to be examined in detail.

[686] See EU, 1979. Commission notice of 18 December 1978 concerning its assessment of certain subcontracting agreements in relation to Article 85 (1) of the EEC Treaty. Offical Journal of the European Union C 1, 3.1.1979: 2-3, No. 2/3 at the end; Article 5 section 1 lit. c) of EU, 2000. Commission Regulation (EC) No 2658/2000 of 29 November 2000 on the application of Article 81(3) of the Treaty to categories of specialisation agreements. Official Journal of the European Union L 304, 5.12.2000: 3-6.

[687] Horizontal Guidelines, mn. 176 *et seq.*.

[688] Horizontal Guidelines, mn. 181 *et seq.*

[689] Horizontal Guidelines, mn. 185.

the market and to understand where they need to improve their performance. In addition, such cooperation also serves to better inform wholesalers, retailers and/or end-consumers.[690] Generally speaking such cooperation is suited to improve competition among market participants.[691] On the other hand, the direct or indirect exchange of information among competitors can violate the confidentiality of competitively relevant company information.[692] Yet not every contact between competitors is suited to create anti-competitive effects. The mere exchange of *opinions* and *experiences* on/with objective circumstances that do not concern the individual market conduct of the participants is permissible.[693] This is because such an exchange of opinions or actual information, e.g. on technical, economic or legal features of the market, is usually not characterised by an exchange of the kind of details that must remain confidential.[694] Whether a specific piece of information is of relevance to the competitive process or not can only be assessed in each and every individual case.[695] Prices, product specifics or product launches

[690] For details see Stancke, F., 2009. Marktinformation, Benchmarking und Statistiken - Neue Anforderungen an Kartellrechts-Compliance. Betriebs-Berater 2009: 912; Voet van Vormizeele, P., 2009. Möglichkeit und Grenzen von Benchmarking nach europäischem und deutschem Kartellrecht. Wirtschaft und Wettbewerb 2009: 143-154; Whish, R., 2006. Information agreements. In: The pros and cons of information sharing, Swedish Competition Authority, Stockholm, Sweden, pp. 19-42 with further information (Available at: http://www.kkv.se/upload/filer/trycksaker/rapporter/pros&cons/rap_pros_and_cons_information_sharing.pdf); Lübbig, T., 2008. In: Wiedemann, G. (ed.) Handbuch des Kartellrechts, 2nd edition, C.H. Beck Verlag, Munchen, Germany, § 8, mn. 240.

[691] Bechtold, R., 2010. In: Bechtold, R. (ed.) GWB, Kartellgesetz, Gesetz gegen Wettbewerbsbeschränkungen, 6th edition, C.H. Beck Verlag, Munchen, Germany, § 1, mn. 85; OLG Düsseldorf, Wirtschaft und Wettbewerb/Entscheidungssammlung DE-R 949, 950 – Transportbeton Sachsen; EU, VII. Bericht über die Wettbewerbspolitik 1978: 19; ECJ Case C-194/99 P Thyssen v Commission ECR 2003: I-10821, mn. 84 and ECJ, 2004. Vom vorgeschriebenen zum freiwilligen Informationsaustausch. Wirtschaft und Wettbewerb 2004: 75-80; ECJ Case C-7/95 P John Deere v Commission ECR 1998: I-3111, mn. 88 et seqq. and ECJ, 1998. Wettbewerbsbeschränkung durch Informationsaustauschsysteme. Wirtschaft und Wettbewerb 1998: 747-752.

[692] Federal Supreme Court, Wirtschaft und Wettbewerb/Entscheidungssammlung 2006: 1337, 1342 - Aluminium-Halbzeug; Federal Supreme Court, Wirtschaft und Wettbewerb/Entscheidungssammlung: 2313, 2315 et seq. - Baumarkt-Statistik; Schröter, H. and Haag, M., 2003. In: Schröter, H., Jakob, T. and Mederer, W. (eds.) Kommentar zum Europäischen Wettbewerbsrecht, 1st edition, Nomos Verlagsgesellschaft, Baden-Baden, Germany, Article 81, mn. 66 with further information; Wagner-von Papp, F. 2004. Marktinformationsverfahren – Grenzen der Information im Wettbewerb. Nomos Verlagsgesellschaft, Baden-Baden, Germany, p. 164 et seq. with further information.

[693] EU, 1968. Notice concerning agreements, decisions and concerted practices in the field of cooperation between Enterprises. Official Journal of the European Union C 075, 29/07/1968, p. 4; Federal Ministry of Economic Affairs, Kooperationsfibel 'Zwischenbetriebliche Zusammenarbeit im Rahmen des Gesetzes über Wettbewerbsbeschränkungen', section II subsection 1.1.

[694] Correspondingly: Karenfort, J., 2008. Der Informationsaustausch zwischen Wettbewerbern - kompetitiv oder konspirativ? Wirtschaft und Wettbewerb 2008: 1154, 1165. (1154-1166).

[695] Haag, M., 2003. In: Schröter, H., Jakob, T. and Mederer, W. (eds.) Kommentar zum Europäischen Wettbewerbsrecht, 1st edition, Nomos Verlagsgesellschaft, Baden-Baden, Germany, Article 81, mn. 65; FCO, Tätigkeitsbericht 1976: 11 et seq., 217 (Annex 1); ECJ Case C-238/05 Asnef-Equifax ECR 2006: I-11125, mn. 54 and ECJ, 2007. EG-kartellrechtliche Grenzen der Zulässigkeit von Kreditinformationssystemen. Wirtschaft und Wettbewerb 2007: 539-544; EU, 1978. VII. Bericht über die Wettbewerbspolitik: 19; Stancke F., 2005. Schadensregulierung und Kartellrecht. Versicherungsrecht 2005: 1324, 1325.

but also billing details must usually be kept confidential.[696] Of relevance can also be information on the functioning of the distribution chain and the internal organisation of a company.[697] According to the FCO the mere exchange of that kind of information constitutes a *per se* infringement, independent of the kind of products concerned and of the market structure.[698]

However, it is well accepted that non-coordinated price announcements in public do not, taken by themselves, infringe the prohibition on anti-competitive conduct.[699] This holds true even if competitors become aware of the announcement regarding prices, as long as the publication of prices is not used to circumvent the prohibition on direct price fixing.

If cooperation with regard to market information or benchmarking is meant to comprise information that is relevant to competition and refers to confidential company data, this will only be permissible if the exchanged information is anonymised.[700] In special cases, the exchange of non-anonymised information may be exempt from the prohibition on anti-competitive agreements and will thus be permissible.[701] Whether the prerequisites for such an exemption are fulfilled must be carefully assessed by the companies concerned.

17.6.14 Information exchanges between distributors and producers

As in every other line of business, food producers and distributors are interested in exchanging certain types of information before concluding contracts. Supply

[696] Wagner-von Papp, F., 2004. Marktinformationsverfahren - Grenzen der Information im Wettbewerb. Nomos Verlagsgesellschaft, Baden-Baden, Germany, p. 219 et seq.; Gehring, S., 2006. In: Mäger, T. (ed.), 2006. Europäisches Kartellrecht. Nomos Verlagsgesellschaft, Baden-Baden, Germany, 1st edition, p. 60, mn. 78 with further information; Carle, M. and Johnsson, J., 1998. Benchmarking and E.C. Competition law. ECLR 19: 74, 78; see also FCO, press release of 20/2/2008 on information exchanges with regard to demands for rebates and separate agreements at the occasion of annual meetings - 'Drogerieartikel'. Available at: http://www.bundeskartellamt.de.

[697] Lübbig, T., 2008. In: Wiedemann, G. (ed.) Handbuch des Kartellrechts, 2nd edition, C.H. Beck Verlag, Munchen, Germany, § 8. mn. 240; with regard to the kind of information exchanged see also Voet van Vormizeele, P., 2009. Möglichkeit und Grenzen von Benchmarking nach europäischem und deutschem Kartellrecht. Wirtschaft und Wettbewerb 2009: 143, 150 et seq. with further information.

[698] FCO, press release of 10/7/2008 - 'Luxuskosmetik'. Available at: http://www.bundeskartellamt.de. The fining decisions are still appealable. On current developments in France and Italy which are similar see Möhlenkamp, A., 2008. Verbandskartellrecht - trittfeste Pfade in unsicherem Gelände. Wirtschaft und Wettbewerb: 428, 433, footnote 29 with further information. (428-440).

[699] ECJ joined Cases C-89/85 *et al.* Ahlströhm *et al.* v Commission, ECR 1993: I-1307, mn. 59-65 and Neue Juristische Wochenschrift 1988: 3086; but see Federal Supreme Court, Wirtschaft und Wettbewerb/ Entscheidungssammlung 2182 - Altölpreise.

[700] For details see Stancke, F., 2009. Marktinformation, Benchmarking und Statistiken - Neue Anforderungen an Kartellrechts-Compliance. Betriebs-Berater 2009: 912, 914 et seq.

[701] For the specific requirements see Stancke, F., 2009. Marktinformation, Benchmarking und Statistiken - Neue Anforderungen an Kartellrechts-Compliance. Betriebs-Berater 2009: 912, 916 with further information.

agreements may provide for an obligation to report on sales, stocks and the market situation in the allocated area. Such agreements raise concerns if the supplier and the distributor are active in the same market, e.g. because the supplier also distributes its goods directly to consumers. In these cases, the considerations described above regarding the exchange of information between competitors apply. In so far as the exchanged information, e.g. the concrete purchase price for the delivery, is immediately connected to the supply agreement and in so far as the flow of information is necessary for the execution of the contract, the information exchange will not raise competitive concerns because it is 'inherent' in the contract.[702] Information exchanges beyond what is necessary, e.g. generally on prices and rebates, sales or contractual relations to third parties should be avoided. It should also be observed that the exchange of information on sales, prices demanded by customers etcetera must not serve to enforce compliance[703] with recommended prices or orientation prices and that the exchange of price-relevant information must not lead to a competition law infringement in the hub and spoke sense (see above).[704]

17.6.15 Abusive conduct

As mentioned above, competition law is particularly strict on companies which have a strong position in their market(s). To begin with, it should be noted that a strong market position in itself is not problematic. It is only the abuse of that position by the company holding it which is prohibited. The difficulties in determining what an 'abuse' is have led to a number of case groups, the most relevant of which shall now be discussed in more detail.

a. Refusals to deal

That a refusal to deal or a discontinuation of supplies can constitute an abuse may surprise. After all, the freedom of contract is a fundamental pillar of our economic order. However, it is well accepted today that the strict enforcement of

[702] On the German 'Immanenztheorie' see Bunte, H.-J., 2010. In: Langen, E. and Bunte, H.J. (eds.) Kommentar zum deutschen und europäischen Kartellrecht Bd. 1, 11th edition, Carl Heymanns Verlag, Koln, Germany, § 1 GWB, mn. 134 et seq.

[703] See section 17.6.4.

[704] FCO, 2010. Vorläufige Bewertung von Verhaltensweisen in Verhandlungen zwischen Herstellern von Markenartikeln, Großhändlern und Einzelhandelsunternehmen zum Zwecke der Konkretisierung der Kooperationspflichten. Wirtschaft und Wettbewerb 2010: 786, 790. (786-791).

this principle can sometimes have market foreclosing effects.[705] For instance, it can be problematic if a dominant producer refuses to supply a distributor because it wants to enter into the distribution of its products itself.[706] The use of a cut-off of supply as a sanctioning mechanism against buyers that export products to other European countries raises similar competitive concerns in the absence of a permissible sales-restriction. A non-permissible refusal to deal may also occur where a company arbitrarily discriminates against certain competitors or in the case of a particular dependency on the products of the supplier.[707]

b. Discrimination

Both European and German Competition Law enumerate the discrimination of trading partners without justification as an example of abusive conduct by a dominant company.[708] Such discrimination may for instance occur with regard to marketing opportunities, prices and other supply terms. Different prices and terms will, however, be justified if they reflect actually existing differences in costs, i.e. if an actual reason for the difference exists.[709] It is also increasingly accepted that in certain cases an efficient allocation of resources is only possible through the use of discriminatory pricing.[710]

[705] Wiedemann, G., 2008. In Wiedemann, G. (ed.) Handbuch des Kartellrechts, 2nd edition, C.H. Beck Verlag, Munchen, Germany, § 23, mn. 49; Federal Supreme Court, Wirtschaft und Wettbewerb/ Entscheidungssammlung 886, 892 – Jägermeister; ECJ Case 6/73 & 7/73 Commercial Solvents, ECR 1974: 223, mn. 25; EU, Wirtschaft und Wettbewerb/Entscheidungssammlung EV 1265 and Wirtschaft und Wettbewerb 1988: 257 – Boosey & Hawkes; EU, 1996. 92/213/EEC: Commission Decision of 26 February 1992 relating to a procedure pursuant to Articles 85 and 86 of the EEC Treaty (IV/33.544, British Midland v. Aer Lingus). Official Journal of the European Union L 96, 10.4.1992: 34-45, mn. 24 et seq. The same reasoning is at the core of the so-called 'essential facilities' – doctrine pursuant to which companies must, under narrow circumstances, afford access to infrastructure and intellectual property. See § 19 IV No. 4 ARC and Federal Supreme Court, Wirtschaft und Wettbewerb/Entscheidungssammlung DE-R 977, 982 – Fährhafen Puttgarden; EU, Wirtschaft und Wettbewerb/Entscheidungssammlung EU-V 931 and Wirtschaft und Wettbewerb 2004: 673 – Microsoft (mainly confirmed by CFI Case T-201/04 Microsoft v Commission, ECR 2007: II-3601 and Europäisches Wirtschafts- und Steuerrecht 2005: 75).

[706] See Federal Supreme Court, Wirtschaft und Wettbewerb/E 2479, 2483 – Reparaturbetrieb; ECJ Case 6/73 & 7/73 Commercial Solvents, ECR 1974: 223, mn. 25.

[707] For details see. Bulst, F.W., 2010. In: Langen, E. and Bunte, H.J. (eds.) Kommentar zum deutschen und europäischen Kartellrecht Bd. 2, 11th edition, Carl Heymanns Verlag, Koln, Germany, Art. 82 EG, mn. 253 et seq.

[708] Article 102 letter c) TFEU, §§ 19 section 4 No. 3, 20 section 1 ARC.

[709] Wiedemann, G., 2008. In: Wiedemann, G. (ed.) Handbuch des Kartellrechts, 2nd edition, C.H. Beck Verlag, Munchen, Germany, § 23, mn. 62.

[710] CFI Case T-168/01 GlaxoSmithKline v Commission, ECR 2006: II-2969, mn. 271 and CFI, 2007. Nichtigerklärung der Versagung der Freistellung des Preisdifferenzierungssystems eines Pharmaherstellers. Wirtschaft und Wettbewerb 2007: 93-108.

c. Loyalty and target rebates

Complex case law has developed on the question under which circumstances dominant companies abuse their position through affording rebates to their customers.[711] Rebates are not generally illegal. There is a justified concern, however, that the design of rebates can have a similarly market foreclosing effect as an exclusive supply obligation.[712] To a particular extent, this risk occurs where rebates are openly linked to the buyer sourcing exclusively or mainly from a specific supplier over long periods of time.[713] But also 'covered' loyalty rebates can raise competitive concerns, for instance if a rebate on the total volume of sales retroactively increases when certain sales thresholds are crossed.[714] Similar problems are raised by target rebates which are linked to the achievement of a sales volume that corresponds to a buyer's foreseeable annual total demand.[715] Against the background of the existing case law, dominant companies should structure their rebates with short reference periods and not retroactively on the total volume of sales.

d. Predatory pricing and similar conduct

The imposition of unfairly high or low prices by a dominant company when buying or selling[716] constitutes a classic abuse of a dominant position. Accordingly both German and European competition law enumerate this conduct as an example of abuse.[717] Quite naturally, it is often highly contentious whether a price is unfairly

---

[711] Federal Supreme Court, Wirtschaft und Wettbewerb/Entscheidungssammlung 2755, 2758 – Aktionsbeiträge; ECJ Case 43/73 *et al.* Société anonyme Générale Sucrière and others v Commission, ECR 1975: 1663 and Wirtschaft und Wettbewerb 1976: 185; ECJ Case 85/76 Hoffmann LaRoche v Commission, ECR 1979: 461, mn. 89 and Neue Juristische Wochenschrift 1979: 2460; ECJ Case 322/81 Michelin I, ECR 1983: 3461, mn. 71 et seq.; CFI T-203/01 Michelin II, ECR 2003: II-4071, mn. 56 et seq. and CFI, 2003. Rabattgewährung durch marktbeherrschendes Unternehmen. Wirtschaft und Wettbewerb 2003: 1331-1346; ECJ Case C-95/04 P British Airways v Commission, ECR 2007: I-2331, mn. 61 et seq. and ECJ, 2007. Missbräuchliche Prämienregelung beim Vertrieb von Flugscheinen. Wirtschaft und Wettbewerb 2007: 821-823.

[712] ECJ Case 85/76 Hoffmann LaRoche v Commission, ECR 1979: 461, mn. 89 and Neue Juristische Wochenschrift 1979: 2460; see also Wiedemann, G., 2008. In: Wiedemann, G. (ed.) Handbuch des Kartellrechts, 2nd edition, C.H. Beck Verlag, Munchen, Germany, § 23, mn. 47 et seq.

[713] The European Commission fined Intel with more than € 1 billion for such (and other) practices in 2009; see EU, 2009. 1,06 Mrd. € Geldbuße wegen schuldhaft missbräuchlicher Ausnutzung einer marktbeherrschenden Stellung. Wirtschaft und Wettbewerb 2009: 1201-1216 - Intel, mn. 323.

[714] Kammergericht, Wirtschaft und Wettbewerb/Entscheidungssammlung 2403, 2410 – Fertigfutter; CFI T-203/01 Michelin II, ECR 2003: II-4071, mn. 95 and CFI, 2003. Rabattgewährung durch marktbeherrschendes Unternehmen. Wirtschaft und Wettbewerb 2003: 1331-1346.

[715] ECJ, Case T-228/97 Irish Sugar, ECR 1999: II-2969, mn. 212. See also European Commission, Decision of 29/3/2006: COMP/E-1/38.113 Prokent-Tomra, mn. 314 et seq. Available at: http://ec.europa.eu/competition/index_en.html.

[716] Wiedemann, G., 2008. In: Wiedemann, G. (ed.) Handbuch des Kartellrechts, 2nd edition, C.H. Beck Verlag, Munchen, Germany, § 23, mn. 52.

[717] Article 102 letter a) TFEU and § 19 IV No. 2 ARC.

low or high. In accordance with the existing case law, the first step is to establish a price that would occur in a situation of regular competition. This is done by looking at the price of the same product on comparable geographical markets. The price thus detected will then be compared to the actually demanded price. For an abuse, the German Federal Supreme Court's case-law requires there to be a 'significant' deviation from the established price;[718] the test under European law is whether there is an 'exaggerated disproportionateness' or an inappropriateness in comparison to the established price.[719]

e. Tying

A further example of an abuse of a dominant position pursuant to Article 102 TFEU is the 'conclusion of contracts subject to acceptance by the other party of supplementary obligations which, by their nature or according to commercial usage, have no connection with the subject of such contracts'. A comparable provision in the ARC was deleted in the past. This does not affect the illegality of certain tying practices also under German law, however.[720] The primary concern with tying practices is that a dominant company uses its position in one market in order to gain a strong position or even dominance in another market.[721] An example is the *Microsoft* case. *Microsoft* was fined almost € 500 million in 2004 by the European Commission for an alleged use of its power in the market for computer operating systems in order to gain a dominant position in the not yet dominated market for multi-media applications.[722] A further example is the German *Soda Club 2* case, in which the tying of drinks makers to the hire of gas cylinders was held to infringe competition law.[723]

[718] Federal Supreme Court, Wirtschaft und Wettbewerb/Entscheidungssammlung 1445, 1454 – Valium; Bechtold, R., 2010. In: Bechtold, R. (ed.) GWB, Kartellgesetz, Gesetz gegen Wettbewerbsbeschränkungen, 6th edition, C.H. Beck Verlag, Munchen, Germany, § 19, mn. 74.

[719] ECJ Case 27/76 United Brands, ECR 1978: 207, mn. 248/257 and Neue Juristische Wochenschrift 1978: 2439; see also EU, Wirtschaft und Wettbewerb/Entscheidungssammlung EU-V 1097 and EU, 2005. Zurückweisung eines Vorwurfs von Preismissbrauch. Wirtschaft und Wettbewerb 2005: 1179-1186 – Port of Helsingborg, mn. 98 et seq.

[720] Wiedemann, G., 2008. In: Wiedemann, G. (ed.) Handbuch des Kartellrechts, 2nd edition, C.H. Beck Verlag, Munchen, Germany, § 23, mn. 48.

[721] See Möschel, W., 2007. In: Immenga, U. and Mestmäcker, E.J. (eds.) Wettbewerbsrecht Band 2: GWB. Kommentar zum Deutschen Kartellrecht, 4th edition, C.H. Beck Verlag, Munchen, Germany, § 19, mn. 133.

[722] EU, Wirtschaft und Wettbewerb/Entscheidungssammlung EU-V 931 and EU, 2004. Kartellrechtswidrige Ausdehnung von Marktmacht. Wirtschaft und Wettbewerb 2004: 673 – Microsoft (mainly confirmed by CFI Case T-201/04 Microsof v. Commission, ECR 2007: II-3601 and Europäisches Wirtschafts- und Steuerrecht 2005: 75). Microsoft then received fines worth more than € 1 billion for alleged non-compliance with the Commission's decision.

[723] FCO, Wirtschaft und Wettbewerb/Entscheidungssammlung DE-V 1177 et seq. – Soda Club; Higher Regional Court Düsseldorf, Wirtschaft und Wettbewerb/Entscheidungssammlung DE-R 1935 et seq. – Soda Club; Federal Supreme Court, 2008. Betriebs-Berater: 970 – Soda Club with case note: Zimmer, D. and Werner, J., 2008. Marktabgrenzung und Verhältnis von Eigentums- und Kartellrecht. Entscheidung des BGH, 4.3.2008 – KVR 21/07. Juristenzeitung 2008: 897, 901. (897-904).

17.6.16 Unfair practices by competitors

Companies may have claims under competition law and the German law of unfair competition more specifically against their competitor's conduct if the purpose of that conduct is merely to impede the process of competition, if the claiming company's business is unduly hindered or if wholesalers and retailers are incited to boycotts.[724] Examples of such behaviour would be agreements among food producers with distributors on exclusive supply and a parallel delisting of all other products if kick-backs, listing fees or other payments of an unusual amount are agreed on. It is also problematic if producers buy up the stock of their competitors from distributors at the regular price in order to then sell those products at a fraction of the usual price and in a way which is detrimental to the reputation of the product.[725]

17.7 Concluding remarks

The case groups discussed above show that many forms of cooperation which are of particular importance to daily business in the food sector can become the focus of prosecutorial concern by competition authorities. The increasing importance of competition law means that market participants must, in times of intense competition, assert themselves not only towards their competitors and their counterparties in the market. If they want to comply with the demands of competition authorities and courts, they are also severely constrained in the design of their supply and cooperation agreements. In order to avoid daily interference between business and legal considerations, it is highly recommendable to take into account competition law concerns already at the stage of planning cooperation and sales and trade strategies. This guarantees that legal risks are minimised and that attractive projects can be implemented in an environment of sufficient legal certainty.

[724] See § 4 No. 10 German Act against Unfair Competition (UWG), §§ 20 section 1, 21 section 1 und § 1 ARC, Article 102, 101 section 1 TFEU. See also Bechtold, R., 2010. In: Bechtold, R. (ed.) GWB, Kartellgesetz, Gesetz gegen Wettbewerbsbeschränkungen, 6th edition, C.H. Beck Verlag, Munchen, Germany, § 21, mn. 3; on boycotts see Federal Supreme Court, Wirtschaft und Wettbewerb/Entscheidungssammlung 2372, 2379 et seq., - Importierte Fertigarzneimittel; Federal Supreme Court, Wettbewerb in Recht und Praxis 1999: 1283, 1287 - Kartenlesegerät; Federal Supreme Court, Gewerblicher Rechtsschutz und Urheberrecht 1999: 1031, 1033 - Sitzender Krankentransport and on purposeful hindrance of competitors Federal Supreme Court, Gewerblicher Rechtsschutz und Urheberrecht 2001: 1061, 1062 - Mitwohnzentrale.de; Federal Supreme Court, Gewerblicher Rechtsschutz und Urheberrecht 2004: 877, 879 - Werbeblocker; Federal Supreme Court, Wettbewerb in Recht und Praxis 2005: 881, 884 - The Colour of Elégance; Federal Supreme Court, Gewerblicher Rechtsschutz und Urheberrecht 200: 8081 - ad-hoc-Mitteilung. On price policies on the basis of an unfair intention to 'squeeze out' competitors from the market Federal Supreme Court, Gewerblicher Rechtsschutz und Urheberrecht 199: 685, 686 - Anzeigenpreis I; Köhler, H., 2010. In: Köhler, H. and Bornkamm, J. (eds.) Gesetz gegen den unlauteren Wettbewerb UWG. 28th edition, C.H. Beck Verlag, Munchen, Germany, § 4 UWG, mn. 10.189.

[725] See Köhler, H., 2010. In: Köhler, H. and Bornkamm, J. (eds.) Gesetz gegen den unlauteren Wettbewerb UWG. 28th edition, C.H. Beck Verlag, Munchen, Germany, § 4 UWG, mn. 10.49.

References

Alexander, C., 2010. Privatrechtliche Durchsetzung des Verbots von Verkäufen unter Einstandspreis. WRP 2010: 727-733.

Bechtold, R. (ed.), 2006. GWB, Kartellgesetz, Gesetz gegen Wettbewerbsbeschränkungen, 6th edition, C.H. Beck Verlag, Munchen, Germany.

Besen, M. and Jorias, R., 2010. Kartellrechtliche Grenzen des Category Managements unter Berücksichtigung der neuen EU-Leitlinien für vertikale Beschränkungen. Betriebs-Berater 2010: 1099, 1100.

Carle, M. and Johnsson, J., 1998. Benchmarking and E.C. Competition law. European Competition Law Review 19: 74-84.

CFI, 1999. Umfang von Verteidigungsrechten. Wirtschaft und Wettbewerb 1999: 623-631.

CFI, 2003. Rabattgewährung durch marktbeherrschendes Unternehmen. Wirtschaft und Wettbewerb 2003: 1331-1346.

CFI, 2007. Nichtigerklärung der Versagung der Freistellung des Preisdifferenzierungssystems eines Pharmaherstellers. Wirtschaft und Wettbewerb 2007: 93-108.

ECJ, 1998. Wettbewerbsbeschränkung durch Informationsaustauschsysteme. Wirtschaft und Wettbewerb 1998: 747-752.

ECJ, 1999. Bestätigung der Polypropylen-Entscheidung durch den Gerichtshof. Wirtschaft und Wettbewerb 1999: 902-907.

ECJ, 2002. Folgen der Nichtigerklärung einer Geldbußenentscheidung. Wirtschaft und Wettbewerb 2002: 1257-1266.

ECJ, 2004. Vom vorgeschriebenen zum freiwilligen Informationsaustausch. Wirtschaft und Wettbewerb 2004: 75-80.

ECJ, 2007. EG-kartellrechtliche Grenzen der Zulässigkeit von Kreditinformationssystemen. Wirtschaft und Wettbewerb 2007: 539-544.

ECJ, 2007. Missbräuchliche Prämienregelung beim Vertrieb von Flugscheinen. Wirtschaft und Wettbewerb 2007: 821-823.

EU, 1968. Notice concerning agreements, decisions and concerted practices in the field of cooperation between enterprises. Official Journal of the European Union C 075, 29/07/1968: 3-6.

EU, 1978. VII. Bericht über die Wettbewerbspolitik.

EU, 1979. Commission notice of 18 December 1978 concerning its assessment of certain subcontracting agreements in relation to Article 85 (1) of the EEC Treaty. Offical Journal of the European Union C 1, 3.1.1979: 2-3.

EU, 1992. 92/204/EEC: Commission Decision of 5 February 1992 relating to a proceeding pursuant to Article 85 of the EEC Treaty (IV/31.572 and 32.571 - Building and construction industry in the Netherlands). Official Journal of the European Union L 92, 7.4.1992: 1-30.

EU, 1994. 94/599/EC: Commission Decision of 27 July 1994 relating to a proceeding pursuant to Article 85 of the EC Treaty (IV/31.865 – PVC). Official Journal of the European Union L 239, 14.9.1994: 14-35.

EU, 1996. 92/213/EEC: Commission Decision of 26 February 1992 relating to a procedure pursuant to Articles 85 and 86 of the EEC Treaty (IV/33.544, British Midland v. Aer Lingus). Official Journal of the European Union L 96, 10.4.1992: 34-45.

EU, 2000. Commission Regulation (EC) No 2658/2000 of 29 November 2000 on the application of Article 81(3) of the Treaty to categories of specialisation agreements. Official Journal of the European Union L 304, 5.12.2000: 3-6.

EU, 2003. Council Regulation (EC) No. 1/2003 of 16 December 2002 on the implementation of the rules on competition laid down in Articles 81 and 82 of the Treaty. Official Journal of the European Union L 1 of 4/1/2003: 1-25.

EU, 2005. Prior notification of a concentration (Case COMP/M.3732 – Procter & Gamble/ Gillette). Official Journal of the European Union C 139, 08/06/2005: 35.

EU, 2004. Kartellrechtswidrige Ausdehnung von Marktmacht. Wirtschaft und Wettbewerb 2004: 673.

EU, 2005. Zurückweisung eines Vorwurfs von Preismissbrauch. Wirtschaft und Wettbewerb 2005: 1179-118.

EU, 2006. Decision of 29/3/2006: COMP/E-1/38.113 Prokent-Tomra. Available at: http:// ec.europa.eu/competition/index_en.html.

EU, 2009. 1,06 Mrd. € Geldbuße wegen schuldhaft missbräuchlicher Ausnutzung einer marktbeherrschenden Stellung. Wirtschaft und Wettbewerb 2009: 1201-1216.

EU, 2009. SEC(2009) 1449 Competition in the food supply chain. Available at: http://ec.europa. eu/economy_finance/publications/publication16065_en.pdf.

EU, 2010. Commission Regulation (EU) No 330/2010 of 20 April 2010 on the application of Article 101(3) of the Treaty on the Functioning of the European Union to categories of vertical agreements and concerted practices. Official Journal of the European Union L 102, 23.4.2010: 1-7.

EU, 2010. Guidelines on Vertical Restraints. Official Journal of the European Union C 130: 1-46.

EU, 2011. Communication from the Commission. Guidelines on the applicability of Article 101 of the Treaty on the Functioning of the European Union to horizontal co-operation agreements (Horizontal Guidelines). Official Journal of the European Union C 11, 14/1/2011: 1-72. Available at: http://eur-lex.europa.eu/LexUriServ/LexUriServ.do?uri=OJ:C:2011:011:0001 :0072:EN:PDF.

FCO, 2008. Kartellbehördlich angeordnete Cluster-Strategie. Wirtschaft und Wettbewerb 2008: 1119-1128.

FCO, 2010. Bußgeld wegen Beschränkung des Internetvertriebs. Wirtschaft und Wettbewerb 2010: 91-97.

FCO, 2010. Freigabe eines Zusammenschlusses in Schreibgerätemärkten. Wirtschaft und Wettbewerb 2010: 568-577.

FCO, 2010. Vorläufige Bewertung von Verhaltensweisen in Verhandlungen zwischen Herstellern von Markenartikeln, Großhändlern und Einzelhandelsunternehmen zum Zwecke der Konkretisierung der Kooperationspflichten. Wirtschaft und Wettbewerb 2010: 786-791.

FCO, 2010. Vorläufige Bewertung von Verhaltensweisen in Verhandlungen zwischen Herstellern von Markenartikeln, Großhändlern und Einzelhandelsunternehmen zum Zwecke der Konkretisierung der Kooperationspflichten. Wirtschaft und Wettbewerb 2010: 786-791.

Hasse, S., 2010. Kartellrecht für Privilegierte (commentary). LebensmittelZeitung, 16/04/2010, p. 2.

Hirsch, G. Montag, F. and Säcker, F.J. (eds.), 2008. Münchener Kommentar zum Europäischen und Deutschen Wettbewerbsrecht (Kartellrecht) Band 2: Gesetz gegen Wettbewerbsbeschränkungen: GWB, 1st edition, C.H. Beck Verlag, Munchen, Germany.

Holzmüller, T. and von Köckritz C., 2009. Zur Kartellrechtswidrigkeit langfristiger Bezugsbindungen und ihrer prozessualen Geltendmachung. Betriebs-Berater 2009: 1712, 1713.

Immenga, U. and Mestmäcker, E.J. (eds.), 2007. Wettbewerbsrecht Band 1: EG. Kommentar zum Europäischen Kartellrecht, 4th edition, C.H. Beck Verlag, Munchen, Germany.

Immenga, U. and Mestmäcker, E.J. (eds.), 2007. Wettbewerbsrecht Band 2: GWB. Kommentar zum Deutschen Kartellrecht, 4th edition, C.H. Beck Verlag, Munchen, Germany.

Immenga, U., 1989. Grenzen des kartellrechtlichen Ausnahmebereichs Arbeitsmarkt. C.H. Beck Verlag, Munchen, Germany.

Kamann, H.-G. and Bergmann, E., 2003. Die neue EG-Kartellverfahrensverordnung - Auswirkungen auf die unternehmerische Vertragspraxis. Betriebs-Berater 2003: 1743, 1745 et seq.

Karenfort, J., 2008. Der Informationsaustausch zwischen Wettbewerbern - kompetitiv oder konspirativ? Wirtschaft und Wettbewerb 2008: 1154-1166.

Karl, M. and Reichelt, D., 2005. Die Änderungen des Gesetzes gegen Wettbewerbsbeschränkungen durch die 7. GWB-Novelle. Der Betrieb 2005: 1436-1437.

Kirchhain, S, 2008. Die Gestaltung von innerstaatlich wirkenden Vertriebsverträgen nach der 7. GWB-Novelle. Wirtschaft und Wettbewerb 2008: 167-178.

Köhler, H. and Bornkamm, J. (eds.), 2010. Gesetz gegen den unlauteren Wettbewerb UWG. 28th edition, C.H. Beck Verlag, Munchen, Germany.

Langen, E. and Bunte, H.J. (eds.), 2010. Kommentar zum deutschen und europäischen Kartellrecht Bd. 1, 11th edition, Carl Heymanns Verlag, Koln, Germany.

Langen, E. and Bunte, H.J. (eds.), 2010. Kommentar zum deutschen und europäischen Kartellrecht Bd. 2, 11th edition, Carl Heymanns Verlag, Koln, Germany.

Loest, T., 2004. 'Category Captaincies' im Ordnungsrahmen des Kartellrechts. Wettbewerb in Recht und Praxis 2004: 454-460.

Loewenheim, U., Meessen, K.M. and Riesenkampff, A. (eds.), 2009. Kartellrecht, 2nd edition, C.H. Beck Verlag, Munchen, Germany.

Mäger, T. (ed.), 2006. Europäisches Kartellrecht. 1st edition, Nomos Verlagsgesellschaft, Baden-Baden, Germany.

Möhlenkamp, A., 2008. Verbandskartellrecht - trittfeste Pfade in unsicherem Gelände. Wirtschaft und Wettbewerb: 428-440.

Roth, P. and Rose, V., 2008. Bellamy & Child: European Community Law of Competition, 6th edition, Oxford University press, Oxford, UK.

Schröter, H., Jakob, T. and Mederer, W. (eds.), 2003. Kommentar zum Europäischen Wettbewerbsrecht, 1st edition, Nomos Verlagsgesellschaft, Baden-Baden, Germany.

Schütze, J. and Beck, R.S., 2010. Drogerie Rossmann darf pauschale Werbekostenzuschüsse auf einzelne Produkte umlegen – Kein Verkauf unter Einstandspreis. Gewerblicher Rechtsschutz und Urheberrecht – Praxis 2010: 208.

Schwintowski, H.-P. and Klaue, S., 2005. Kartellrechtliche und gesellschaftsrechtliche Konsequenzen des Systems der Legalausnahme für die Kooperationspraxis der Unternehmen. Wirtschaft und Wettbewerb 2005: 370-378.

Senge, L. (ed.), 2006. Karlsruher Kommentar zum OWiG, 3rd edition, C.H. Beck Verlag, Munchen, Germany.

Siemes, J., Mager, G., Gerling, M. and Vogell, K., 2010. KPMG, Trends im Handel 2010, p. 51. Available at: http://www.kpmg.de/docs/trends_im_handel_2010_de.pdf.

Stancke F., 2005. Schadensregulierung und Kartellrecht. Versicherungsrecht 2005: 1324, 1325.

Stancke, F., 2009. Marktinformation, Benchmarking und Statistiken - Neue Anforderungen an Kartellrechts-Compliance. Betriebs-Berater 2009: 912.

Stancke, F., 2010. Das Kartellrecht der Lebensmittelbranche. Zeitschrift für das gesamte Lebensmittelrecht 2010: 543-565.

Voet van Vormizeele, P., 2009. Möglichkeit und Grenzen von Benchmarking nach europäischem und deutschem Kartellrecht. Wirtschaft und Wettbewerb 2009: 143-154.

Wagner-von Papp, F., 2004. Marktinformationsverfahren – Grenzen der Information im Wettbewerb: Die Herstellung praktischer Konkordanz zwischen legitimen Informationsbedürfnissen and Geheimwettbewerb. Wirtschaftsrecht und Wirtschaftspolitik: Vol.191. Nomos Verlagsgesellschaft, Baden-Baden, Germany.

Whish, R., 2006. Information agreements. In: The pros and cons of information sharing, Swedish Competition Authority, Stockholm, Sweden. Available at: http://www.kkv.se/upload/filer/trycksaker/rapporter/pros&cons/rap_pros_and_cons_information_sharing.pdf.

Wiedemann, G. (ed.), 2008. Handbuch des Kartellrechts, 2nd edition, C.H. Beck Verlag, Munchen, Germany.

Wiring, R., 2010. Sortimentsplanung im Supermarktregal – Welchen Spielraum lässt das Kartellrecht? Gewerblicher Rechtsschutz und Urheberrecht. Praxis 2010: 332.

Zimmer, D. and Werner, J., 2008. Marktabgrenzung und Verhältnis von Eigentums- und Kartellrecht. Entscheidung des BGH, 4.3.2008 – KVR 21/07. Juristenzeitung 2008: 897-904.

18. EU 'new approach' also for food law?

Nicole Coutrelis

18.1 What is the 'new approach'

The 'new approach' is a legislative technique put in place in the EU in the 80's to harmonise the legislations of the Member States regarding products, where such harmonisation proved to be necessary in order to achieve one of the fundamental objectives of the Common Market, i.e. the free movement of goods.

This approach and the reasons why it was put in place cannot be understood without a little bit of history, related both to the jurisprudence of the EC Court of Justice and the Policy of the Commission, which have continuously supported each other to enforce the free movement of goods principle (originally Articles 30-36, then 28-30 after the Amsterdam Treaty, and now 34-36 under the Lisbon Treaty). Indeed, we must always keep in mind that one of the basic purposes of EU law, from the beginning and, hopefully, still now, is the realisation of an Internal Market without borders within the Union.

18.1.1 A little bit of history

In order to achieve an internal market where products circulate freely, when those products are defined and regulated in very different ways in the various Member States, two avenues can be taken: either harmonise the regulations, to have them all identical, or at least compatible, in all the Member States, or decide that products can circulate freely, despite their differences. The entire history of food regulation in the EU has been dominated by the balance between these two ways – or a combination of both.

Until the 1970s (with only 6 founding Member States), the main tool was harmonisation, giving birth to some so called 'recipe laws', through directives related to products such as chocolate, fruit juice, etc. Obviously, it was not possible to harmonise that way the definitions and regulations of every single food product sold in the EU. In this context, came the famous 'Cassis de Dijon' case,[726] in which the Court of Justice said that in the absence of harmonisation, the Member States are free to set up national rules for products sold on their territory, and can ban the importation of products which are not in conformity with national rules, if those national rules are justified by 'mandatory requirements', such as, for instance, the protection of the consumer or fair trade. On the other hand, if the national rules are not justified by such 'mandatory requirements', they cannot

[726] ECJ, 1979. Case 120/78, February 20, 1979. Rewe-Zentral AG v Bundesmonopolverwaltung für Branntwein (Cassis de Dijon). ECR 1979: 649.

justify restrictions to the importation of products legally produced and marketed in other Member States.

On this basis, the Commission issued a Communication on October 3, 1980, on the consequences of the 'Cassis de Dijon' case, putting the emphasis on the principle of free movement of products legally produced and marketed in one Member State. In the Commission's view, this was the basic remedy to what was called 'technical barriers to trade', i.e. obstacles due to the existence of different rules and specifications of products among the Member States. As a consequence, the Commission announced its intention to limit harmonisation to what was strictly necessary to achieve the Single Market, where the mere enforcement of the principle of free movement of goods was not sufficient, i.e. where 'mandatory requirements' could justify national measures entailing barriers to trade.

Further to this first Communication, the year 1985 brought essential developments in view of the complete Achievement of the Single Market, which was to be reached in 1992. On the basis of a Communication issued by the Commission ('the white paper'), the Council took a Resolution on the 'new approach',[727] valid for all products except foodstuffs. As to foodstuffs, the Commission issued the same year a specific Communication called 'White Paper bis'.

18.1.2 What is the 'new approach'

The 'new approach' for the regulation of products specifications is based on the four following principles:

- EU legislation should be limited to the adoption of essential requirements, regarding specifically safety or other requirements of general interest.
- The task of drawing up the technical specifications of products, in conformity with the essential requirements laid down in the legislation, is entrusted to organisations which are competent in the standardisation area.
- These technical specifications are not mandatory.
- However, products conform to the standards are presumed to conform to the essential requirements.

This means that products may conform to the essential requirements by other means than the conformity to standards, a system which leaves space for individual initiatives and innovation. This legal technique has been enforced and has become quite common in a lot of industrial sectors such as toys, electricity, vehicles, etc.

[727] EU, 1985. Council Resolution of 7 May 1985 on a new approach to technical harmonization and standards. Official Journal of the European Union C 136, 04/06/1985: 1-9.

18.1.3 Scope and limitation of the new approach for food

The purpose of the 'white paper bis' issued in 1985 and specifically devoted to food is similar to the New Approach, i.e. to achieve the Single Market. However, it has been considered at that time that for food, the approach should be slightly different. The main difference is that for foodstuffs, there is no reference to standardisation.

Despite this basic difference, we can nevertheless consider that the orientation taken in 1985 for food is also a 'new approach' because full harmonisation ('recipe laws') is abandoned: harmonisation of the specification of products will be limited to the 'essential requirements', i.e. the basic requirements which, according to the case-law of the ECJ regarding the free movement of goods, allow the Member States to keep barriers to trade. As to other specifications of products, which are not considered as essential, the principle of Mutual Recognition of national rules should apply.

This position was solemnly reiterated by the Commission in 1989 in its Communication 'On the Free Movement of Foodstuffs in the Community',[728] where the two basic pillars are put forward: Minimal Harmonisation, and Mutual Recognition. As to standardisation, it was still totally absent from this text.

Does it mean that the 'new approach' as defined above is totally unknown in EU Food Law?

18.2 Is the 'new approach' unknown in EU Food Law?

Many references to private rules or standards are found in regulations regarding Food Law, but this sole reference to standards does not correspond to the real 'new approach' as defined above for industrial products, because they do not meet the four criteria. For example:

- The specifications of products which benefit from Geographic Indications and specificities are mostly derived from professional rules. However, abiding by these rules is not optional, but compulsory in order to benefit from the protected appellation.
- In the 'labelling directive' (2000/13)[729], the name under which the product is sold can be, in the absence of a legal name defined by laws or regulations, the 'customary' name, which may be ruled by private or professional standards.

[728] EU, 1989. Communication on the free movement of foodstuffs within the Community. Official Journal of the European Union C 271, 24/10/1989: 3.

[729] EU, 2000. Directive 2000/13/EC of the European Parliament and of the Council of 20 March 2000 on the approximation of the laws of the Member States relating to the labelling, presentation and advertising of foodstuffs. Official Journal of the European Union L 109, 6/5/2000: 29-42.

However, in such a case, the name defined by private bodies in a specific Member State may not be valid all over the EU.

Closer to the real 'new approach' technique is the case of accredited laboratories and some methods of analysis, which are often established by standardisation bodies. Also, the regulation on hygiene recognises the relevance of GMP codes as presumptions of conformity to the rules. However, all this cannot be really qualified as the enforcement of the 'new approach', which is not only a reference to standards. As explained above, the 'new approach' is a complete legal system in which essential requirements are compulsory, and where conformity to standards are not compulsory, but provides a presumption of conformity to the essential requirements. As a consequence, conformity to the essential requirements can be established by other means than by conformity to the standards, and the presumption of conformity to the essential requirements provided by the conformity to standards can be overturned is specific circumstances.

The closest situation to that system in the Food sector is not within the EU, but at the International level, in WTO, where SPS and TBT agreements[730] refer to Codex standards. Codex standards are not compulsory, but conformity to Codex standards provides a presumption of conformity to SPS/TBT principles, i.e. national measures justified by what we can call, by analogy with the EU system, 'mandatory requirements'. However, the legal consequences in this context are very specific, limited to the scope and objective of WTO, i.e. international trade: conformity to Codex rules helps Member States of the WTO justify barriers to trade, since their national rules are presumed to conform to the essential requirements accepted as barriers in SPS and TBT agreements. However, this mechanism is limited to international trade, and is not supposed to replace national rules.

Also, and in any case, Codex rules are set up by representatives of the Governments, and not by private bodies such as standardisation bodies in the EU 'new approach'.

18.3 Public/private – regulation/standards: present situation and questions

Before answering, or trying to answer, the question which is the topic of this discussion, it is essential to keep in mind a few essential issues. The basic question is, to our view: do the interests and goals of Public Authorities (who regulate) and of Private Bodies (who can set up all sorts of 'standards') converge? Probably yes, sometimes, but certainly up to a limit. Indeed, both sets of rules are, by their very nature, fundamentally different. The 'new approach' is just a technique to

[730] Agreement on the application of Sanitary and Phytosanitary Measures (WTO). Available at: http://www.wto.org/english/docs_e/legal_e/15-sps.pdf; Agreement on Technical Barriers to Trade (WTO). Available at: http://www.wto.org/english/docs_e/legal_e/17-tbt.pdf.

delegate the setting up of technical rules, the purpose of this technique being to achieve *compliance with public rules*.

On the other hand, private standards taken at the initiative of private bodies, or referred to at their initiative, are set up in the interest of those private bodies, and their very nature is of *contractual relationship*. One should also not forget that private standards may raise competition law issues: if standards are set up by associations of undertakings with a view to, or the effect of, restricting access to the market, they may be considered as decisions of associations caught by Article 101 of the Treaty on the Functioning of the European Union[731]. Likewise, standards imposed to their counterparts by a company, or by a group of companies, holding a dominant position on the market, might, under certain circumstances, constitute an abuse of a dominant position caught by Article 102.

Under this very general background, several specific issues should be kept in mind, such as (among others), depending on the situations:

- Reference to private standards or codes of practices in national law, to assess the conformity of products to those national standards, may contradict the principle of mutual recognition between Member States, which should be applied in the Single Market.
- Various optional standards granting a 'plus' to products or to processes (for instance, 'eco labels' or labels referring to fair trade), are purely private and the consumer should not be misled as to what they do represent, a requirement of transparency which is obviously of public interest.
- Some purely technical standards (such as, for example, set up for logistic considerations) may have nothing to do with legal requirements, and may raise competition issues by constituting barriers to enter or to remain on the market if such standards become *de facto* compulsory.
- Similar problems may derive from insurance companies requirements.

These general questions have been recently addressed by the EU Commission, who has published a detailed Communication on December 16, 2010, which deserves careful consideration.[732] In the introduction of this Communication, the Commission underlines that:

> 'Clearly, private certification is not needed to show compliance with legal requirements. Any private certification scheme for the agricultural and food sector must remain voluntary. Where operators employ certification of compliance with basic requirements in order to facilitate transactions

[731] EU, 2010. Consolidated versions of the Treaty on European Union and the Treaty on the Functioning of the European Union. Official Journal of the European Union C 83, 30/03/2010: 1-388.
[732] Commission Communication, 2010. EU best practice guidelines for voluntary certification schemes for agricultural products and foodstuffs. Official Journal of the European Union C 341, 16/12/2010, p. 5. See appendix 1.

> with other actors along the food chain, it should be clear that this practice cannot be used to differentiate products in the market'.

In such a context, would the 'new approach' now be possible/desirable in EU Food Law?

18.4 Is the 'new approach' now possible/desirable in EU Food Law?

It seems obvious that all the above mentioned questions could not be solved by the mere virtue of applying now the 'new approach' in the food sector. However, some of these questions might be solved, at least partially, and the changes occurred since 1985 should also be taken into account. Indeed, many things have changed since 1985. The expectations are not the same as they used to be, and different new procedures have been put in place.

As to the expectations, the main trauma was the 'mad cow' crisis in 1996, which entailed a dramatic shift of priorities. Since then, after the publication of the 'Green Paper on Food Law' issued by the Commission in 1997,[733] and up to now, the emphasis has been put on safety rather than on the internal market - even if the internal market is still the necessary legal basis for regulating at the EU level. In parallel - and this is also a clear symptom of that shift - the entire area of food law has been taken out from the DG 'Enterprise' (at that time DG III, in charge of Industry and of Internal Market at the same time) and placed under the responsibility of DG SANCO (Health and protection of the Consumer).

During the same period of time, and parallel to that shift of priorities, the procedures for setting up EU food law have changed too. Harmonisation is easier than it used to be in the past, since it does not require any more unanimity at the Council as was the case with the old 'Article 100' of the Treaty of Rome. All basic texts now require a qualified majority at the Council. But at the same time, the decision making process is more complex, involving the Parliament in co-decision. Likewise, the Commission has more powers since many texts can now be taken in 'comitology', but the Parliament comes now into play with the procedure of 'scrutiny'. Experience shows that the Parliament, in these matters, has become a key actor of the protection of the consumer's interests, particularly, but not exclusively, as regards food safety.

Since 1985, the principle of subsidiarity has also been introduced in EU Law, but at the same time the decision making process in Food Law has withdrawn powers from the Member States, by replacing systematically directives by regulations -

[733] The General Principles of Food Law in the European Union. Commission Green Paper. COM (97) 176 final, 30 April 1997.

which facilitates uniform implementation and, consequently, the materialisation of the single market. Also, crisis management is less and less only left to national authorities, and has more and more become an EU competence with the Rapid Alert System and obligations deriving from it.

Last but not least, the legal frame of Food Law has totally changed with the so-called 'General Food Law' Regulation (Regulation 178/2002)[734]. In the system put in place by this regulation, the share of competences between the authorities is between Risk Assessors (the EFSA) and Risk Managers (the EU Institutions), according to the well-known principles of Risk Analysis. This share of competences has nothing to do with the 'new approach' system, where the line is drawn between essential requirements and technical standards. In such a frame, there seems to be very little room for the 'new approach'.

However, the situation is not totally clear cut: as we have seen before, the general safety requirements which lay upon the food operators can be enforced through compliance with standards, such as the GMP codes of hygiene referred to in the hygiene regulation. There is probably other room for other private standards in some specific fields, such as codes of ethics, 'self-regulation' in the publicity sector, or for very specific ranges of products (kosher, hallal, etc.). Does this mean that we could now envisage the implementation of the 'new approach', and if yes, to what extent?

We do not see why some questions could not, or could not have been, solved by delegating to standardisation the setting up of technical requirements, in order to implement the essential requirements laid down in the Regulation. As a practical existing example, we can refer to the definition of yoghourt, which is found in an AFNOR standard in France. Why not delegate to the CEN some topics such as the purity criteria for additives, or the definition of what is a 'medical' claim, or even nutritional profiles, or some claims for which no harmonisation seems to be foreseen, even if it would be most useful, such as 'natural'? Once the essential requirement of not misleading the consumer and ensuring its safety are laid down in the regulation, there is no theoretical obstacle to delegating the practical implementation of such principles to standardisation. It could also be envisaged that, in some cases, conformity to those standards would be optional, constituting only a presumption of conformity.

We are well aware that some of the examples given above may be considered to be provocative, because the idea of acting that way has not been seriously examined,

[734] EU, 2002. Regulation (EC) No 178/2002 of the European Parliament and of the Council of 28 January 2002 laying down the general principles and requirements of food law, establishing the European Food Safety Authority and laying down procedures in matters of food safety. Official Journal of the European Union L 31, 1/2/2002: 1-24.

at least as far as we know. However, such an avenue would certainly deserve to be explored in some specific cases.

This having been said, we do not go to the conclusion that the 'new approach' could now be applicable to the entire Food Law. It is also clear in our mind that, in any case, the role of EFSA is not questioned, and should not be questioned. We do not advocate for attributing the EFSA's competences to standardisation bodies. As opposed to industrial sectors where the 'new approach' has been used to ensure the mere safety of the products, food is also closely related to health, which goes far beyond safety, and which necessitates unquestionable independent scientific expertise. In these matters too, technical specifications cannot be always optional.

Also, and in any case, the central objectives of EU Food Law – as well as any EU Law – should not be forgotten: Market access and free trade should remain the basic goals, while ensuring the protection of the consumer through safety and correct information, which should not be misleading and should be relevant. To achieve these goals, a broader use of the 'new approach' technique might well help better monitor the ever expanding flow of private standards, by providing legitimacy to those standards which really deserve it, and only to them.

So, as a conclusion and in an attempt to answer the question: 'EU new approach also for Food Law?', we tend to answer: 'why not?'. The question does deserve to be seriously addressed, 25 years after the birth of the concept. It could certainly be useful in some cases ... but totally useless or unrealistic in other cases: pragmatism is of the essence. But at least, the 'new approach' is a tool, among others, the existence of which regulators should not ignore when they face new or unsolved questions.

References

EU, 1985. Council Resolution of 7 May 1985 on a new approach to technical harmonization and standards. Official Journal of the European Union C 136, 04/06/1985: 1-9.

EU, 1997. The General Principles of Food Law in the European Union. Commission Green Paper. COM (97) 176 final, 30 April 1997.

EU, 1989. Communication on the free movement of foodstuffs within the Community. Official Journal of the European Union C 271, 24/10/1989: 3.

EU, 2000. Directive 2000/13/EC of the European Parliament and of the Council of 20 March 2000 on the approximation of the laws of the Member States relating to the labelling, presentation and advertising of foodstuffs. Official Journal of the European Union L 109, 6/5/2000: 29-42.

EU, 2002. Regulation (EC) No 178/2002 of the European Parliament and of the Council of 28 January 2002 laying down the general principles and requirements of food law, establishing the European Food Safety Authority and laying down procedures in matters of food safety. Official Journal of the European Union L 31, 1/2/2002: 1-24.

EU, 2010. Commission Communication – EU best practice guidelines for voluntary certification schemes for agricultural products and foodstuffs. Official Journal of the European Union C 341, 16/12/2010: 5-11.

EU, 2010. Consolidated versions of the Treaty on European Union and the Treaty on the Functioning of the European Union. Official Journal of the European Union C 83, 30/03/2010: 1-388.

EUJ, 1979. Case 120/78, February 20, 1979. Rewe-Zentral AG v Bundesmonopolverwaltung für Branntwein (Cassis de Dijon). ECR 1979: 649.

WTO, undated. Agreement on Technical Barriers to Trade (WTO). Available at: http://www.wto.org/english/docs_e/legal_e/17-tbt.pdf.

WTO, undated. Agreement on the application of Sanitary and Phytosanitary Measures (WTO). Available at: http://www.wto.org/english/docs_e/legal_e/15-sps.pdf.

Appendix 1.
Commission Communication – EU best practice guidelines for voluntary certification schemes for agricultural products and foodstuffs[735]

[735] Official text available at: http://eur-lex.europa.eu/LexUriServ/LexUriServ.do?uri=OJ:C:2010:341:0005:0011:EN:PDF.

Commission Communication — EU best practice guidelines for voluntary certification schemes for agricultural products and foodstuffs

(2010/C 341/04)

1. INTRODUCTION

nt years have seen substantial growth in voluntary certifi- ɔn schemes for agricultural products and foodstuffs. An ntory compiled for the Commission in 2010 ([1]) lists more 440 different schemes, most of which were established ng the last decade.

ification schemes for agricultural products and foodstuffs ride assurance (through a certification mechanism) that ain characteristics or attributes of the product or its luction method or system, laid down in specifications, been observed. They cover a wide range of different atives that function at different stages of the food supply n (pre- or post-farm gate; covering all or part of the food ly chain; affecting all sectors or just one market segment, They can operate at business-to-business (B2B) level (where supermarket or processing business is the intended final ient of the information) or at business-to-consumer (B2C) . They can use logos although many, especially the B2B mes, do not.

e certification schemes by definition employ third-party tation, there are other schemes in the market which ate on the basis of a label or logo (often registered as a mark) without involving any certification mechanism. erence to these schemes is done by self-declaration or ugh selection by the scheme owner. In line with the defi- ns provided in Section 2, these schemes will be referred to elf-declaration schemes'. The use of certification is most opriate when the undertakings made are complex, laid n in detailed specifications and checked periodically. Self- aration is more appropriate for relatively straightforward le-issue) claims.

development of certification schemes is driven mainly by rs such as societal demands for certain characteristics ([2]) of product or its production process on the one hand (mostly B2C schemes), and operators' desire to ensure that their liers meet specified requirements, on the other hand tly for B2B schemes). In the area of food safety, Regulation No 178/2002 laying down general principles and irements of food law ([3]) puts the primary responsibility ensuring that food and feed satisfy the requirements of law and for verifying that such requirements are met, on ood and feed business operator. Large players in the food ly chain in particular often rely on certification schemes in r to satisfy themselves that a product meets the irements and to protect their reputation and liability in vent of a food safety incident.

Clearly, private certification is not needed to show compliance with legal requirements. Any private certification scheme for the agricultural and food sector must remain voluntary. Where operators employ certification of compliance with basic requirements in order to facilitate transactions with other actors along the food chain, it should be clear that this practice cannot be used to differentiate products in the market.

Certification schemes can bring benefits:

— to intermediate actors in the food supply chain, by assuring standards and thereby protecting liability and reputation for product and label claims,

— to producers, by increasing market access, market share and product margins for certified products and also, potentially, by increasing efficiency and reducing transaction costs, and

— to consumers, by providing reliable and trustworthy information on product and process attributes.

Some stakeholders have argued that certification schemes can have drawbacks:

— threats to the single market ([4]),

— questions relating to the transparency of scheme requirements and the credibility of claims particularly for schemes that certify compliance with baseline requirements,

— potential for misleading consumers,

— costs and burdens on farmers, particularly where they have to join several schemes to meet demands from their buyers,

udy conducted by Areté for DG AGRI; see http://ec.europa.eu/ riculture/quality/index_en.htm

r example: animal welfare; environmental sustainability; fair trade.

J L 31, 1.2.2002, p. 1.

([4]) In its Communication 'A better functioning food supply chain in Europe' (COM(2009) 591), the Commission stated its intention to review selected environmental standards and origin-labelling schemes that may impede cross-border trade.

— risk of rejection from the market of producers not participating in key certification schemes, and

— impacts on international trade, especially with developing countries ([1]).

The Commission has noted that the issue of consumer confusion arising from different schemes with similar objectives is being taken up by private initiatives ([2]) aiming to create 'codes of good practice' for private standard-setting organisations mainly in the social and environmental field. Moreover, certain proponents of existing schemes have already taken steps to align requirements with similar schemes and some existing certification schemes (mostly at B2B level) have emerged from a harmonisation process of various individual standards.

1.1. **Types of scheme**

There is a great diversity of schemes in terms of their scope, their objectives, their structure and their operational methods. As mentioned earlier, one important distinction between schemes is whether or not they rely on a third-party attestation procedure, thereby grouping them into self-declaration schemes on the one hand and certification schemes on the other. Ce cation schemes can be further distinguished based on whe they operate at business-to-business (B2B) level or whether aim to provide information from the business chain to consumer (B2C).

Another important classification criterion pertains to whe the scheme assesses products and processes (mostly B2C management systems (mostly B2B). In terms of spec requirements, schemes may attest compliance with provis laid down by governmental authorities (baseline) or they add criteria which go beyond the legal requirements (a baseline). Distinction between the two is not always eas make: on the one hand, schemes often combine bas criteria in some areas with higher requirements in others the other hand, certain baseline requirements particularly in environmental and farming area require operators to use and best practice, and make value-judgment about due car that the concrete actions to be taken can differ between a and between Member States. Indeed, the technical requirem of some certification schemes are used by operators to inte and make concrete these general obligations.

The following table illustrates this classification:

Classification of schemes

Type of attestation:	Self-declaration	Certification (third-party attestation)	
Audience:	B2C	B2C	B2B
Objects of specified requirements:	Products and processes	Mostly products (including services) and processes	Mostly management systems
Content of requirements:	Mostly above baseline	Mostly above baseline	Baseline and above baseline

The guidelines will focus on certification schemes as outlined in the right-hand side of the table above.

1.2. **Purpose of the guidelines**

In its Communication on agricultural product quality policy ([3]), the Commission stated that in the light of developments and initiatives in the private sector, legislative action was not warranted to address the potential drawbacks in certification schemes at this stage ([4]). Instead, drawing on comments from stakeholders, the Commission undertook to develop guidelines for certification schemes for agricultural products and food in consultation with the Advisory Group on Quality ([5]).

These guidelines are designed to describe the existing framework and to help improving the transparency, credi and effectiveness of voluntary certification schemes ensuring that they do not conflict with regulatory requirem They highlight best practice in the operation of such sche thereby offering guidance on how to:

— avoid consumer confusion and increase the transpa and clarity of the scheme requirements,

([1]) The issue of private standards has been discussed in the SPS Committee of the WTO.

([2]) E.g. the ISEAL Alliance (http://www.isealalliance.org).

([3]) COM(2009) 234.

([4]) This conclusion was based on a thorough impact assessment that explored different options for the way forward (see 'Certification schemes for agricultural products and foodstuffs'; http://ec.europa.eu/agriculture/quality/policy/com2009_234/ia_annex_d_en.pdf).

([5]) Advisory Group on 'Quality of Agricultural Production', estab under Commission Decision 2004/391/EC (OJ L 120, 24.4. p. 50).

reduce the administrative and financial burden on farmers and producers, including those in developing countries, and

ensure compliance with EU internal market rules and principles on certification.

e guidelines are directed primarily to scheme developers and erators.

take of the guidelines is voluntary. Adherence to these delines does not mean that the Commission has endorsed requirements set up by these schemes. The present delines neither have a legal status in the EU nor are they ended to alter requirements under EU legislation.

ally, these guidelines should not be considered as a legal erpretation of the EU legislation as such interpretations are exclusive competence of the Court of Justice of the opean Union.

2. SCOPE AND DEFINITIONS

2.1. Scope

e guidelines are applicable to voluntary certification schemes ering:

agricultural products, whether or not intended for human consumption (including feed),

foodstuffs covered by Article 2 of Regulation (EC) No 178/2002, and

processes and management systems related to the production and processing of agricultural products and foodstuffs.

guidelines do not apply to official controls carried out by lic authorities.

2.2. Definition of terms [1]

Specified requirement: need or expectation that is stated.

2. Conformity assessment: demonstration that specified requirements relating to a product, process, system, person or body are fulfilled.

3. Review: verification of the suitability, adequacy and effectiveness of selection and determination activities, and the results of these activities, with regard to fulfilment of specified requirements.

4. Attestation: issue of a statement, based on a decision following review that fulfilment of specified requirements has been demonstrated.

5. Declaration: first-party attestation. For the purpose of these guidelines, the term 'self-declaration schemes' is used for collective schemes and label claims that are not certified, and which rely on the producer's self-declaration.

6. Certification: third-party attestation related to products, processes, systems or persons.

7. Accreditation: third-party attestation related to a body conveying formal demonstration of its competence to carry out specific tasks. In the EU [2], accreditation shall mean an attestation by a national accreditation body that a conformity assessment body meets the requirements set by harmonised standards and, where applicable, any additional requirements including those set out in relevant sectoral schemes, to carry out a specific conformity assessment activity.

8. Inspection: examination of a product design, product, process or installation and determination of its conformity with specific requirements or, on the basis of professional judgement, with general requirements.

9. Audit: systematic, independent, documented process for obtaining records, statements of fact or other relevant information and assessing them objectively to determine the extent to which specified requirements are fulfilled.

3. EXISTING LEGAL PROVISIONS AT EU LEVEL

3.1. Rules related to the operation of schemes

Certification schemes operating in the EU are subject to the following basic EU provisions:

[1] Based on EN ISO/IEC 17000 'Conformity assessment — Vocabulary and general principles'.

[2] Article 2(10) of Regulation (EC) No 765/2008 setting out the requirements for accreditation and market surveillance relating to the marketing of products (OJ L 218, 13.8.2008, p. 30).

— Rules on the internal market. Certification service-providers may benefit from the freedom of establishment and freedom to provide services as enshrined in Articles 49 and 56 of the Treaty on the Functioning of the European Union (TFEU) and relevant provisions of the Directive on Services [1]. They shall face no unjustified restrictions when establishing in another Member State. Equally, they should face no unjustified restrictions when providing the services across the borders. Certification schemes must also not result in de facto barriers to trade in goods in the internal market.

— Rules on State involvement in schemes. Certification schemes supported by public bodies, such as regional or national authorities, may not lead to restrictions based on the national origin of producers or otherwise impede the single market. Any support for certification schemes granted by a Member State or through State resources within the meaning of Article 107 of the TFEU, must comply with State aid rules.

— Rules on competition. Certification schemes may not lead to anticompetitive behaviour, including in particular on a non-exhaustive basis:

  — horizontal or vertical agreements restricting competition,

  — foreclosure of competing undertakings by one ore more undertakings with significant market power (such as preventing access of competing buyers to supplies and/or access of competing suppliers to distribution channels),

  — preventing access to the certification scheme by market operators that comply with the applicable pre-requisites,

  — preventing the parties to the scheme or other third parties from developing, producing and marketing alternative products which do not comply with the specifications laid down in the scheme.

— Consumer information and labelling requirements [2]. The labelling, advertising and presentation of food must not be such as it could mislead a purchaser to a material degree, particularly:

  — as to the characteristics of the foodstuff and, particular, as to its nature, identity, propert composition, quantity, durability, origin or provena method of manufacture or production,

  — by attributing to the foodstuff effects or proper which it does not possess,

  — by suggesting that the foodstuff possesses special ch acteristics when in fact all similar foodstuffs possess s characteristics.

Schemes certifying only compliance with legal requireme may not lead to any suggestion that the certified prod possess special characteristics which are different from th of similar products. Nor should the effect of the schemes to discredit or tend to discredit other products on the mar nor the reliability of official controls.

Moreover, labelling, advertising and presentation of food m not be such as it could mislead consumers according to provisions of the Directive in Unfair Commercial Practices

— The EU takes into account its international obligations particular the requirements set out in the WTO Agreem on Technical Barriers to Trade, when it introduce conformity assessment procedure in a given piece of le lation.

3.2. **Rules related to the content of schemes**

In addition, specific legislation exists on many subjects cove by the requirements of certification schemes (e.g. regulat requirements for food safety and hygiene [4]; organic farm animal welfare; environmental protection; marketing stand for specific products).

In areas where relevant standards or legislation exist, cla must take into account and be consistent with such stand or legislation and make reference to them in the specificati (e.g. if a scheme is making organic farming claims, it must based on Regulation (EC) No 834/2007 about org production and labelling of organic products [5]; sche making claims about nutrition and health must be accordance with Regulation (EC) No 1924/2006 [6], and through the required scientific assessment by EFSA).

(1) Directive 2006/123/EC of 12 December 2006 on services in the internal market (OJ L 376, 27.12.2006, p. 36).

(2) Article 2(1)(a) of Directive 2000/13/EC on the approximation of the laws of the Member States relating to the labelling, presentation and advertising of foodstuffs (OJ L 109, 6.5.2000, p. 29).

(3) Directive 2005/29/EC concerning unfair business-to-consu commercial practices in the internal market (OJ L 149, 11.6.2 p. 22) and guidance for its implementation: SEC(2009) 1666.

(4) Regulation (EC) No 852/2004 of 29 April 2004 on the hygien foodstuffs; Regulation (EC) No 853/2004 of 29 April 2004 la down specific hygiene rules for the hygiene of foodstuffs and R lation (EC) No 854/2004 laying down specific rules for the org sation of official controls on products of animal origin intendec human consumption (OJ L 139, 30.4.2004, p. 1).

(5) OJ L 189, 20.7.2007, p. 1.

(6) OJ L 404, 30.12.2006, p. 9.

articular, with regard to food safety and hygiene:

schemes may not prejudice or aim to replace existing official standards and/or requirements, nor should they purport to substitute for official controls carried out by competent authorities for the purposes of official verification of compliance with official obligatory standards and requirements,

product marketed under schemes which set safety and hygiene standards beyond legal requirements may not be advertised or promoted in a way that would discredit or tend to discredit the safety of other products on the market or the reliability of official controls.

Rules governing conformity assessment, certification and accreditation

s on the organisation and operation of accreditation of es performing conformity assessment activities in the lated area have been laid down in Regulation (EC) No /2008. While this Regulation does not contain a irement for conformity assessment bodies to become edited, such a requirement is part of some other EU legis- n [1].

ddition, the internationally recognised rules for operating luct/process or system certification schemes are set out in International Standards Organisation (ISO) Guide 65 (EN 11) or ISO 17021, respectively. While product/process or m certification schemes are voluntary initiatives, to deliver uct/process or system certificates under accreditation, fication bodies have to be accredited against EN 11/ISO 65 or ISO 17021.

ever, the above is without prejudice to all applicable EU law requirements, including the general objectives laid n in Article 5(1) of Regulation (EC) No 178/2002:

od law shall pursue one or more of the general objectives a high level of protection of human life and health and the otection of consumers' interests, including fair practices in od trade, taking account of, where appropriate, the otection of animal health and welfare, plant health and e environment'.

in this framework, Regulation (EC) No 882/2004 [2] of the pean Parliament and of the Council on official controls performed to ensure the verification of compliance with feed and food law, animal health and animal welfare rules includes certain rules for delegation by competent authorities of official control tasks to independent third parties (including accreditation and reporting obligations).

The guarantees given by the official control activities are the baseline, on top of which specific certification schemes may operate on a voluntary basis, bearing in mind that any breach is liable to food law. Assessment of conformity with baseline requirements through certification schemes does not exempt the official control authorities from their responsibility.

4. RECOMMENDATIONS REGARDING SCHEME PARTICIPATION AND DEVELOPMENT

1. Schemes should be open under transparent and non-discriminatory criteria to all participants willing and able to comply with the specifications.

2. Schemes should have a supervisory structure which allows for the contribution of all concerned stakeholders in the food chain (farmers and their organisations [3], agricultural and agri-food traders, food industry, wholesalers, retailers and consumers, as appropriate) in the development of the scheme and in decision-making in a representative and balanced way. Mechanisms for participation by stakeholders and the organisations involved should be documented and publicly available.

3. Managers of schemes operating in different countries and regions should facilitate the participation of all concerned stakeholders from those regions in scheme development.

4. Scheme requirements should be developed by technical committees of experts and submitted to a broader group of stakeholders for inputs.

5. Managers of schemes should ensure the participation of concerned stakeholders in the development of inspection criteria and checklists, as well as in the design and determination of thresholds for sanctions.

6. Managers of schemes should adopt a continuous development approach where feedback mechanisms exist to regularly review rules and requirements in a participatory manner. In particular, scheme participants should be involved in the future development of the scheme.

g. Article 11(3) of Regulation (EC) No 510/2006 on the protection f geographical indications and designations of origin for agri- iltural products and foodstuffs requires that 'The product certifi- ation bodies referred to in paragraphs 1 and 2 shall comply with id, from 1 May 2010 be accredited in accordance with European andard EN 45011 or ISO/IEC Guide 65 (General requirements for odies operating product certification systems)'.

J L 165, 30.4.2004, p. 1.

(3) E.g. cooperatives.

7. Changes to scheme requirements must be made only when justified, so as to avoid unnecessary adaptation costs for scheme participants. Scheme participants must be given appropriate notice of any change to the scheme requirements.

8. Schemes should include contact information on all documentation associated with the scheme (including on a website) and establish a process to receive and reply to comments on the scheme.

5. RECOMMENDATIONS REGARDING SCHEME REQUIREMENTS AND CORRESPONDING CLAIMS

5.1. Clarity and transparency of scheme requirements and claims made

1. Schemes should clearly state the social, environmental, economic and/or legal objectives.

2. Claims and requirements should be clearly linked to the objectives of the scheme.

3. The scope of the scheme in terms of products and/or processes should be clearly defined.

4. Scheme specifications [1], including a public summary, should be freely available (e.g. on a website).

5. Schemes operating in different countries should provide translations of the specifications if a duly justified request is made by potential participants or certification bodies.

6. Scheme specifications should be clear, sufficiently detailed and easily understandable.

7. Schemes using logos or labels should provide information about where consumers can find further details on the scheme, such as a website address, either on the product packaging or at the point of sale.

8. Schemes should clearly state (e.g. on their website) that they require certification by an independent body and provide contact details of certification bodies which provide this service.

(1) Exceptions may be needed where scheme specifications are based on standards which are not freely available (e.g. ISO and EN standards).

5.2. Evidence base of scheme claims and requiremen

1. All claims should be based on objective and verif evidence and scientifically sound documentation. T documents should be freely available, e.g. on a websit

2. Schemes operating in different countries and regions sh adapt their requirements in line with the relevant local a ecological, socio-economic and legal conditions and cultural practices, while ensuring consistent results ac different contexts.

3. Schemes should clearly indicate (e.g. on a website) whe where and to what extent their specifications go beyonc relevant legal requirements, including in the areas reporting and inspections, if applicable.

6. RECOMMENDATIONS REGARDING CERTIFICATION INSPECTIONS

6.1. Impartiality and independence of certification

1. Certification of compliance with the scheme requiren should be carried out by an independent body accred

— by the national accreditation body appointed by Me States according to Regulation (EC) No 765/2008 accordance with relevant European or internati standards and guides setting out general requiren for bodies operating product certification systems, c

— by an accreditation body signatory to the multila recognition arrangement (MLA) for product certific of the International Accreditation Forum (IAF).

2. Schemes should be open to certification by any qualified accredited certification body, without the impositio geographical restrictions.

6.2. Inspections

As a general principle, inspections should be effective, c transparent, based on documented procedures and rela verifiable criteria underlying the claims made by the certific scheme. Unsatisfactory inspection results should lead to ap priate action.

1. Regular inspections of scheme participants should be ca out. There should be clear and documented procedure inspections, including frequency, sampling and labo ry/analytical tests in parameters related to the scope o certification scheme.

(2) An exception should be made for confidential and/or propr information, which should be clearly indicated.

The frequency of inspections should take into account previous inspection results, inherent risks posed by the product or process or management system, as well as the existence of internal audits in collective producer organisations which can complement third-party inspections. A minimum inspection frequency for all scheme participants should be determined by the scheme supervisor.

There should be a systematic evaluation of the results of inspections.

Unannounced inspections and inspections at short notice should be used as a general rule (e.g. within 48 hours).

Inspections and audits should be based on publicly available guidelines, checklists and plans. The inspection criteria should be closely linked to the requirements of the scheme and the corresponding claims.

There should be clear and documented procedures for dealing with non-compliance which are effectively implemented. Knock-out criteria should be defined which could lead to:

— non-issue or withdrawal of the certificate,

— withdrawal of membership, or

— reporting to the relevant official enforcement body.

These knock-out criteria should include at least non-fulfilment of basic legal requirements in the area covered by the certification. Cases of non-compliance with adverse implications for health protection should be notified to the relevant authorities in accordance with regulatory requirements.

Inspections should focus on analysing the verifiable criteria which underlie claims made by certification schemes.

6.3. **Costs**

Scheme managers should make public the membership fees (if any) and require their certification bodies to publish the costs associated with certification and inspection for different types of scheme participants.

Possible discrepancies in fees charged to different scheme participants should be justified and proportionate. They should not serve to deter certain groups of potential participants, e.g. from other countries, to join the scheme concerned.

3. Any cost savings arising form mutual recognition and benchmarking should be passed on to the operators subject to inspections and audits.

6.4. **Qualification of auditors/inspectors**

As a general principle, auditors/inspectors should be impartial, qualified and competent.

Auditors carrying out the certification audits should have the relevant knowledge in the specific sector and should work for certification bodies that are accredited under the relevant European or international standards and guides for product certification schemes and for management system certification schemes. The required auditor skills should be described in the scheme specifications.

6.5. **Provisions for small-scale producers**

Schemes should include provisions enabling and promoting the participation of small-scale producers (especially from developing countries, if relevant) in the scheme.

7. RECOMMENDATIONS REGARDING MUTUAL RECOGNITION AND BENCHMARKING/OVERLAP WITH OTHER SCHEMES

1. Where schemes are entering a new sector and/or expanding in scope, the need for the scheme should be justified. Where possible, scheme managers should make explicit reference (e.g. on their website) to other relevant schemes operating in the same sector, policy area and geographical region and identify where approaches converge and agree. They should actively explore possibilities for mutual recognition for parts or all of the scheme requirements.

2. In areas where schemes have been identified to have partial or total overlap with the requirements of other schemes, schemes should include recognition or acceptance partially or totally of inspections and audits already carried out under those schemes (aiming to not re-audit the same requirements).

3. If mutual acceptance cannot be achieved, scheme managers should promote combined audits based on combined audit checklists (i.e. one combined checklist and one combined audit for two or more different schemes).

4. Managers of schemes that overlap in their requirements should as much as practically and legally possible also harmonise their auditing protocols and documentation requirements.

Appendix 2.
Communication from the Commission to the Council, the European Parliament and the European Economic and Social Committee[736]

Contributing to Sustainable Development: The role of Fair Trade and nongovernmental trade-related sustainability assurance schemes

[736] Official text available at:
http://eur-lex.europa.eu/LexUriServ/LexUriServ.do?uri=COM:2009:0215:FIN:EN:PDF

COMMISSION OF THE EUROPEAN COMMUNITIES

Brussels, 5.5.2009
COM(2009) 215 final

COMMUNICATION FROM THE COMMISSION TO THE COUNCIL, THE EUROPEAN PARLIAMENT AND THE EUROPEAN ECONOMIC AND SOCIAL COMMITTEE

Contributing to Sustainable Development: The role of Fair Trade and non-governmental trade-related sustainability assurance schemes

COMMUNICATION FROM THE COMMISSION TO THE COUNCIL, THE EUROPEAN PARLIAMENT AND THE EUROPEAN ECONOMIC AND SOCIAL COMMITTEE

Contributing to Sustainable Development: The role of Fair Trade and non-governmental trade-related sustainability assurance schemes

TABLE OF CONTENTS

1. INTRODUCTION

This Communication examines the current situation of Fair Trade and other non-governmental (i.e. private) trade-related sustainability assurance schemes. The Commission has long recognised that consumers can support sustainable development objectives by purchasing decisions. This Communication responds to the growing interests that have been articulated, at political level as well as in the growing level of purchases by EU consumers. At political level the European Parliament adopted a report in 2006 on Fair Trade and Development[1]. The report points out the need for raising awareness among consumers, and the risk of abuse by companies that enter the Fair Trade market without complying with certification criteria. Additionally, it recognises that Fair Trade is an essentially voluntary, private sector phenomenon, and that too heavy regulatory embrace could prove damaging rather than beneficial.

The 2005 exploratory opinion of the European Economic and Social Committee (EESC) looked at "consumer assurance schemes". Key findings were to identify the need for authoritative quality assessment of consumer assurance schemes and to fix central definitions. In June 2006 the European Council adopted its renewed sustainable development strategy and encouraged Member States to promote sustainable products, including Fair Trade[2].

EU consumers each year purchase Fair Trade certified products for approximately €1.5 billion; which is 70 times more than in 1999 when the Commission adopted a communication on this topic. This success underlines the need for consumers public authorities and other stakeholders, including producer organisations in developing countries to measure the real impact of Fair Trade.

In this communication the term "Fair Trade" is used in conformity with standards established by the international standard setting and conformity assessment organisations, that are members of the ISEAL[3], and as applied by the Fair Trade organisations. The term "other private sustainability assurance schemes" is used to describe other labelling schemes that aim to inform consumers about the sustainability of the production of the product. (A brief overview of terms and organizations is appended in Annex I).

This Communication provides an up-date on developments arising since the 1999 Commission Communication on fair trade[4] and suggests preliminary considerations on the role of public authorities and stakeholders in the field of Fair Trade and other private sustainability assurance schemes. Issues to be addressed are relevant for several EU policy areas, e.g. consumer protection, economic and social development, trade, corporate social responsibility, environment and the EU internal market. Where appropriate, this Communication may be followed by more targeted initiatives in one or more policy fields.

---

1 European Parliament Report on Fair Trade and Development (2005/2245(INI) "The Schmidt Report".

2 "Member States should promote sustainable products that stem for organic farming and fair trade as well as environmentally sound products" http://register.consilium.europa.eu/pdf/en/06/st10/st10117.en06.pdf, page 13.

3 International Social And Environment Accreditations and Labelling

4 COM(1999)619 of 29-11-1999. Information on the 1999 Commission Communication is appended in Annex II.

This Communication does not cover sustainability and labelling schemes established by public authorities (such as the EU eco-label).

2. FAIR TRADE DEVELOPMENTS SINCE 1999

The most striking developments since 1999 have taken place in national markets where certified Fair Trade products were already present. Answering the 1999 Communication's call for a single label and the need for independent verification and control, the "Fairtrade Certification Mark" has been successfully implemented[5].

The consumer recognition level for the Fair Trade mark in the UK was above 70% in 2008 (compared to 12% in 2000)[6] and in France 74% in 2005 (compared to 9% in 2000)[7]. Worldwide sales of certified Fair Trade goods exceeded €2.3 billion by the end of 2007,[8] (but still an order of magnitude behind organic food sales and still less than 1% of total trade)[9]. Europe is Fair Trade's home: between 60% and 70% of global sales take place here, with large variations between its fastest growing market, Sweden, and newer Member States where the concept is still relatively young.

Fair Trade has played a pioneering role in illuminating issues of responsibility and solidarity, which has impacted other operators and prompted the emergence of other sustainability regimes. Trade-related private sustainability initiatives use various social or environmental auditing standards[10], which have grown in number and market share. The best known social standard is perhaps SA8000, initiated by Social Accountability International (SAI) in 1997[11]. Assurances that extend into broader issues, including both social and environmental criteria, are for example Utz certified and the Rainforest Alliance (RA).

Multi-enterprise sustainability trade initiatives, in different parts of Europe, range from national arrangements to pool the results of social audits to transnational initiatives with some government backing, such as the Ethical Trading Initiative (ETI)[12]. The background work by operators to fulfil and audit standards need not be transmitted by means of a certification and label for consumers; fulfilling and auditing standards can count as a company's efforts of corporate social responsibility (CSR)[13], which is not always indicated on the product. CSR activities can be reinforced by a company committing to a recognised set of criteria or objectives, such as through the UN Global Compact[14].

---

5 See further information on definitions in Annex I.

6 Fairtrade Foundation, 2008.

7 OECD, Trade Policy Working Paper No. 47. Part 1; Jan 10, 2007.

8 Fairtrade Labelling Organizations International, 2008.

9 Land, P. & Andersen, M, "What is the world market for certified products", Commodities and Trade Technical Paper, OECD.

10 See also the Portal for Responsible Supply-Chain Management, established as part of the European Alliance on CSR; www.csr-supplychain.org.

11 SAI claims that "retailers, brand companies and other employers worldwide with annual sales over USD175 billion are using SA8000"; www.sa-intl.org.

12 Other initiatives to mention in this context include the Business Social Compliance Initiative (http://www.bsci-eu.com/), and the Global Social Compliance Programme (http://www.ciesnet.com/2-wwedo/2.2-programmes/2.2.gscp.background.asp).

13 Communication (2006) 136 of 22 March 2006 on "Making Europe a pole of excellence on Corporate Social Responsibility".

14 www.unglobalcompact.org.

Private labelling markets can be divided between;

(1) Fair Trade proper;

(2) other "niche" certified products not participating formally in Fair Trade but targeting consumers aware of sustainability issues (Rainforest Alliance, Utz Certified);

(3) products covered by baseline standards that aspire to be "industry-wide" (e.g. Code for the Coffee Community (4C's); Ethical Tea Partnership);

(4) the rest ("no name" commodity supplies).

A single producer may sell into all four of these categories. It can be tricky for the consumer to assess the significance of various sustainability schemes. It is against this complex and evolving backdrop that political and institutional developments should be assessed.

3. Sustainability Criteria Applied

Private trade-related private sustainability schemes use a set of criteria to assess and/or guarantee the sustainability of the products. Criteria often build on one or more of the three pillars of sustainable development; economic, environmental and social development, sometimes linking into international standards and agreements. Some schemes focus on a particular issue and objective (e.g. carbon footprint for climate change mitigation) whereas others rely on criteria in a wider sustainable development context.

This section describes the first category – Fair Trade – referred to above[15] which achieved significant levels of consumer recognition in those markets where it is operating. Recognition goes with a good measure of understanding of the issues that Fair Trade promotes. The criteria and standards applied by Fair Trade are among the most comprehensive and ambitious in terms of addressing a broad set of issues and conditions that impact the producers in developing countries, including in particular a minimum price for the producer and a premium paid to the community of the producer.

Fair Trade criteria

The criteria, as defined by the Fair Trade movement and recalled in the 2006 European Parliament report are;

- **a fair producer price**, guaranteeing a fair wage, covering the costs of sustainable production and living. This price needs to be at least as high as the Fair Trade minimum price and premium where they have been defined by the international Fair Trade associations;

- **part payments to be made in advance** if so requested by the producer;

15 Appended in Annex III is a presentation of the additional private sustainability schemes; referred to in this section.

- **long-term, stable relations** with producers and producers' involvement in Fair Trade standard-setting;
- **transparency and traceability** throughout the supply chain to guarantee appropriate consumer information;
- conditions of production **respecting** the eight International Labour Organization (ILO) **Core Conventions**;
- respect for the environment, protection of human rights and in particular women's and children's rights and **respect for traditional production** methods which promote economic and social development;
- **capacity building and empowerment** for producers, particularly small-scale and marginalised producers and workers in developing countries, their organisations as well as the respective communities, in order to ensure the sustainability of Fair Trade;
- support for production and **market access** for the producer organisations;
- **awareness-raising activities** about Fair Trade production and trading relationships, the mission and aims of Fair Trade and about the prevailing injustice of international trade rules;
- **monitoring and verification** of compliance with these criteria, in which southern organisations must play a greater role, leading to reduced costs and increased local participation in the certification process;
- regular **impact assessments** of the Fair Trade activities.

4. POLICY CONSIDERATIONS

4.1. Contribution to Sustainable Development

One of the particular features of Fair Trade and other private sustainability assurance schemes is that it is an essentially voluntary, dynamic mechanism that develops along with societal and consumer awareness and demands. As the understanding of sustainability challenges develops, private trade-related sustainability assurance schemes tend to follow. In some cases, they are at the forefront of issues; raising awareness and pushing consumer interest and understanding of new and emerging sustainable development challenges. Niche markets and schemes can influence mainstream business and government policy making.

The Commission considers that it should not take a role in ranking or regulating criteria related to private trade-related sustainability assurance schemes, and their relevance in relation to sustainable development objectives. Regulating criteria and standards would limit a dynamic element of private initiatives in this field and could stand in the way of the further development of Fair Trade and other private schemes and their standards.

Sustainable development can be served by schemes that prioritise environmental, social, or economic elements.. It is important for good market functioning that consumers and producers have access to reliable information on the schemes. Here, it is possible to indicate some elements that are relevant in assessing good practice that operators should undertake according to the Commission:

Standards and criteria should be objective and non-discriminatory to avoid any (unintended) negative impact on, in particular, producers in developing countries. The Commission welcomes efforts under way towards greater definitional clarity, such as the publication of a Fair Trade Charter. To allow consumers to make their choices in a well informed manner, standards and criteria should be applied in a transparent manner. Part of the information which consumers and producers may require to maintain confidence in the market is the proportion of the extra price which is transmitted to producers[16].

Ideally, there should be independent monitoring to guarantee that the products are the result of practices carried out according to a specific set of criteria balancing ecological, economic and social considerations. The nature and results of the auditing process should be available for inspection[17]. The Commission therefore encourages relevant parties to improve their evaluation methodology so as to allow consumers to make informed choices.

Further clarity and understanding is needed of the actual impact of the private sustainability schemes on producers in developing countries and also on their environment in a broader sense. Consumers should ideally be offered some element of objective assessment of the impact of schemes. In this area the Commission expects improvements given the work already under way and looks forward to progress which could form the basis for further policy considerations[18].

Annex IV contains a list of process issues relating to consumer assurance schemes identified by the European Economic and Social Committee. The Commission encourages further work towards a common understanding of what basic process requirements it is reasonable to expect schemes to meet, while continuing to avoid entering into defining appropriate sustainability standards for private schemes.

---

16 The U.K. House of Commons report "Fair Trade and Development, June 2007, suggested a label to inidiciate the percentage of the price received by the producer.

17 Appended in Annex IV is a list of issues relating to consumer assurance schemes identified by the EESC.

18 The ISEAL Alliance is undertaking a project of writing to examine good practice for measuring the impact of standards and certification.

Principles for maximising the impact of private trade-related sustainability assurance schemes;

- Maintaining the non-governmental nature of private schemes throughout the EU.
- Exploring the scope for possible synergies between schemes and enhancing clarity for the consumer and producers.
- Achieving a common understanding of reasonable basic process requirements.
- Establishing objective facts on the relative impacts of different private trade-related sustainability assurance schemes.

4.2. Private Trade-related Sustainability Assurance Schemes and the WTO

Trade liberalisation can offer opportunities for economic growth and sustainable development. Development and the integration of developing countries into the global economy, especially the least developed, are key objectives of the WTO and of EU trade policy.

Multilateral trade liberalisation through the WTO system is the most effective way to expand and manage world trade, and may help to create opportunities for economic growth and sustainable development. However, trade liberalisation is not sufficient; impact of trade policies on growth, development and sustainability is in part framed by regulation and policies in a wide range of other areas that impact on growth and sustainable development.

Private initiatives that operate through essentially voluntary participation are consistent with a non-discriminatory multilateral trading system. Any government intervention or regulatory mechanisms relating to such labelling schemes, while not problematic per se, need to take account of WTO obligations, in particular to ensure their transparent and non-discriminatory functioning.

Principle in relation to WTO;

- Ensuring transparent and non-discriminatory functioning of labelling schemes.

4.3. Public procurement

A field in which important developments have been taking place is public procurement. Public authorities spend the equivalent of 16% of the EU GDP and therefore constitute a key strategic market.

In order to better respond to the contracting authorities' need for guidance to implement sustainable public procurement, the Commission has recently adopted a Communication on

8

public procurement for a better environment[19] (complementing the Commission's Green Procurement Guide) and is currently working on publishing a parallel guide on social procurement. Together, these guides constitute a comprehensive guide to sustainable public procurement.

Many authorities are calling for tenders including sustainable objectives or "fair trade" in their procurement policies. Some Member States have gone further and require specific "Fair Trade label or equivalent". According to European public procurement rules, contracting authorities that wish to purchase fair trade goods, cannot require specific labels because this would limit the access to the contract of products which are not so certified but meet similar sustainable trade standards.

If a contracting authority intends to purchase Fair Trade goods, it can define in the technical specifications of the goods the relevant sustainable criteria, that must be linked to the subject-matter of the contract and comply with the other relevant EU public procurement rules, including the basic principles of equal treatment and transparency. These criteria must relate to the characteristics or performance of the products (e.g. glasses made out of recycled material) or the production process of the products (e.g. organically grown).

Contracting authorities that intend to purchase sustainability assurance goods should not simply take the concept of a particular label and include it in the technical specifications of their purchases. They ought instead look at the sub-criteria underlying, for example, the Fair Trade label and use only those which are relevant to the subject matter of their purchase. Contracting authorities must always allow bidders to prove compliance with these standards by using Fair Trade labels or by other means of proof.

Environmental and social criteria may also be incorporated in the execution clauses, provided these criteria are linked to the execution of the contract in question (e.g. minimum salary for the workers involved in the performance of the contract) and comply *mutatis mutandis* with the other requirements mentioned above in relation to the technical specifications.

Principles to help realise the potential contribution to sustainable development from public purchasing decisions:

- Secure that appropriate guidelines are available on how to implement sustainable public procurement

4.4. EU support

The Commission has provided financial support for Fair Trade and other sustainable trade related activities essentially through its development cooperation instruments (budget chapter 19), through co-financing actions with NGO's. Between 2007 and 2008, € 19.466 million were allocated for various NGO implemented and co-financed actions. The majority of these actions were in the field of awareness raising within the EU.

Actions financed within the framework of multiannual Country Strategy Papers and Indicative Programmes, covering agricultural and rural sectors, include activities that contribute to facilitating Fair Trade. The Special Framework of Assistance for Traditional ACP Suppliers of Bananas and the Accompanying Measures for Sugar Protocol have also contributed to

19 Commission Communication on public procurement for a better environment: COM (2008)400 of 16 July 2008.

helping farmers to sell in the Fair Trade niche. On the other side of the chain, projects in support of trade and private sector development may also contribute to facilitating trade activities, including Fair Trade.

For the budget years 2008 and 2009, additional credits of €1 million each year have been included specifically for actions related to Fair Trade in the credits for trade budget (chapter 20). These credits will be used to top up the financing under the development instruments[20].

The EC has provided support to "fair trade related projects" mainly on a demand-driven basis, responding to grant requests from NGOs for co-financing actions in this area, mostly related to awareness raising within the EU. The EU Commission considers paying more attention to supporting impact assessments, market transparency efforts and assessing difficulties in implementing schemes and obtaining certification. This could be further supported by similar action by EU member states to finance studies on the impact of Fair Trade.

A Commission project taken forward by UNCTAD is to develop an internet portal on sustainability claims schemes. The project aims to provide comparable information on the content and processes of the range of existing schemes, to the benefit of both consumers and producers. The intention is increase transparency on how different schemes tackle the various relevant criteria and to allow stakeholder exchanges on this.

Principles to help the EU to use its direct support to schemes optimally;

- Identifying target areas under existing budget provisions such as studies clarifying the impacts of different schemes, supporting market transparency efforts and cost-benefit analyses of support given.

5. CONCLUSIONS: THE ROLE OF PUBLIC AUTHORITIES AND OF OPERATORS IN RELATION TO FAIR TRADE AND OTHER PRIVATE TRADE-RELATED SUSTAINABILITY ASSURANCE SCHEMES

Given the potential contribution of Fair Trade and other trade-related sustainability assurance schemes to sustainable development, the Commission intends to stay engaged and further support such schemes. Where appropriate, this Communication may be followed by additional initiatives in one or more policy fields. At this stage, the Commission;

- ***Reiterates*** the importance of maintaining the non-governmental nature of Fair Trade and other similar sustainability schemes throughout the EU. Public regulation could interfere with the workings of dynamic private schemes.

- ***Observe*** that Fair Trade has a significant presence in much of the EU market and a high level of consumer recognition linked to the development and transparency of standards and principles underlying the system.

- ***Observe*** that many different types private schemes can contribute towards sustainability objectives, but their multiplicity can carry risks of consumer confusion. The Commission sees scope for further reflection around the principles for maximising the impact of private trade-related sustainability assurance schemes, while avoiding entering into defining what are the appropriate sustainability standards to be followed by these private schemes: This

20 Appended in Annex V are examples of current financing.

10

is, however, without prejudice to compliance with relevant sustainability-related standards and legislation set by public authorities.

In this context the Commission;

- ***Recalls*** that transparency and adequacy of information to consumers about standards of private sustainability schemes are key, and that there could be benefit from arriving at a common understanding of what basic process requirements, such as independent monitoring, are reasonable to expect.

- ***Recalls*** that further assessment of the impact of private sustainability schemes could be a key step forward.

- ***Intends*** to explore the scope for further dialogue, co-operation and, where appropriate, convergence between different private labelling schemes to promote possible synergies and enhance clarity for the consumer.

In the context of public purchasing, the Commission;

- ***Underlines*** the interest of providing guidance to public purchasing authorities help realise the full potential contribution to sustainable development from their decisions.

- ***Underlines*** that a contracting authority that intends to purchase sustainability assurance goods should use only criteria linked to the subject matter of their purchase and comply with the other relevant EU public procurement rules. Contracting authorities must always allow bidders to prove compliance with these standards by using Fair Trade labels or by other means of proof.

In the context of financing, the Commission;

- ***Intends*** to continue funding for relevant Fair Trade and other sustainable trade related activities in accordance with its practice to date. This does not exclude the possibility of financing also more targeted actions in order to pursue priorities identified.

- ***Recalls*** the need to assess the results of analyses of the impact of private sustainability assurance scheme on sustainable development parameters, including the implications for economic, social and developmental criteria in producing countries. Given the focus of private sustainability assurance scheme on the working and living conditions for producers in developing countries, the Commission considers that particular attention should be given to this aspect. Analysis should compare the impact of various private schemes so as to provide a basis for possible further initiatives in this field.

ANNEX I

FAIR TRADE DEFINITION

Fair Trade standards are the result of consultation of stakeholders and experts and are set in accordance with the requirements of *the International Social and Environment Accreditations and Labelling* Alliance (ISEAL). The alliance is a formal collaboration of leading international standard-setting and conformity assessment organizations focused on social and environmental issues.

There are two international Fair Trade standard setters that certify Fair Trade Organizations across the world, according to ISEAL principles; the *Fairtrade Labelling Organizations* (FLO) and the *World Fair Trade Organization* (WFTO) (previously the International Fair Trade Association, IFAT). The WFTO is an associate member of ISEAL. These two standard setters have produced the "*Charter of Fair Trade principles*".

In accordance with the "*Charter of Fair Trade principles*" (January 2009) Fair Trade is defined as (based on the FINE definition in 2001):

"*Fair Trade is a trading partnership, based on dialogue, transparency and respect, that seeks greater equity in international trade. It contributes to sustainable development by offering better trading conditions to, and securing the rights of, marginalized producers and workers – especially in the South. Fair Trade Organizations, backed by consumers, are engaged actively in supporting producers, awareness raising and in campaigning for changes in the rules and practice of conventional international trade*".

This above definition is used in this Communication. .

The Fairtrade Labelling Organizations (FLO) is a multi-stakeholder association involving 23 member organizations, traders and external expert. The organisation develops and reviews Fairtrade standards and provides support to Fairtrade Certified producer by assisting them in gaining and maintaining Fairtrade certifications and capitalizing on market opportunities. For example the Fairtrade Labelling Organisation (FLO) sets the standards, and a separate international certification company - FLO-CERT - regularly inspects and certifies producers against these standards, and audits the flow of goods between producers and importers.

Furthermore, *the World Fair Trade Organization (WFTO)* has developed an independent third party certification system: the sustainable fair trade management system.

A distinction not easy to make is that between NGO-initiated goal-driven operations, i.e. the primary objective is to contribute to sustainable development, and mainstream initiatives that are foremost business-oriented but seek to contribute to sustainability objectives. For example supermarkets propose their own fair trade brands together with other Fair Trade-certified products.

ANNEX II

THE 1999 COMMUNICATION ON FAIR TRADE

The issues identified in the Communication of 1999 have been addressed in different instances. At a European level, the 2006 report of the European Parliament (the "Schmidt Report") and the 2005 exploratory opinion of the European Economic and Social Committee (EESC) (rapporteur Richard Adams) presented considerations relating to Fair Trade and similar private sustainability schemes. In June 2006 the European Council adopted its renewed sustainable development strategy and included fair trade in the call to Member States to promote sustainable products[21].

Issues of relevance to sustainability labelling have also been referred to in many EC policy documents; the Communication on Agricultural Commodity chains, poverty and dependence; the EU Policy for Africa; the Action Plan on Cotton; the Aid for Trade Strategy adopted by the council in October 2007) and the Commission's Green Paper on agriculture product quality (October 2008)[22]. Although the Commission's 1999 Communication on *"fair trade"* remains the most comprehensive statement of the Commission's stance towards what was then called "fair trade".

The Communication pointed out three key issues; (i) the development of Fair Trade and "ethical trade" need to be dealt with in a coherent manner; (ii) Fair Trade should contribute to sustainable development through voluntary participation, and EC involvement should take WTO obligations into account; and (iii) schemes must satisfy the needs of producers from developing countries and allow consumers to make properly informed choices.

---

21 http://register.consilium.europa.eu/pdf/en/06/st10/st10117.en06.pdf, p. 13.

22 The Commission's Green Paper on agricultural product quality of October 2008 addresses the issue of fair trade in the context of food quality certification schemes. A Commission Communication (forthcoming) on the same subject is planned. Paper on agricultural product quality: product standards, farming requirements and quality schemes COM (2008) 641 final of 15 October 2008.

ANNEX III

CRITERIA RELATED TO GOOD AGRICULTURAL AND BUSSINESS PRACTICE AS WELL AS SOCIAL AND ENVIRONMENTTAL CRITERIA

This part refers to the section three in the Communication and provides examples for certified products targeting consumers awareness of sustainability issues.

It is common for certification schemes to include criteria related to good agricultural and business practice as well as social and environmental criteria. The Utz Certified Code of Conduct (which currently applies to coffee and is due to be extended to cocoa, tea and palm oil) includes elements such as standards for record-keeping, minimised and documented use of agrochemicals for crop protection, protection of labour rights and access to health care and education for employees and their families. In the social field, workers' protection is based on both national laws and ILO conventions but also relate to housing, clean drinking water and training for workers. Environmental criteria relate to the prevention of soil erosion, water usage, energy use and sustainable energy sources as well as deforestation.

Other private schemes have a more environmental focus: it is evident from the name that the Rainforest Alliance is one of these, although in practice the RA certification scheme combines both environmental and social concerns:

- Social and Environmental Management System
- Ecosystem Conservation
- Wildlife Protection
- Water Conservation
- Fair Treatment and Good Working Conditions for Workers
- Occupational Health and Safety
- Community Relations
- Integrated Crop Management
- Soil Management and Conservation
- Integrated Waste Management

A third type listed in the report, section 3, is standards that have been set up with the intention that they should apply "industry-wide" rather than to cater for a niche market of discriminating consumers. One example of this type of initiative is the Common Code for the Coffee Community (4C) Association, which has worked over the past five years to set the baseline for sustainable development within the mainstream coffee sector. The 4C Association standards build on the Millennium Development Goals of the United Nations and exclude the worst forms of social, environmental and economic practices in the production, post-harvest processing and trading of green coffee. Definitions are primarily based on the UN Human Rights Declaration as well as existing UN conventions and standards and, usually, national legislation. Once the ten worst practices have been eliminated participants have to continuously improve on the other parameters set out in the Code.

15

ANNEX IV

Range of process issues relating to consumer assurance schemes identified by the European Economic and Social Committee:

a) Scheme Governance

Where does ultimate control of the scheme lie?

b) Scheme Goals

Are the goals clearly defined?

c) Scheme scope

Does the scheme address the "problem" as normally defined?

d) Scheme standards or terms

Do the standards set and monitored by the scheme express the goals?

e) Impact assessment

Is there credible assessment of the impact of the scheme on the goals?

f) Independent review

Is there any independent review of the scheme's operation?

g) Cost-benefit analysis

Is there any process to monitor and evaluate the costs of the scheme borne by suppliers, traders and consumers in comparison to the progress made to achieve the goals?

h) Public claims

Do the public claims by certified companies or suppliers match the goals, standards and outcomes of the scheme?

16

ANNEX V

THEMATIC PROGRAMMES "CO-FINANCING WITH EUROPEAN NGOs" and "NON STATE ACTORS AND LOCAL AUTHORITIES IN DEVELOPMENT"

2007 and 2008 Projects including the promotion of Fair Trade among its objectives and / or activities

Contract year	Title	Contracting party	Nationality	Budget	DAC Code
2008	*ESTRATEGIA PARA EL FORTALECIMIENTO DE INICIATIVAS COMUNITARIAS PRODUCTIVAS ACORDES CON EL DESARROLLO SUSTENTABLE*	FUNDACION TIERRA VIVA	Venezuela	49968	15150
2008	*COMPETITIVIDAD PRODUCTIVA Y COMERCIAL DE LA RED DEPARTAMENTAL DE PEQUEÑOS PRODUCTORES DE HABA CONVENCIONAL Y ORGÁNICA (ASOHABA) EN EL MERCADO COMUNITARIO Y DE COMERCIO JUSTO*	CREDI FUTURO ASOCIACION		380000	43040
2008	*Cafe amigable con la naturaleza Santa Cruz - Bolivia*	ASOCIACION CENTRO DE PROMOCION AGROPECUARIA CAMPESINA	Bolivia	515267	43040
2008	*OntunLan, N"do Botor - Turismo Socialmente responsavel no sector de Quinhamel*	INSTITUTO MARQUES DE VALLE FLOR FUNDACAO	Portugal	496389,32	33210
2008	*Espaço por um Comércio Justo: alternativas em rede*	CENTRO DE INFORMACAO E DOCUMENTACAOAMILCAR CABRAL ASSOCIACAO	Consortium	370011,99	15150
2008	*Decent Life - decent work. Enhancing international strategies and policies of trade unions*	SUDWIND DIE AGENTUR FUR SUD NORD BILDUNGS UND OFFENTLICHKEITSARBEIT GMBH	Consortium	662264	99820
2008	*Mobilizing for a sector dialogue for the improvement of working conditions in the globalized toy industry*	SUDWIND DIE AGENTUR FUR SUD NORD BILDUNGS UND OFFENTLICHKEITSARBEIT GMBH	Consortium	929043	99820
2008	*F.R.A.M.E. (Fair and Responsible Action in MEditerranean area)*	CONSORZIO CTM-ALTROMERCATO SOCIETACOOPERATIVA	Consortium	494821	99820
2008	*Network Sustainable Consumption*	SUDWIND DIE AGENTUR FUR SUD NORD BILDUNGS UND OFFENTLICHKEITSARBEIT GMBH	Consortium	647023	99820
2008	*Creating Coherence. Trade for Development: Development Aid for Trade*	MOVIMENTO PER L AUTOSVILUPPOL INTERSCAMBIO E LA SOLIDARIETA	Consortium	968233	99820

17

2008	*A NETWORK FOR THE DEVELOPMENT*	PROVINCE OF PESARO AND URBINO	Consortium	115621	99820
2008	*Frauen und Globalisierung: Decent work for ALL! – Informations- und Mobilisierungskampagne für menschenwürdige Arbeit für Frauen in der globalen Exportindustrie am Beispiel Bekleidung*	CHRISTLICHE INITIATIVE ROMERO EV	Consortium	720446	99820
2008	*Verantwortliche Öffentliche Beschaffung und Menschenwürdige Arbeit JETZT!– Öffentlichkeits-, Bewusstseins- und Lobbykampagne zur Durchsetzung sozialer und ökologischer Beschaffung von Öffentlicher Hand und privaten Institutionen*	CHRISTLICHE INITIATIVE ROMERO EV	Consortium	701163	99820
2008	*Local capacity building for Fairtrade in Sweden, Finland and Estonia*	FORENINGEN FOR RATTVISEMARKT SVERIGE	Consortium	823148	99820
2008	*Network of Schools and Local Communities contributing to the achievement of the MDGs*	POLSKA AKCJA HUMANITARNA	Consortium	999000	99820
2008	*Fair Flowers - a gift to all involved. Raising the awareness of local authorities, consumers and traders on the production of cut flowers in developing countries*	FIAN FOODFIRST INFORMATIONS & AKTIONS NETWERK SEKTION DER BUNDESREPUBLIK DEUTSCHLAND EV	Consortium	669087	99820
2008	*A case for poverty reduction: Consumer awareness and action in 6 EU member states*	CONSUMERS INTERNATIONAL LBG	Consortium	857713	99820
2008	*Introducing the MADE-BY label for sustainable fashion in the United Kingdom*	STICHTING INTERKERKELIJKE AKTIE VOOR LATIJNS AMERIKA SOLIDARIDAD	Consortium	770000	99820
2008	*La sensibilisation sur les interdépendances entre Nord et Sud : un enjeu pour la mobilisation des citoyens européens en faveur du développement.*	ASSOCIATION FRERES DES HOMMES	France	789205,5	99820
2007	*Export Trade from Kenya – Enabling the poor to share the fruits.*	AFRICA NOW LBG	Royaume-Uni	408000	31191
2007	*Appui aux familles vulnérables dans deux zones cotonnières du Mali*	ASSOCIATION VETERINAIRES SANS FRONTIERES - CENTRE INTERNATIONAL DE COOPERATION POUR LE DEVELOPPEMENT AGRICOLE VSF CICDA	France	740614,53	43040
2007	*Empowering Emerging Farmers through fair trade development in South Africa*	STICHTING INTERKERKELIJKE AKTIE VOOR LATIJNS AMERIKA SOLIDARIDAD	Pays-Bas	1000000	33120
2007	*PUBLIC AFFAIRS - Mobilising action for Fair Trade Public Procurement*	STICHTING EUROPEAN FAIR TRADE ASSOCIATION	The Netherlands	568200	99820
2007	*Campaign for sustainable purchasing of computers: Making public purchasing in Europe work for development by raising awareness ot the working conditions and environmental issues in the global supply chain of computers.*	WELTWIRTSCHAFT, OKOLOGIE & ENTWICKLUNG - WEED EV	Germany	1038334,5	99820

2007	*Enlarging FAIR*	PANGEA - NIENTE TROPPO SOCIETA COOPERATIVA SOCIALE SCSARL	Italy	448198,2	99820
2007	*Expanding Fair Trade Awareness in Slovakia and the Czech Republic*	NADACIA INTEGRA	Slovakia	202779	99820
2007	*Fair consumption*	MAGOSFA KORNYEZETI NEVELESI ES OKOTURISZTIKAI ALAPITVANY		99880,26	99820
2007	*Supermarkets, supply chains and poverty reduction*	WAR ON WANT	United Kingdom	360000	99820
2007	*Decent work, trade and development: raising awareness among trade unions and women´s groups of the employment implications of international trade relations*	WAR ON WANT	United Kingdom	720000	99820
2007	*Education for Global Sustainability, Responsible Consumption and Fair Trade*	UUSI TUULI RY		496579,78	99820
2007	*FEEDING AND FUELLING EUROPE*	MAGYAR TERMESZETVEDOK SZOVETSEGE	Hungary	1078521,66	99820
2007	*Médiatiser la face invisible du développement*	ASSOCIATION COMITE FRANCAIS POUR LASOLIDARITE INTERNATIONALE	France	346591,06	99820
				19.466.102,80	

19

Keyword index

A

B

D

Q

R

S

www.ingramcontent.com/pod-product-compliance
Lightning Source LLC
Chambersburg PA
CBHW061154240926
51846CB00006B/1009
9789086861767